VOLUME SIX HUNDRED AND ONE

METHODS IN ENZYMOLOGY

Mechanisms of DNA Recombination and Genome Rearrangements: Intersection between Homologous Recombination, DNA Replication and DNA Repair

METHODS IN ENZYMOLOGY

Editors-in-Chief

ANNA MARIE PYLE
Departments of Molecular, Cellular and Developmental Biology and Department of Chemistry
Investigator, Howard Hughes Medical Institute
Yale University

DAVID W. CHRISTIANSON
Roy and Diana Vagelos Laboratories
Department of Chemistry
University of Pennsylvania
Philadelphia, PA

Founding Editors

SIDNEY P. COLOWICK and NATHAN O. KAPLAN

VOLUME SIX HUNDRED AND ONE

METHODS IN ENZYMOLOGY

Mechanisms of DNA Recombination and Genome Rearrangements: Intersection between Homologous Recombination, DNA Replication and DNA Repair

Edited by

MARIA SPIES

University of Iowa Carver College of Medicine, Iowa City, IA, United States

ANNA MALKOVA

University of Iowa, Iowa City, IA, United States

ELSEVIER

ACADEMIC PRESS

An imprint of Elsevier

Academic Press is an imprint of Elsevier
50 Hampshire Street, 5th Floor, Cambridge, MA 02139, United States
525 B Street, Suite 1800, San Diego, CA 92101–4495, United States
The Boulevard, Langford Lane, Kidlington, Oxford OX5 1GB, United Kingdom
125 London Wall, London, EC2Y 5AS, United Kingdom

First edition 2018

ISBN: 978-0-12-813979-0
ISSN: 0076-6879

For information on all Academic Press publications
visit our website at https://www.elsevier.com/books-and-journals

Publisher: Zoe Kruze
Acquisition Editor: Zoe Kruze
Editorial Project Manager: Alina Cleju
Production Project Manager: Divya KrishnaKumar
Cover Designer: Matthew Limbert

Typeset by SPi Global, India

CONTENTS

CONTRIBUTORS

Jasvinder S. Ahuja
Laboratory of Biochemistry and Molecular Biology, Center for Cancer Research, National Cancer Institute, Bethesda, MD, United States

Aneesa T. Al-Soodani
School of Molecular Biosciences, College of Veterinary Medicine, Washington State University, Pullman, WA, United States

Frank Alber
University of Southern California, Los Angeles, CA, United States

Andrew S. Arvai
The Scripps Research Institute, La Jolla, CA, United States

Rahul Bhowmick
Center for Chromosome Stability, University of Copenhagen, Copenhagen, Denmark

Lotte Bjergbæk
Aarhus University, Aarhus, Denmark

Kevin Brick
Genetics & Biochemistry Branch, NIDDK, National Institutes of Health, Bethesda, MD, United States

Chris A. Brosey
The University of Texas, M.D. Anderson Cancer Center, Houston, TX, United States

Alexander J. Brown
School of Molecular Biosciences, College of Veterinary Medicine, Washington State University, Pullman, WA, United States

Rafael D. Camerini-Otero
Genetics & Biochemistry Branch, NIDDK, National Institutes of Health, Bethesda, MD, United States

Christopher P. Caridi
University of Southern California, Los Angeles, CA, United States

Irene Chiolo
University of Southern California, Los Angeles, CA, United States

Devika Das
University of Southern California, Los Angeles, CA, United States

Luther Davis
University of Washington, Seattle, WA, United States

Laetitia Delabaere
University of Southern California, Los Angeles, CA, United States

Gajendrahar Dwivedi
Rosenstiel Basic Medical Sciences Research Center, Brandeis University, Waltham, MA, United States

Rajula Elango
University of Iowa, Iowa City, IA, United States

Laura Folly da Silva Constantino
The University of Iowa, Iowa City, IA, United States

Juan C. Garcia
Lineberger Comprehensive Cancer Center, Curriculum in Genetics and Molecular Biology, University of North Carolina at Chapel Hill, Chapel Hill, NC, United States

Lorenza Garribba
Center for Chromosome Stability, University of Copenhagen, Copenhagen, Denmark

James E. Haber
Rosenstiel Basic Medical Sciences Research Center, Brandeis University, Waltham, MA, United States

Terrence Hanscom
Tufts University, Medford, MA, United States

Chengtao Her
School of Molecular Biosciences, College of Veterinary Medicine, Washington State University, Pullman, WA, United States

Stephanie Her
Dartmouth College, Hanover, NH, United States

Wolf-Dietrich Heyer
University of California, Davis, Davis, CA, United States

Ian D. Hickson
Center for Chromosome Stability; Center for Healthy Aging, University of Copenhagen, Copenhagen, Denmark

Hannah Hopp
University of Southern California, Los Angeles, CA, United States

Neil Hunter
Howard Hughes Medical Institute, University of California, Davis, Davis, CA, United States

Kristoffer P. Jakobsen
Aarhus University, Aarhus, Denmark

Hardeep Kaur
Laboratory of Biochemistry and Molecular Biology, Center for Cancer Research, National Cancer Institute, Bethesda, MD, United States

Scott Keeney
Molecular Biology Program; Howard Hughes Medical Institute, Memorial Sloan Kettering Cancer Center, New York, NY, United States

Havva Keskin
Georgia Institute of Technology, Atlanta, GA, United States

Varandt Y. Khodaverdian
Tufts University, Medford, MA, United States

Zachary Kockler
University of Iowa, Iowa City, IA, United States

Romain Koszul
Institut Pasteur, Paris, France

Quinn Li
The University of Iowa, Iowa City, IA, United States

Michael Lichten
Laboratory of Biochemistry and Molecular Biology, Center for Cancer Research, National Cancer Institute, Bethesda, MD, United States

Liping Liu
University of Iowa, Iowa City, IA, United States

Ying Liu
Center for Chromosome Stability, University of Copenhagen, Copenhagen, Denmark

Nancy Maizels
University of Washington, Seattle, WA, United States

Anna Malkova
University of Iowa, Iowa City, IA, United States

Jean-Yves Masson
Genome Stability Laboratory, CHU de Québec Research Center; Laval University Cancer Research Center, Québec City, QC, Canada

Mitch McVey
Tufts University, Medford, MA, United States

Eleni P. Mimitou
Molecular Biology Program, Memorial Sloan Kettering Cancer Center, New York, NY, United States

Davide Moiani
The University of Texas, M.D. Anderson Cancer Center, Houston, TX, United States

Shannon Owens
Howard Hughes Medical Institute; Biochemistry, Molecular, Cellular and Developmental Biology Graduate Group, University of California, Davis, Davis, CA, United States

Özgün Özer
Center for Chromosome Stability, University of Copenhagen, Copenhagen, Denmark

Elena Petricci
University of Siena, Siena, Italy

Galina Petukhova
Uniformed Services University of the Health Sciences, Bethesda, MD, United States

Aurèle Piazza
University of California, Davis, Davis, CA, United States; Institut Pasteur, Paris, France

Florencia Pratto
Genetics & Biochemistry Branch, NIDDK, National Institutes of Health, Bethesda, MD, United States

Dale A. Ramsden
Lineberger Comprehensive Cancer Center, Curriculum in Genetics and Molecular Biology, University of North Carolina at Chapel Hill, Chapel Hill, NC, United States

Steven A. Roberts
School of Molecular Biosciences; Center for Reproductive Biology, College of Veterinary Medicine, Washington State University, Pullman, WA, United States

Daryl A. Ronato
Genome Stability Laboratory, CHU de Québec Research Center; Laval University Cancer Research Center, Québec City, QC, Canada

Miles Saul
University of Washington, Seattle, WA, United States

M. Ashley Spies
The University of Iowa, Iowa City, IA, United States

Francesca Storici
Georgia Institute of Technology, Atlanta, GA, United States

Chi-Yu Sun
Uniformed Services University of the Health Sciences, Bethesda, MD, United States

Aleem Syed
The University of Texas, M.D. Anderson Cancer Center, Houston, TX, United States

John A. Tainer
The University of Texas, M.D. Anderson Cancer Center, Houston, TX; Lawrence Berkeley National Laboratory, Berkeley, CA, United States

Shangming Tang
Howard Hughes Medical Institute; Biochemistry, Molecular, Cellular and Developmental Biology Graduate Group, University of California, Davis, Davis, CA, United States

Harianto Tjong
University of Southern California, Los Angeles, CA, United States

Wei Wu
Center for Chromosome Stability, University of Copenhagen, Copenhagen, Denmark

Yinbo Zhang
University of Washington, Seattle, WA, United States

PREFACE

Homologous genetic recombination remains the most enigmatic process in DNA metabolism. The molecular machines of recombination preserve the integrity of the genetic material in all organisms and generate genetic diversity in evolution. The same molecular machines that support genetic integrity by orchestrating accurate repair of the most deleterious DNA lesions, however, also promote survival of cancerous cells and emergence of radiation and chemotherapy resistance. The machinery and the cellular process of recombination are intrinsically interconnected with those of DNA replication and repair. Recent work from many labs has revealed new and deeper connections between these 3Rs of genome maintenance.

These two volumes in this "Mechanisms of DNA Recombination and Genome Rearrangements" set offer a comprehensive set of cutting-edge methods to study various aspects of homologous recombination and cellular processes that utilize the enzymatic machinery of recombination. The two volumes feature contributions by the leading experts in the field of DNA repair, recombination, replication, and genome stability. The chapters within document methods are presently cutting edge, but will become a standard laboratory practice in the years to come. The first volume in the set (Volume 600, "Mechanisms of DNA Recombination and Genome Rearrangements: Methods to Study Homologous Recombination") concerns with the methods to study the initiation of homologous recombination, the central step in recombination, which involves formation of the recombinase nucleoprotein filament, homology search and DNA strand exchange, and the methods to study the processing of the recombination intermediates through resolution or dissolution. The second volume in the series (Volume 601, "Mechanisms of DNA Recombination and Genome Rearrangements: Intersection between Homologous Recombination, DNA Replication, and DNA Repair") focuses on the methods to study the interface between recombination and replication, selection of recombination as homology-directed DNA repair pathway, homologous recombination machinery as a target for anticancer drug discovery, and meiotic recombination. In addition, the second volume covers a number of topics that recently emerged in the field of recombination, including mutagenesis associated with recombination, recombination between RNA and DNA, recombination initiated by DNA nicks, and recombination in a context of heterochromatin.

We selected the topics and contributors with a goal that these volumes to become a useful reference for the researchers engaged in the field of DNA repair, DNA recombination, DNA replication, and genome stability. The techniques described in the chapters range from cellular imaging to biochemical, structural, and single-molecule approaches, to drug discovery and next-generation sequencing. We hope to see this two-volume set as a well-used reference in the labs actively studying these processes and those contemplating entering the field. The chapters are written in a way to make the methodology accessible not only to the established experts but to a broader group of researchers, from undergraduates and rotation students to postdoctoral fellows and faculty who wish to expand their research programs.

Maria Spies and Anna Malkova
University of Iowa

CHAPTER ONE

Using the Flp Recombinase to Induce Site-Specific Protein–DNA Nicks

Kristoffer P. Jakobsen, Lotte Bjergbæk[1]
Aarhus University, Aarhus, Denmark
[1]Corresponding author: e-mail address: lbj@mbg.au.dk

Contents

Methods in Enzymology, Volume 601
ISSN 0076-6879
https://doi.org/10.1016/bs.mie.2017.11.023

Abstract

A natural and frequent occurring replication insult is generated by the action of DNA Topoisomerase I (Top1). When Top1 gets trapped in a cleavage complex on the DNA, a protein-linked DNA nick (PDN) is generated. Today it is known that PDNs are generated at a high incidence in the cell. If not rapidly removed, PDNs can have a profound impact on cell destiny, as a nick in proliferating cells is passively transformed into a single-ended DSB, when encountered by the replication machinery. A DSB can in turn lead to chromosomal rearrangements and thus jeopardize genome stability if not appropriately repaired.

In order to study repair pathways associated with PDNs, we have developed a cellular system (Flp-nick), where we can generate a single PDN at a specific genomic site in the model organism *Saccharomyces cerevisiae*. The system takes advantages of the Flp recombinase, which catalytically operates like Top1 by generating a nick in the DNA backbone and during this process becomes covalently linked to the DNA. Flp cleaves at well-defined target sites. Thus, a target site has been inserted in the genome and a mutant Flp, which cleaves but do not religate, is expressed. In this way, a single PDN mimicking the one generated by Top1 is induced at a known genomic site. The Flp-nick system allows detailed molecular analysis of repair pathways associated with this type of damage and can be designed to study repair at any genomic context.

1. INTRODUCTION

During DNA replication topological stress is building up in front of the replication fork, which needs to be removed in order to prevent excess supercoiling with the risk of replication fork breakdown (Wang, 2002). Topoisomerase I (Top1) is a ubiquitous enzyme capable of cleaving and rapidly religating one nucleic acid strand through a transesterification reaction. This reaction generates a transient nick in the DNA backbone through which topological changes can occur (Schoeffler & Berger, 2008; Wang, 2002). However, the catalytic mode of Top1 also poses a threat for the cells as it generates an intermediate in which Top1 becomes covalently linked to the 3′ end of the nick. Failure to reseal the nick, which is caused by misalignment of the 5′-hydroxyl DNA end, results in the trapping of Top1 on the DNA, thereby generating a long-lived protein-linked DNA nick (PDN). Misalignment can occur under conditions of damage, e.g., abasic sites or long-lived PDNs obtained upon treatment of cells with agents that block the religation step, e.g., the drug camptothecin (Pommier, 2006; Pommier et al., 2006).

Top1-generated PDNs are a widespread problem for proliferating cells. When encountered by the replication machinery, they are passively converted into double-stranded breaks (DSBs) (see Fig. 1B). The generated DSB is single ended and thus repair of this will likely dependent on some type of break-induced replication (BIR) pathway (Anand, Lovett, & Haber, 2013). However, for this to be initiated, removal of the protein–DNA adduct is also needed.

To study the cellular responses to Top1-generated PDNs, we have developed the cellular Flp-nick system (Nielsen et al., 2009). The system takes advantage of the site-specific Flp recombinase encoded by the 2 μ plasmid found naturally in *Saccharomyces cerevisiae*. Flp recombinase is able to catalyze a site-specific recombination reaction between two Flp recognition targets (FRT), which are also found on the 2 μ plasmid (Chan, Liu, Ma, Jayaram, & Sau, 2013). A FRT site consists of three 13-bp long symmetric elements, two of which are present in inverse orientation and flank an 8 bp spacer (see Fig. 1A). The third element present in tandem with one of the other repeats can be deleted without significant effects on target-site function (Andrews, Proteau, Beatty, & Sadowski, 1985). During a normal recombination reaction catalyzed by Flp, four monomers of the enzyme are needed along with two FRT sites. A monomer of Flp recombinase harbors an incomplete active site that is made functionally complete by association with another monomer, thus sharing catalytic residues between them (Chen, Lee, & Jayaram, 1992; Chen, Yang, & Jayaram, 1993). Recombination is initiated by strand cleavage by the active-site tyrosine in Flp (Tyr343) at one position of the FRT site (see Fig. 1A). This results in a nick with a free 5′-hydroxyl group and a 3′-phosphotyrosine linkage between Flp recombinase and the 3′-end of the DNA (Chan et al., 2013). The trans-esterification reaction as well as the established transient protein–DNA adduct are identical to what is seen during the catalytic mechanism of Top1.

During the religation step of the catalytic cycle, the 5′-hydroxyl group attacks the phosphotyrosine bonds across substrates (in a second FRT substrate), which establishes a Holliday junction. Next strand cleavage on the other end of the spacer occurs and the religation reaction is repeated, which resolves the Holliday junction into two reciprocal recombinant DNA products. Cleavage by Flp follows a *trans*-horizontal mode, thus strand cleavage can be initiated when one FRT site is occupied by two Flp monomers (Lee, Jayaram, & Grainge, 1999). Although such a complex can organize two shared active sites, in reality only one of the two sites is organized at any

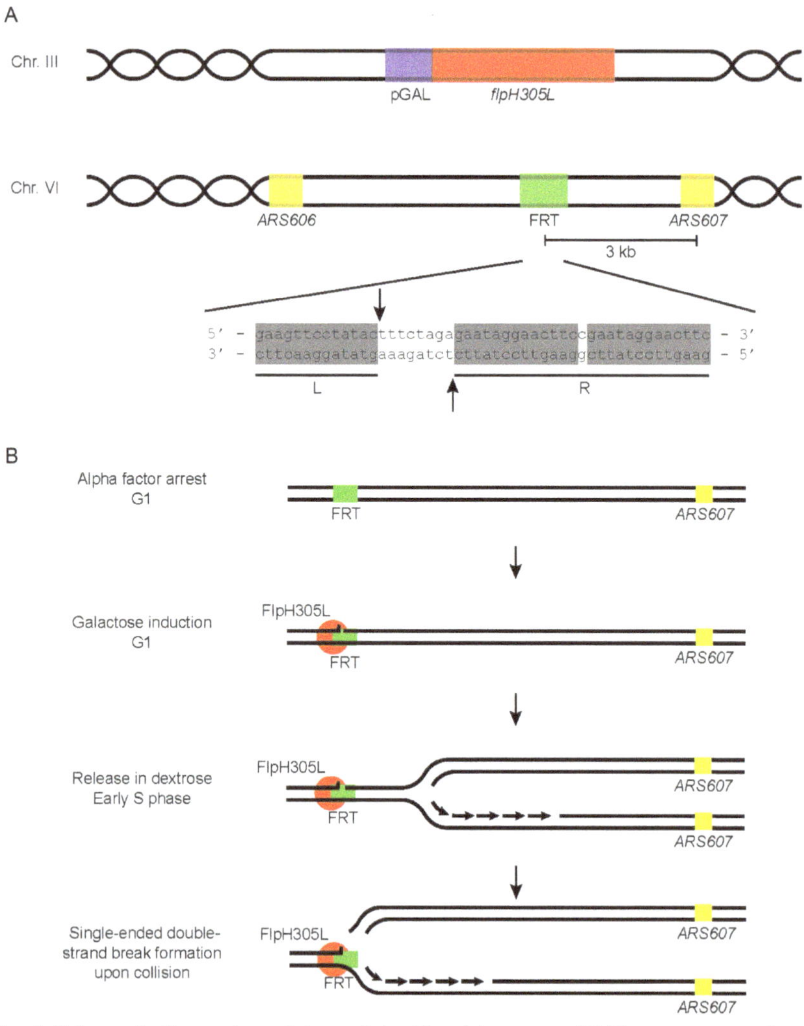

Fig. 1 Schematic illustration of the cellular Flp-nick system. (A) The gene encoding a mutant form of the Flp recombinase (*flpH305L*) is integrated on Chr. III and expressed from the GAL promoter. The Flp recognition target (FRT) is in our system integrated in a noncoding region on Chr. VI 3 kb from *ARS607*. The FRT consists of three DNA elements (shown as *gray areas*) and an 8 bp DNA region, where nicking takes place. *Vertical black arrows* indicate the positions for Flp-mediated cleavage. Cleavage only takes place at one of the two positions at a time. L and R show the *left* and the *right side* of the FRT site, respectively. (B) Illustrates events at the FRT site upon induction of *flpH305L* and when cells progress into S phase.

given time (Qian, Inman, & Cox, 1990; Voziyanov, Lee, Whang, Lee, & Jayaram, 1996). Hence, only single-strand breaks are made within a substrate.

In the Flp-nick system, we mimic a situation, where Top1 gets trapped on the DNA and thus form a long-lived PDN. To obtain a long-lived PDN we use a "step-arrest" mutant of Flp, which carries a mutation at position 305, changing a histidine in the active site to a leucine (Parsons, Prasad, Harshey, & Jayaram, 1988). This mutant form of Flp (*flpH305L*) cleaves the DNA but is unable to accomplish the religation reaction. With the use of an integration plasmid, the *flpH305L* gene expressed behind the *GAL10* promoter is recombined into the *S. cerevisiae* genome (Tsalik & Gartenberg, 1998). This generates a controlled system, where high expression of the *flpH305L* is obtained by addition of galactose to the yeast culture under nonfermenting conditions (see Fig. 1).

A series of FRT integration plasmids with different selection markers based on the pFA6a plasmids have been generated in the lab. This allows us to integrate the FRT site at any desired genomic locus in *S. cerevisae* by using PCR-directed homologous recombination (Nielsen et al., 2009). Design of the primers for the PCR reaction will determine, where integration takes place and in which orientation the FRT is inserted with respect to, e.g., lagging and leading strand synthesis. The Flp-nick system has the advantage that it can easily be modified to study PDNs in any chromosomal context. Furthermore, the system is ideal to investigate the destiny of a replication-generated DSB, as the single-ended DSB is entirely dependent on replication (Mayle et al., 2015; Nielsen et al., 2009). This creates a unique possibility to study BIR-related repair during S phase. The system also offers the option for studying repair of PDNs, without formation of the single-ended DSB by keeping the cells in a G1 arrest.

In the following we describe how to make a basic yeast strain with the Flp-nick system integrated. We furthermore describe our normal growth protocol and standard techniques we are using to monitor nick formation and repair, DSB generation and repair as well as a modified chromatin immunoprecipitation (ChIP) protocol to monitor removal of the Flp protein from the DNA.

2. GENERATION OF A BASIC FLP-NICK YEAST STRAIN

To generate the cellular Flp-nick system in a specific yeast strain several modifications are needed. First, integration of the *flpH305L* gene under control of the *GAL10* promoter into the *LEU2* locus of the genome is carried out by one-step gene replacement using a plasmid containing the *GAL10* promoter and the *flpH305L* gene (Tsalik & Gartenberg, 1998). Second, the generated yeast strain is cured from the natural existing 2 μ plasmid. Finally, the FRT site, where the nick is generated upon induction of *flpH305L*, needs to be integrated at a genomic locus of interest. This is done by transformation of a PCR-generated fragment containing the FRT site into the yeast strain, which requires plasmids containing the FRT site together with at selectable marker. Together these steps result in a basic Flp-nick yeast strain, which can be further modified (additional gene deletions, gene tagging, etc.) according to wishes.

2.1 Integration of the *flpH305L*-Containing Plasmid and Cure of the 2 μ Plasmid

Integration of *flpH305L* in the yeast genome is an easy step, which requires linearization of plasmid pFV17Δ-H305L, and transformation of the yeast strain with the plasmid. When integration has been confirmed the yeast strain is next cured from the 2 μ plasmid. The 2 μ plasmid is found naturally in most *S. cerevisiae* yeast strains with an average copy number of 20–40. It does not seem to provide any advantage or disadvantage to its host, when present at its normal copy number. The plasmid encodes four genes, one of them being the *FLP* gene. Furthermore, it also contains two FRT sites (Chan et al., 2013). To avoid recombination between the inserted FRT site and FRT sites on the 2 μ plasmid as well as wild-type Flp activity, the *flpH305L*-containing yeast strain is cured for the 2 μ plasmid before the FRT site is inserted.

2.1.1 Equipment

- PCR machine
- Equipment to run agarose gels and visualize DNA
- Heating block for enzyme digestion
- Microcentrifuge and centrifuge for 50-mL tubes
- 30°C incubator

2.1.2 Buffers and Reagents

- pFV17Δ-H305L plasmid (see note 1)
- BstEII (10,000 U/mL)
- Reaction buffer for the enzyme (supplied by distributor of enzyme)
- Commercial available DNA purification kit (we use GFX kit from GE Healthcare)
- 1 × TE buffer (10 m*M* Tris–HCL pH 7.5, 1 m*M* Na_2EDTA pH 8.0)
- W303 yeast strain
- Reagents for standard yeast transformation (LiAC method): 10 × TEL: 0.1 *M* Tris–HCl pH 7.5, 10 m*M* Na_2EDTA pH 8.0, and 10.2 g lithium acetate per 100 mL buffer. 40% PEG4000 in TEL: 40% polyethylene

glycol 4000 dissolved in 1 × TEL. Carrier DNA from salmon sperm (10 mg/mL)

- SC-leucine selection plates
- 20% galactose: Dissolve galactose in water at room temperature. Filter sterilize. Do not autoclave. Store at 4°C
- YP + 3% galactose plates: 2% bacto peptone, 1% yeast extract, 0.0025% adenine hemisulfate salt, 2% agar, 1 NaOH pellet/L. Autoclave. When cooled down add galactose from a 20% solution to 3% and pour the plates
- Primers to test for *flpH305L* integration: Dissolve to 100 μ*M* in H_2O
- Primers to test for absence of 2 μm plasmid: Dissolve to 100 μ*M* in H_2O
- Taq PCR master mix kit from QIAGEN or equivalent
- Reagents for running agarose gels: 10 × DNA-loading buffer: 20% ficoll (type 400), 0.1 *M* Na_2EDTA pH 8.0, 1% sodium dodecyl sulfate (SDS), 0.25% bromophenol blue, 0.25% xylene cyanol. 10 × TBE buffer: 890 m*M* Trizma base, 890 m*M* boric acid, 20 m*M* Na_2EDTA pH 8.0. Agarose. Ethidium bromide

2.1.3 Procedure

1. To linearize the *flpH305L*-containing plasmid pFV17Δ-H305L (see note 1), which contains the *flpH305L* gene behind a *GAL10* promoter and a functional *LEU2* gene, use restriction enzyme BstEII (2 μL) in the presence of appropriate 1 × reaction buffer in a total volume of 20 μL. Digest at the optimal temperature for the enzyme for 2 h. BstEII will linearize the plasmid by cleaving in the *LEU2* gene. We use 5 μg of pFV17Δ-H305L.
2. Purify the digested plasmid using the GFX kit and elute in 30 μL of 1 × TE buffer.
3. Transform the purified linearized plasmid into the *leu2-3,-112* locus of a W303 *S. cerevisiae* yeast strain (see note 2) using the LiAc method for transformation (Gietz & Woods, 2002). Use different amounts of the linearized pFV17Δ-H305L plasmid for the transformation. Plate on SC-leucine plates as correct integration of pFV17Δ-H305L plasmid will result in a functional *LEU2* gene. Incubate the plates for 2–3 days at 30°C.
4. Restreak yeast colonies to a new SC-leucine plate and test for correct integration by standard colony PCR using primers 5′CGTTT CTATTATGAATTTCA and 5′GGTGGTTAGCAATCGTCTTA (see note 3). These primers anneal in the *flpH305L* gene and in the genome upstream of the wild-type *LEU2* gene and will result in a

PCR product of 1250 bp if *flpH305L* has been correctly integrated. Run the products on a 1% agarose gel with ethidium bromide in 1 × TBE.

5. To cure the *flpH305L*-containing yeast strain from the natural occurring 2 μ plasmid, restreak single colonies on YP + 3% galactose plates. Incubate at 30°C for 1 day and restreak single colonies on a new YP + 3% galactose plate. Repeat this once more.
6. Test the colonies for presence (cir^+) or absence (cir^0) of the 2 μ plasmid by standard colony PCR using primers 5′ CCATCCACCTTT CGCTCCTT and 5′ CGAGAGACTGCTTGCTTGTA located in the *REP1* gene of the 2 μ plasmid. This will result in a PCR product of around 1000 bp only if the cells are still Cir^+. Alternatively, perform a Southern blot to confirm lack of 2 μ plasmid.

2.1.4 Notes

1. Upon request plasmid pFV17Δ–H305L can be obtained from our laboratory. This plasmid was generated by Marc Gartenberg and first published in Tsalik and Gartenberg (1998).
2. Our yeast strains are derivatives of W303-1A from R. Rothstein.
3. We first perform colony PCR, and if this does not give positive hits, we make genomic DNA and perform PCR on this. We often see that although colony PCR does not work, we get positive results with PCR on genomic DNA.

2.2 Generation of FRT-Containing Plasmids and Insertion of the FRT Site Into the Genome

We have generated a series of FRT integration plasmids with different selection markers based on the pFA6a plasmid (Longtine et al., 1998), which enables us to integrate the FRT site at any given location in the yeast genome by PCR-directed homologous recombination. The primers used for the PCR reaction should contain approximately 40 bp of DNA in their 5′ end, which are homologous to the genomic region, where you wish to insert the FRT site. Furthermore, depending on the design of the primers, the FRT site can be inserted in any of the two orientations.

The FRT integration plasmids can be obtained from our laboratory upon request (see note 1). Alternatively pFA6a or pAG plasmids containing the FRT site can be constructed as briefly described later. To make FRT-containing pFA6a plasmids, two oligos corresponding to the upper and lower strand of a full-length FRT sequence are annealed and cloned into a pFA6a plasmid.

2.2.1 Equipment

- PCR machine
- Equipment to run agarose gels and visualize DNA
- Heating block for enzyme digestion, ligation, etc.
- Microcentrifuge and centrifuge for 50-mL tubes
- 30°C and 37°C incubator

2.2.2 Buffers and Reagents

- SacII (20,000 U/mL), SpeI (10,000 U/mL), XbaI (20,000 U/mL)
- Oligos corresponding to the FRT site: Dissolve to 1 μg/μL
- 10 m*M* ATP
- T4 polynucleotide kinase (10,000 U/mL, buffer supplied from distributor)
- 2 *M* KCl
- 1 × TE buffer (10 m*M* Tris–HCL pH 7.5, 1 m*M* Na_2EDTA pH 8.0)
- pFA6a vector
- Calf intestinal alkaline phosphatase (CIP, 10,000 U/mL)
- T4 DNA ligase (400,000 U/mL, buffer supplied from distributor)
- Reagents for standard *E. coli* transformation
- FlpH305L Cir^0 strain
- Reagents for standard yeast transformation (see Section 2.1.2)
- SC selection plates (selection depends on the chosen selection marker)
- Primers to integrate FRT: Dissolve to 100 μ*M* in H_2O
- Primers to test for integration of FRT: Dissolve to 100 μ*M* in H_2O
- Taq PCR master mix kit from QIAGEN or equivalent
- Reagents for running agarose gels (see Section 2.1.2)

2.2.3 Procedure

1. Two oligos corresponding to the upper and lower strand of the FRT site are synthesized. We used oligos containing a full FRT site and recognition sites for SacII and SpeI in the ends.

   Oligo A: CTAGTGAAGTTCCTATACTTTCTAGAGAATAGG AACTTCCGAATAGGAACTTCCCGC.

   Oligo B: GGGAAGTTCCTATTCGGAAGTTCCTATTCTCT AGAAAGTATAGGAACTTCA.
2. Dissolve each oligo to a concentration of 1 μg/μL in H_2O.
3. For each oligo make a kinase reaction consisting of: 1 μg oligo, 1 × kinase buffer, 1 m*M* ATP, and 1 μL T4 polynucleotide kinase. Mix in a volume of 20 μL and leave at 37°C for 30–45 min.

4. Annealing of oligo A and B: Heat water in a beaker to 80°C. Mix the two kinase reactions and add 2 μL 2 *M* KCl. Place the eppendorf tube in a polystyrene rack so that the reaction is in the water. Cool down to 30°C.
5. Approximately 3–5 μg of the vector (pFA6a vector with the wanted selection marker) is digested with 1 μL SacII and 1 μL SpeI in a reaction containing 1 × reaction buffer in a volume of 20 μL. Leave at 37°C for 2 h, add 1 μL CIP, and leave for 1 more hour at 37°C. Run the digested plasmid on a 1% agarose gel with ethidium bromide in 1 × TBE and purify using a commercial available DNA purification kit. Elute with 10 μL 1 × TE buffer. Set up standard ligation reaction using purified vector and annealed oligos. Transform into competent *E. coli* strain by standard procedures.
6. Plasmids from individual *E. coli* colonies are isolated by standard procedures and the product is digested to confirm insertion of the FRT site. The FRT site contains an XbaI restriction site and this is used together with a BglII site in the vector, resulting in products of 1500 and 2480 bp if cloning was successful (see note 2).
7. To insert the FRT site in the yeast genome we use PCR-directed homologous recombination using standard protocol. Make a PCR product using primers with homologous ends to the genomic region where you wish to insert the FRT site. To insert FRT on Chr. VI approximately 3 kb from *ARS607* we use the following primers: 5′ CCTGATTCATCTATTAGCCGTATAGAGAATATACCGCATC ACGACTCACTATAGGGAGAC and 5′ GCTTTCATGGATCGA GTATTTGTTTGTCGAAAAAGGATCAGAACGCGGCCGCCA GCTGAA.

   The underlined sequence is homologous to the genomic region, whereas the rest anneals to the pFA6a vectors.
8. Transform the generated PCR product into the FlpH305L Cir^0 strain. Plate on appropriate selection plates and incubate at 30°C for 2–3 days to obtain colonies. Restreak the colonies on selection plates.
9. Confirm correct integration of the FRT site by either colony PCR or PCR on genomic DNA. We use the following primers, when we confirm integration at 3 kb from *ARS607*: 5′GGTGATGTGATCG TATTCAA and 5′CCGTGCGGCCATCAAAATG.

2.2.4 Notes

1. pLB116 (FRT with KanMX6 marker), pLB122 (FRT with NatMx4), and pLB133 (FRT with URA3 marker)

2. To verify correct insertion and sequence of the FRT site the plasmid is sequenced over the inserted FRT site using primer: 5′CCTCG ACATCATCTGCCC

3. GROWTH OF FLP-NICK YEAST STRAINS AND INDUCTION OF *FLPH305L*

Expression of the *flpH305L* gene is under control of the *GAL10*, thus expression is induced upon growth in the presence of galactose, whereas growth in the presence of glucose will suppress expression. This enables a tight regulation of nick formation. Together with the possibility of arresting cells in a specific cell cycle phase, this gives the advantage, that induced nick formation and associated repair or cellular responses can be studied in a specific part of the cell cycle, e.g., the S phase. Here follows a general arrest and induction protocol for the Flp-nick system, used when S phase experiments are carried out. Modifications are made to this protocol depending on which experiment is performed.

3.1 Equipment

- 30°C shaking incubator
- Centrifuge for 50-mL tubes

3.2 Buffers and Reagents

1. YPD: 2% bacto peptone, 1% yeast extract, 2% D-(+)-glucose monohydrate, 0.0025% adenine hemisulfate salt. Autoclave.
2. YP + raffinose: 2% bacto peptone, 1% yeast extract, 2% D-(+)-raffinose pentahydrate, 0.0025% adenine hemisulfate salt. Autoclave.
3. YP + raffinose pH 3.5: As for YP + raffinose but adjust pH to 3.5 with 6 *M* HCl before autoclaving (see note 1).
4. 20% galactose: Dissolve galactose in water at room temperature. Filter sterilize. Do not autoclave. Store at 4°C.
5. Alpha factor: Dissolve to 6 mg/mL in 10 m*M* HCl, 1 m*M* β-mercaptoethanol, and 0.2 m*M* Na_2EDTA pH 8.0. Store aliquots at −20°C.

3.3 Procedure

1. Grow a small culture of cells o/n in YP + raffinose at 30°C.
2. Dilute the cells and let them grow one generation in YP + raffinose pH 3.5.

3. Add alpha factor (1:1000, see note 1) and treat for 60 min at 30°C. Alpha factor is a mating pheromone that affects MATa cells to inhibit DNA synthesis thereby arresting the cells in G1.
4. Add galactose to a final concentration of 3% (see note 2) and additional alpha factor (1:3000, see note 1) and treat for 2 h at 30°C.
5. Wash the cells once in cold distilled water.
6. Release the cells into YPD at 25°C to slow down the replication process and take out samples throughout the S/G2 phase. We also recommend taking out samples for flow cytometry (see note 3).

3.4 Notes

1. Alpha factor concentration depends on yeast strain and supplier. For optimal concentration test the alpha factor, whenever you change yeast strain or buy new alpha factor.
2. We use galactose solutions no older than 14 days.
3. Samples are taken for flow cytometry from the random growing culture, from the G1-arrested culture, and from every time point, where samples are also taken for other analysis. We take out approximately 0.5×10^7 cells and analyze the samples in a BC Cytoflex. Data are further analyzed using Flow JO.

4. ALKALINE DENATURING GEL ANALYSIS TO MEASURE NICKING ACTIVITY

To visualize nicking at the FRT site alkaline gel electrophoresis can be carried out on purified enzymatically digested genomic DNA. After electrophoresis Southern blotting and hybridization are carried out.

4.1 Equipment

- Equipment to run agarose gels and visualize DNA
- Heating block for enzyme digestion
- Microcentrifuge and centrifuge for 50-mL tubes
- Shaking incubator
- Hybridization oven

4.2 Buffers and Reagents

- Media for growth and alpha factor (see Section 3.2)
- QIAGEN Genomic-tip 20/G kit

- Required buffers and reagents used in the QIAGEN Genomic top 20/G protocol: G2, Y1, QBT, QC, and QF buffers. See recipes in QIAGEN handbook. Proteinase K solution (make stock of 40 mg/mL), Zymolyase (20,000 U/g) solution (make stock of 1000 U/mL), RNase A (make stock of 10 mg/mL), isopropanol, 70% ethanol, 1 × TE buffer
- Enzyme for digestion and buffer supplied from distributor
- 3 *M* sodium acetate pH 5.2
- 0.5 *M* Na_2EDTA pH 8.0
- Ethanol: 96% and 70%
- 1 × alkaline loading buffer: 50 m*M* NaOH, 1 m*M* Na_2EDTA pH 8.0, 2.5% ficoll (type 400) in water, bromocresol green, xylene cyanol FF
- Agarose
- 10 *M* NaOH
- 1 × alkaline electrophoresis buffer: 50 m*M* NaOH, 1 m*M* Na_2EDTA pH 8.0
- Standard buffers for Southern blotting and hybridization

4.3 Procedure

4.3.1 Growth of Yeast Cells

Grow the yeast cells as described in Section 3.3 with the following modifications:

1. Grow cells until 2.0×10^7 cells/mL.
2. Take out 2.0×10^9 cells from the G1-arrested culture before galactose induction. This sample is used as a negative control, where no break should be detected.
3. Take out 2.0×10^9 cells from the G1-arrested, galactose-induced culture. We normally take out a sample just prior to release into S phase, which is after 2 h of induction. This sample is used as a positive control to detect the single-strand break.
4. 2.0×10^9 cells are taken out, e.g., 30, 60, and 90 min after release into S phase or if G1 experiments are conducted samples can be collected at different times during a G1 arrest.
5. The cells are pelleted and stored at −80°C until DNA extraction.

4.3.2 Preparation and Enzymatic Digestion of Genomic DNA

1. Thaw cells on ice. Dissolve in 2 mL TE and spin 3000 rpm, 3 min, 4°C. Remove supernatant.
2. Dissolve in 1 mL buffer Y1 and add 100 μL Zymolase (1000 U/mL). Incubate at 30°C for 30 min.

3. Pellet the spheroplast by centrifuging at 3000 rpm, 10 min, 4°C.
4. Resuspend in 1.8 mL G2 (with RNase A and prot K).
5. Incubate (no shaking) at 37°C, 2 h. Add additional 45 μL Prot K after 1 h.
6. Move sample to 2-mL tube. Spin at 7000 rpm, 10 min, 4°C.
7. Equilibrate a Qiagen Genomic-tip 20/G with 2 mL Buffer QBT.
8. Apply sample to the Genomic-tip.
9. Wash the Genomic-tip with 3 mL Buffer QC.
10. Elute the genomic DNA with 2 × 1 mL Buffer QF (50°C) into 2 × 2-mL tubes.
11. Add 0.7 mL isopropanol, mix, and spin 13,000 rpm, 15 min, 4°C.
12. Wash with 1 mL 70% EtOH and spin 13,000 rpm, 10 min, 4°C.
13. Remove supernatant and air dry 5–10 min.
14. Add 50 μL TE (pH 8.0) and shake overnight at 30°C, 300 rpm.
15. Collect in one tube and leave on shaker till DNA is dissolved (1–3 days/ preferable over weekend).
16. Apply 1 μL on a 1% agarose gel with ethidium bromide and run in 1 × TBE buffer to see if the DNA has been dissolved.
17. Digest DNA o/n in a total volume of 200 μL (100 μL of DNA + water, buffer and enzyme 50 units). For detection of a nick at Chr. VI approximately 3 kb from *ARS607* we use HindIII enzyme.
18. Vortex short and add extra enzyme (30 units), digest 3 h.
19. Add: 500 μL 96% EtOH, 20 μL NaAc (3 *M*, pH 5.2), and 2 μL EDTA (0.5 *M*, pH 8).
20. Precipitate 1 h at −80°C and spin (30 min/13 k/4°C).
21. Wash with 70% EtOH, spin (10 min/13 k, 4°C), and dissolve in 30 μL TE. Add 30 μL 6 × alkaline loading buffer. Samples are now ready for loading on gel.

4.3.3 Preparation and Running Alkaline Gel

1. Melt 4 g agarose in 400 mL H_2O and cool to 60°C.
2. Add 2 mL NaOH (10 *M*) and 0.8 mL EDTA (0.5 *M*, pH 8). No EtBr!
3. Pour the gel and add slot former (our gel has a size of: 20 × 23.5 cm).
4. Let the agarose set for 1 h at RT.
5. Fill running chamber with 1 × alkaline running buffer (see note 1).
6. Load marker and samples (e.g., 20 μL) (see note 2).
7. Run 1 h at 40 V (set mA as high as possible). Place glass plate on gel and run ~40 h at 20 V.

8. Set up for Southern blotting (use standard Southern blot protocol) and afterward hybridization with a probe recognizing the genomic place in close proximity to where the FRT site has been inserted.

4.4 Notes

1. Alkaline running buffer should be mixed fresh each time.
2. It is recommended to load a marker composed of different sizes of DNA recognized by the probe on the gel. We use genomic DNA purified from wild-type cells as template in the PCR reaction.

5. DSB ASSAY USING AGAROSE PLUGS TO DETECT DSB FORMATION AND REPAIR

To monitor single-ended DSB formation during DNA replication, we use a slightly modified protocol from Maringele and Lydall (2006). Embedding cells in agarose plugs prior to genomic DNA isolation allow the visualization of DNA intermediates generated during either replication or repair. Furthermore, this procedure also minimizes additional strand breakage at the FRT site, when a nick is already formed.

5.1 Equipment

- Equipment to run agarose gels and visualize these
- Plug molds (Biorad or equivalent)
- Heating block for enzyme digestion
- Microcentrifuge and centrifuge for 50-mL tubes
- Shaking incubator
- Hybridization oven

5.2 Buffers and Reagents

1. Media for growth and alpha factor (see Section 3.2).
2. 0.5 *M* EDTA pH 8.0. Adjust pH with NaOH pellets (~20 g) and autoclave. Make 1000 mL.
3. 0.125 *M* EDTA pH 7.0. Dilute 0.5 *M* EDTA solution and adjust pH to 7.0 with concentrated HCl. Make 500 mL and autoclave.
4. 50 m*M* EDTA pH 8.0. Dilute 0.5 *M* EDTA solution and adjust pH to 7.0. Make 1000 mL and autoclave.
5. SCE solution: 1 *M* sorbitol, 0.1 *M* sodium citrate, and 60 m*M* EDTA. Adjust pH to 7.0 with NaOH. Make 100 mL and autoclave.

6. 1.2% LMP agarose in 125 m*M* EDTA pH 7.0. Make right before use and store at 50°C.
7. Agarose.
8. Zymolyase powder 20T (20,000 U/g). Store at 4°C.
9. Zymolyase solution: 1 mL SCE solution, 20 mg Zymolyase 20T, 50 μL β-mercaptoethanol. Make fresh just prior to use.
10. 1 *M* Tris–HCl pH 8.0. Adjust pH with about 40 mL of concentrated HCl and autoclave.
11. ETB solution: For 42 mL ETB solution: 36 mL 0.5 *M* EDTA pH 8.0, 4 mL 1 *M* Tris–HCl pH 8.0, and 2 mL β-mercaptoethanol. Make fresh just prior to use.
12. Proteinase K powder.
13. RNase A (10 mg/mL).
14. 10% *N*-lauroylsarcosine sodium salt in 0.5 *M* EDTA, pH 8.0. Store at RT and do not autoclave.
15. Proteinase solution: For 40.4 mL: 36 mL 0.5 *M* EDTA, 4 mL 10% *N*-lauroylsarcosine, 40 mg Proteinase K, 4 mg RNase A (400 μL of 10 mg/mL).
16. Tris–EDTA (TE) buffer pH 8.0.
17. Storage solution: For 15 mL: 13.5 mL 0.5 *M* EDTA and 1.5 mL 1 *M* Tris–HCl, pH 8.0.
18. 10 mg/mL EtBr stock. 0.5 μg/mL is used for the staining.
19. Restriction enzyme for digestion of plugs.
20. TBE running buffer.
21. Standard buffers for Southern blotting and hybridization.

5.3 Procedure

5.3.1 Growth of Yeast Cells

Grow the yeast cells as described in Section 3.3 with the following modifications:

1. Dilute to 1.5×10^7 cells/mL in 185 mL YP + raffinose pH 3.5. Let the cells grow for ~2 h at 30°C until the cells start budding (approx. 2.5×10^7 cells/mL). Take out 20 mL random sample.
2. Take out 20 mL cells from the G1-arrested, galactose-induced culture. This sample is used as a negative control, where no break should be detected.
3. Wash the cells once in cold water (spin down 3′ at 3000 rpm) and release into 25 mL preheated YPD at 25°C.

4. Take out 3 mL aliquots at different times after release. Use two eppendorf tubes for each sample.

5.3.2 Treatment of Aliquots/Plugs

1. Spin 30 s at 3000 rpm (or spin down 2′ at 3000 rpm in 50-mL tube for the large samples).
2. Remove supernatant.
3. Add 1 mL of 50 m*M* EDTA pH 8.0, to one of the pellets, resuspend it, and combine it with the cell pellet from the other tube.
4. Spin 30 s at 3000 rpm and remove the supernatant.
5. Add 50 μL of Zymolyase solution (keep cold at 4°C).
6. Add 100 μL of 50 m*M* EDTA pH 8.0 and dissolve the pellet.
7. Incubate at 42°C for 10 s.
8. Add 250 μL LMP agarose and mix by pipetting.
9. Pipet 80 μL into each of the five wells in the plug molds, making five plugs in total pr sample.
10. Put the plug mold into the cold room (4°C) for minimum 15 min.
11. Use the mini spatula to push the five plugs into 2-mL eppendorf tubes (2–3 in each).
12. Add 1 mL ETB solution to each 2-mL tube.
13. Incubate o/n at 37°C.
14. ETB solution. Wash plugs twice with 2 mL of 50 m*M* EDTA.
15. Add 1 mL proteinase solution to each 2-mL eppendorf tube. Incubate for at least 6 h to o/n at 37°C.
16. Remove the Proteinase solution, wash the plugs with 2 mL of 50 m*M* EDTA, and incubate in 2 mL 1 × TE for 1 h at 37°C.
17. Store the plugs in 1 mL storage solution at 4°C. If you want to use the plugs right away continue with the digest. The plugs can be stored for weeks in 1 mL storage solution at 4°C.

5.3.3 Digestion of Plugs

1. Transfer plugs into 2-mL autoclaved eppendorf tubes with 1 mL 1 × TE buffer and incubate at 4°C for 30 min. Repeat this step.
2. Exchange the buffer with 170 μL 1 × restriction buffer compatible with the restriction enzyme to be used. Incubate 30 min at 4°C.
3. Remove the buffer and add 170 μL enzyme mastermix to each plug, so it will contain: 5 μL restriction enzyme (with a conc. of 10 U/μL), 25 μL 10 × restriction buffer, 140 μL autoclaved water.
4. Incubate o/n at 37°C.

5. Add 2 μL enzyme, mix, and incubate at 37°C for 2 h (optional, depends on the efficiency of the restriction enzyme).
6. Remove enzyme mix and add 1 mL 1 × TE buffer and incubate for 1 h at room temperature.
7. Exchange the buffer with 1 mL new 1 × TE buffer and incubate at 4°C for at least 1 h to make the plugs more solid and easier to load on the gel. Keep plugs on ice when loading them into the wells of the gel (see Section 5.3.4 point 5).

5.3.4 Gel Casting and Gel Electrophoresis

1. Make 3 L 1 × TBE buffer and pour into the running chamber placed in the cold room.
2. Put autoclave tape on the gel chamber. Put in the slot former to use for plugs.
3. Make a 400 mL 0.7% agarose gel (can also use 1%). No ethidium bromide added. Our gel has a dimension of 20 × 23.5 cm.
4. Keep 1–2 mL of the melted agarose and keep it in a 60°C heat block in an eppendorf tube. This is used for sealing the wells after the plugs have been inserted.
5. The plugs are gently inserted into the wells with the help of a mini spatula and an inoculating stick. Avoid air bubbles in the wells. The wells are sealed with the agarose that has been kept at 60°C.
6. Run the gel at 4°C for 22 h at 50 V. Load 20–25 ng of a marker with DNA fragments recognized by the probe and a 1 kb marker in the first two wells.
7. Set up for Southern blotting (use standard Southern blot protocol) and afterward hybridization with a probe recognizing the genomic place in close proximity to where the FRT site has been inserted.

6. CHIP WITHOUT CROSSLINKING TO DETECT FlpH305L BINDING AND CLEVAGE AT THE FRT SITE

In order to verify that FlpH305L gets covalently attached to the FRT site, and thus cleavage occurs, ChIP can be performed. In the general ChIP procedure formaldehyde is used to crosslink proteins to DNA, but as FlpH305L forms long-lived covalent protein–DNA complexes upon DNA cleavage, formaldehyde treatment is omitted. We have successfully performed ChIP using an Flp-nick yeast strain where FlpH305L has been HA tagged. The ChIP assay can also be used to analyze enzymes involved in the removal of protein–DNA, as deleting genes encoding enzymes engaged in Flp removal will result in persistent FlpH305L at the FRT site over a time course.

6.1 Equipment

1. Magnetic rack
2. 2-mL tubes with screw cap
3. Glass beads, acid washed
4. Homogenizer to crack cells
5. Microcentrifuge and centrifuge for 50-mL tubes
6. Sonication tubes
7. Needle
8. Bunsen burner
9. Sonicator
10. Thermoshaker
11. Safe-lock tubes
12. Water bath or incubator at 65°C
13. qPCR machine

6.2 Buffers and Reagents

1. Media for growth and alpha factor (see Section 3.2).
2. 10× PBS: 8% NaCl, 0.2% KCl, 1.42% $Na_2HPO_4 \cdot 2H_2O$, 0.2% KH_2PO_4. Adjust pH to 7.0. Autoclave and store at 4°C.
3. Lysis buffer: For 500 mL: 25 mL 1 *M* Hepes pH 7.0, 14 mL 5 *M* NaCl, 1 mL 0.5 *M* Na_2EDTA pH 8.0, 50 mL 10% Triton X-100, and add water to 500 mL. Use autoclaved Hepes, NaCl, and Na_2EDTA. Add 0.025 g sodium deoxycholate and protease inhibitors per 25 mL lysis buffer before the ChIP procedure is started. Store the lysis buffer at 4°C.
4. Wash buffer: For 500 mL: 5 mL 1 *M* Tris–HCl pH 8.0, 50 mL 5 *M* NaCl, 25 mL 10% Igepal, 1 mL 0.5 *M* Na_2EDTA pH 8.0, and add water to 500 mL. Use autoclaved Tris–HCl, NaCl, and Na_2EDTA. Add 0.12 g sodium deoxycholate and protease inhibitors per 25 mL wash buffer before the ChIP procedure is started. Store the wash buffer at 4°C.
5. Protease inhibitor (Pierce Protease Inhibitor from Thermo Fisher or equivalent).
6. Albumin from bovine serum (BSA) powder.

7. Magnetic dynabeads: Sheep antimouse IgG with a concentration of 4×10^8 beads/mL (from Invitrogen).
8. Primary antibody, e.g., HA-probe mouse monoclonal (200 μg/mL).
9. 1 × TE buffer: 10 m*M* Tris–HCl pH 7.5, 1 m*M* Na_2EDTA pH 8.0.
10. 1 × TE + 1% SDS.
11. Proteinase K (20 mg/mL).
12. 5 *M* lithium acetate.
13. Phenol/chloroform/isoamyl alcohol (25:24:1).
14. Ethanol: 96% and 70%.
15. Glycogen solution from Roche or equivalent.
16. 3 *M* sodium acetate pH 5.2.
17. SYBR Green qPCR kit or equivalent.
18. Primers for qPCR.

6.3 Procedure

6.3.1 Growth of Yeast Cells

Grow the cells as described in Section 3.3 with the following modifications:

1. Grow cells until 1.0×10^7 cells/mL.
2. Take out 25 mL from the random growing culture. This is used as a negative control in the ChIP assay. FlpH305L should not be detected at the FRT in these cells.
3. Optional: After induction take out 1 mL of culture for a trichloroacetic acid preparation. Perform Western blotting on this sample to verify induction of FlpH305L-3HA.
4. Take out 25 mL of culture at different time points during the replication process, e.g., every 20 min for up to 160 min.
5. Spin down the samples, wash in 1 mL 1 × PBS, and store at −20°C in 2-mL tubes with screw cap until the ChIP procedure is performed.

6.3.2 Preparation of Buffers for ChIP

1. Add sodium deoxycholate monohydrate and protease inhibitors only to the volume of lysis- and wash buffer that is needed during the experiment (see note 1).
2. Dissolve BSA in 1 × PBS. Usually it is enough to make 10–15 mL.

6.3.3 Coupling of Antibody to Beads

We use magnetic dynabeads. For each sample taken during the time course 100 μL beads are used (e.g., 10 samples = 1 mL of beads).

1. Wash the beads in 1 mL 1 × PBS + 5 mg/mL BSA by pipetting up and down and then placing the tube in a magnetic rack and allowing the beads to move to the magnet before removing the buffer. Repeat this wash three times.
2. After wash dissolve the beads in 1 mL 1 × PBS + 5 mg/mL BSA and divide into two tubes of 500 μL each.
3. Add antibody to one of the tubes. The amount of antibody used will depend on the antibody. We use 5 μL of HA-probe mouse monoclonal/ 50 μL original volume beads (e.g., 10 samples = 20 μL antibody).
4. Leave the two tubes (beads with antibody and beads without antibody) to rotate on a wheel for 2 h at 4°C.
5. After coupling wash beads by pipetting up and down.
6. 1 × wash buffer (0.8 mL).
7. 2 × lysis buffer (0.8 mL).
8. Add lysis buffer to beads (original volume) and add 45 μL to each sample of extract.

6.3.4 Making the Extract

During the preparation of the extract it is important to keep the samples cold. Thus prepare everything on ice and make sure that the buffers are cold before use. To each sample add glass beads (approximately twice as many as the volume of the cells) and 700 μL lysis buffer.

1. Crack the cells. The time needed to break open ~80%–90% of the cells will depend on the machine used for cracking. Breakage can be checked under microscope. Cracked cells will look transparent.
2. Heat up a needle using a bunsen burner and quickly make a hole at the bottom of the tube to spin out the extract. Place the tube in another tube and spin at 3800 × *g* for 5 min at 4°C.
3. Transfer the extract, including the pellet, to an eppendorf tube and spin at 13,000 rpm in a microcentrifuge for 15 min at 4°C.
4. Remove the supernatant. Keep 40 μL of the supernatant in another tube for Western blotting (see note 2).
5. Dissolve the pellet in 1000 μL lysis buffer by pipetting up and down. The pellet is hard so it may take some time. When dissolved transfer the extract to sonication tube.
6. Sonicate the samples to obtain DNA fragments of 500–1000 bp (see note 3).
7. Transfer the extract to a new eppendorf tube and add 400 μL lysis buffer. Spin at 7000 rpm in a microcentrifuge for 2 min at 4°C.

8. From each supernatant take out 40 μL for the Input sample. Optional: 40 μL can also be stored for later Western blotting.
9. Divide the extract from each sample into two eppendorf tubes (2 × 600 μL). The washed beads are now ready to be added. Add 45 μL of beads to the samples. For each sample: one eppendorf tube with extract will get beads coupled with antibody and the other tube beads without antibody.
10. Incubate on a rotating wheel for 2 h at 4°C.
11. After the incubation place the samples into the magnetic rack. Optional: take out 40 μL for Western blotting from selected +antibody and −antibody samples. Wash the samples twice in lysis buffer, once in wash buffer, and once in 1 × TE buffer. Use a volume of 750 μL. Between each washing step incubate the samples for 5 min with shaking of 1000 rpm at 4°C.
12. Disrupt the coupling between beads and antibody by adding 60 μL 1 × TE buffer + 1% SDS and shake at 1100 rpm at 65°C for 10 min.
13. Transfer the supernatant into a safe-lock tube (see note 4) and repeat step 12. Optional: Take 5 μL out of selected +ab and −ab samples for Western blotting.
14. Combine the supernatants from the two rounds of uncoupling (the volume is now ~120 μL).
15. To reverse the crosslink add 210 μL 1 × TE buffer + 1% SDS to the Input samples (these were kept in step 8) and 130 μL 1 × TE buffer + 1% SDS to the +ab and −ab samples and incubate o/n at 65°C, e.g., in a hybridization oven or water bath.
16. Next day treat the samples with proteinase K for 2 h at 37°C by adding 240 μL 1 × TE buffer and 20 μL proteinase K (20 mg/mL) to each sample.
17. Isolate the DNA by adding 50 μL 5 *M* lithium acetate and 450 μL phenol/chloroform.
18. Vortex the samples and spin for 5 min at 13,000 rpm in a microcentrifuge.
19. Transfer the water phase (~480 μL) to a new eppendorf tube and add 1 mL of ethanol, glycogen, and sodium acetate solution. To make a stock for 25 samples mix 25 mL 96% ethanol, 50 μL glycogen, and 1.1 mL 3 *M* sodium acetate pH 5.2.
20. Precipitate at −80°C for at least 2 h or o/n. Precipitation takes longer for small DNA fragments.
21. Spin the samples for 30 min at 13,000 rpm at 4°C in a microcentrifuge and carefully remove the supernatant. The pellet is small and may be very loose.

22. Add 200 μL 70% ethanol to the pellet and spin for 2 min at 13,000 rpm in a microcentrifuge.
23. Dry the samples and dissolve the pellets in 60 μL H_2O.

6.3.5 Quantitative PCR

Recovered DNA can now be analyzed for the enrichment of the FRT site by quantitative PCR. We use two different primer sets, one locates close to the FRT site in the genomic region where FRT has been inserted, and one locates to a telomeric region. The last one serves as the negative control primer in our setup. The primer concentrations are optimized to obtain efficiencies as close to 100% as possible, meaning that the amount of DNA is doubled in every qPCR cycle. We conduct our quantitative PCR using SYBR green, thus primers are designed to give PCR products of equal lengths. It might be necessary to dilute the Input and +ab samples 10–100 times to make the amount of DNA fit in the area of the standard curve where the primers have 100% efficiency. Remember to take the dilution into account when calculating fold increases.

1. We use the following qPCR program using Stratagenes MX3000P with SYBR Green master mix:
   a. Segment 1: 95°C for 15 min.
   b. Segment 2: 95°C for 30 s, 56°C for 60 s, 72°C for 40 s. Repeat this for 40 cycles.
2. The fold increase is calculated by this formula: $2^{-ct\ \text{value}+ab} - 2^{-ct\ \text{value}-ab}/2^{-ct\ \text{value input}}$, assuming that the primer efficiency is 100%.

6.4 Notes

1. Add the protease inhibitors to the wash buffer right before it is required to wash the extract incubated with beads.
2. Only take out samples for Western blotting from one or two samples where presence of FlpH305L-3HA will be expected. The molecular weight of FlpH305L with the 3HA tag will be of the same size as the long IgG chain from the primary antibody, and it can therefore be difficult to see FlpH305L on the Western blot in the samples with antibody.
3. We use an S2 sonicator from Covaris with the following sonication program: Duty 20%, intensity 8, and cycles/burst 400.
4. Use safe-lock tubes as phenol/chloroform extraction will be carried out in these tubes the next day.

REFERENCES

Anand, R. P., Lovett, S. T., & Haber, J. E. (2013). Break-induced DNA replication. *Cold Spring Harbor Perspectives in Biology*, *5*(12), a010397. https://doi.org/10.1101/cshperspect.a010397.

Andrews, B. J., Proteau, G. A., Beatty, L. G., & Sadowski, P. D. (1985). The FLP recombinase of the 2 micron circle DNA of yeast: Interaction with its target sequences. *Cell*, *40*(4), 795–803.

Chan, K. M., Liu, Y. T., Ma, C. H., Jayaram, M., & Sau, S. (2013). The 2 micron plasmid of Saccharomyces cerevisiae: A miniaturized selfish genome with optimized functional competence. *Plasmid*, *70*(1), 2–17. https://doi.org/10.1016/j.plasmid.2013.03.001.

Chen, J. W., Lee, J., & Jayaram, M. (1992). DNA cleavage in trans by the active site tyrosine during Flp recombination: Switching protein partners before exchanging strands. *Cell*, *69*(4), 647–658.

Chen, J. W., Yang, S. H., & Jayaram, M. (1993). Tests for the fractional active-site model in Flp site-specific recombination. Assembly of a functional recombination complex in half-site and full-site strand transfer. *The Journal of Biological Chemistry*, *268*(19), 14417–14425.

Gietz, R. D., & Woods, R. A. (2002). Transformation of yeast by lithium acetate/single-stranded carrier DNA/polyethylene glycol method. *Methods in Enzymology*, *350*, 87–96.

Lee, J., Jayaram, M., & Grainge, I. (1999). Wild-type Flp recombinase cleaves DNA in trans. *The EMBO Journal*, *18*(3), 784–791.

Longtine, M. S., McKenzie, A., 3rd, Demarini, D. J., Shah, N. G., Wach, A., Brachat, A., et al. (1998). Additional modules for versatile and economical PCR-based gene deletion and modification in Saccharomyces cerevisiae. *Yeast*, *14*(10), 953–961. https://doi.org/10.1002/(SICI)1097-0061(199807)14:10<953::AID-YEA293>3.0.CO;2-U.

Maringele, L., & Lydall, D. (2006). Pulsed-field gel electrophoresis of budding yeast chromosomes. *Methods in Molecular Biology*, *313*, 65–73. https://doi.org/10.1385/1-59259-958-3:065.

Mayle, R., Campbell, I. M., Beck, C. R., Yu, Y., Wilson, M., Shaw, C. A., et al. (2015). DNA REPAIR. Mus81 and converging forks limit the mutagenicity of replication fork breakage. *Science*, *349*(6249), 742–747. https://doi.org/10.1126/science.aaa8391.

Nielsen, I., Bentsen, I. B., Lisby, M., Hansen, S., Mundbjerg, K., Andersen, A. H., et al. (2009). A Flp-nick system to study repair of a single protein-bound nick in vivo. *Nature Methods*, *6*(10), 753–757. https://doi.org/10.1038/nmeth.1372.

Parsons, R. L., Prasad, P. V., Harshey, R. M., & Jayaram, M. (1988). Step-arrest mutants of FLP recombinase: Implications for the catalytic mechanism of DNA recombination. *Molecular and Cellular Biology*, *8*(8), 3303–3310.

Pommier, Y. (2006). Topoisomerase I inhibitors: Camptothecins and beyond. *Nature Reviews. Cancer*, *6*(10), 789–802. https://doi.org/10.1038/nrc1977.

Pommier, Y., Barcelo, J. M., Rao, V. A., Sordet, O., Jobson, A. G., Thibaut, L., et al. (2006). Repair of topoisomerase I-mediated DNA damage. *Progress in Nucleic Acid Research and Molecular Biology*, *81*, 179–229. https://doi.org/10.1016/S0079-6603(06)81005-6.

Qian, X. H., Inman, R. B., & Cox, M. M. (1990). Protein-based asymmetry and protein-protein interactions in FLP recombinase-mediated site-specific recombination. *The Journal of Biological Chemistry*, *265*(35), 21779–21788.

Schoeffler, A. J., & Berger, J. M. (2008). DNA topoisomerases: Harnessing and constraining energy to govern chromosome topology. *Quarterly Reviews of Biophysics*, *41*(1), 41–101. https://doi.org/10.1017/S003358350800468X.

Tsalik, E. L., & Gartenberg, M. R. (1998). Curing *Saccharomyces cerevisiae* of the 2 micron plasmid by targeted DNA damage. *Yeast, 14*(9), 847–852. https://doi.org/10.1002/(SICI)1097-0061(19980630)14:9<847::AID-YEA285>3.0.CO;2-9.

Voziyanov, Y., Lee, J., Whang, I., Lee, J., & Jayaram, M. (1996). Analyses of the first chemical step in Flp site-specific recombination: Synapsis may not be a pre-requisite for strand cleavage. *Journal of Molecular Biology, 256*(4), 720–735.

Wang, J. C. (2002). Cellular roles of DNA topoisomerases: A molecular perspective. *Nature Reviews. Molecular Cell Biology, 3*(6), 430–440. https://doi.org/10.1038/nrm831.

CHAPTER TWO

A Proximity Ligation-Based Method for Quantitative Measurement of D-Loop Extension in *S. cerevisiae*

Aurèle Piazza*,†, Romain Koszul†, Wolf-Dietrich Heyer*,1
*University of California, Davis, Davis, CA, United States
†Institut Pasteur, Paris, France
1Corresponding author: e-mail address: wdheyer@ucdavis.edu

Contents

Abstract

Homologous recombination faithfully restores the sequence information interrupted by a DNA double-strand break by referencing an intact DNA molecule as a template for repair DNA synthesis. DNA synthesis is primed from 3′-OH end of the invading DNA strand in the displacement loop (D-loop). Here, we describe a simple and quantitative proximity ligation-based assay to study the initiation of homologous recombination-associated DNA synthesis initiated at the D-loop and final product formation. The D-loop extension assay overcomes the semiquantitative nature and some limitations of the current PCR-based technique and facilitates the study of the recombination-associated DNA synthesis.

Methods in Enzymology, Volume 601
ISSN 0076-6879
https://doi.org/10.1016/bs.mie.2017.11.024

1. INTRODUCTION

1.1 DNA Synthesis in Homologous Recombination

Homologous recombination (HR) faithfully repairs complex DNA damage including DNA double-stranded breaks (DSBs) by referencing an intact donor template in the form of the sister chromatid, a homolog, or an ectopic sequence. DNA strand invasion posits the 3′-OH of the invading DNA strand on the donor template (Fig. 1A). HR-associated DNA synthesis recovers the sequence information disrupted by the DSB. The extended D-loop is the last intermediate common to all HR subpathways (double Holliday Junction, Synthesis-Dependent Strand Annealing, Break-Induced Replication (BIR)) and their associated repair outcome (Heyer, 2015).

1.2 Current Method for Detection of Initiation of Recombination-Associated DNA Synthesis and Its Limitations

The kinetics and regulation of multiple steps of HR up to the repair product can be studied in detail following the controlled and synchronous induction of a site-specific DSB in *Saccharomyces cerevisiae* (White & Haber, 1990). The initiation of DNA synthesis has so far been studied semiquantitatively using PCR-based assays. These assays score the amplification of the unique sequence produced upon DNA synthesis templated from an ectopic sequence, with one primer upstream of the region of homology on the broken molecule and the other downstream of the region of homology on the donor (Aylon, Liefshitz, Bitan-Banin, & Kupiec, 2003; Hicks, Yamaguchi, & Haber, 2011; White & Haber, 1990). This approach has some inherent limitations. First, semiquantitative PCR methods are labor intensive and imprecise. Second, the length of the fragment generated by the PCR imposes a limit on the length of homology that can be provided between the broken molecule and the donor, as long ectopic homologies cause PCR artifacts. Since the length of homology dictates the efficiency of the upstream HR steps such as homology search and D-loop stability (Forget & Kowalczykowski, 2012) with consequences for the crossover/noncrossover HR outcome (Inbar, Liefshitz, Bitan, & Kupiec, 2000; Jinks-Robertson, Michelitch, & Ramcharan, 1993; Mehta, Beach, & Haber, 2017), this experimental approach constrains the study of recombinational DNA synthesis to particular substrates. Third, in the case of BIR, the kinetics of DNA synthesis initiation and final product formation, which

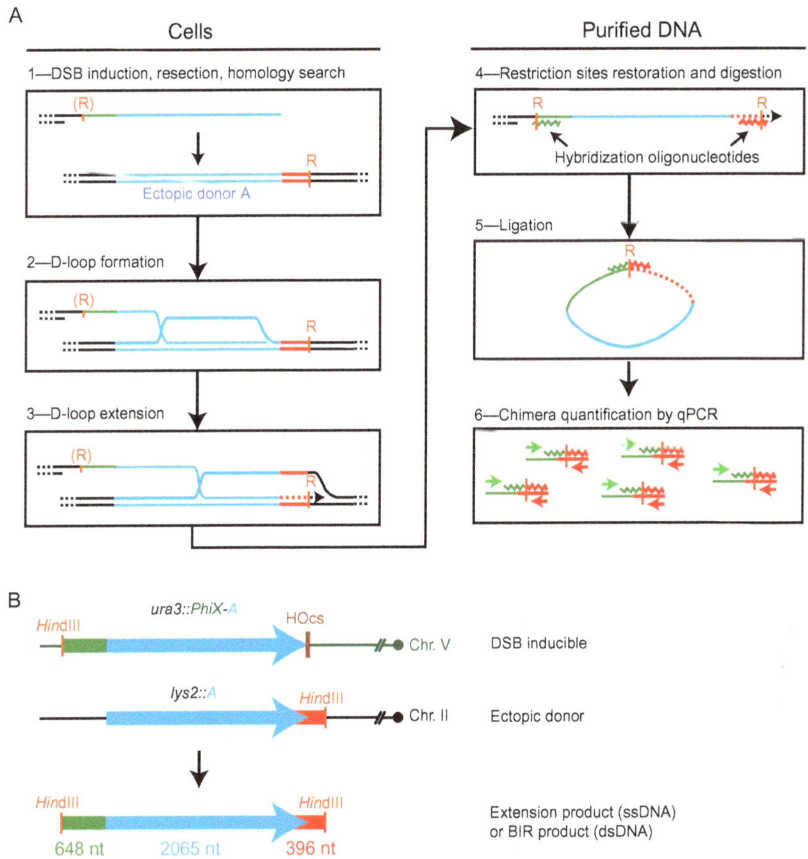

Fig. 1 Rationale of the D-loop extension assay. (A) (1) Site-specific DSB induction and resection enables homology search and (2) DNA strand invasion of an ectopically positioned donor. (3) Priming of DNA synthesis from the 3′-OH end of the invading molecule will lead to the copy of the restriction site located downstream of the donor. The two compatible restriction sites are now present on the same ssDNA molecule. (4) Following DNA extraction and restoration of the restriction sites upon annealing of long oligonucleotides, the extended ssDNA can be digested. The product of this digestion is a ssDNA with short dsDNA extremities. (5) Proximity ligation of these extremities, achieved in dilute conditions, produces a unique chimeric molecule. (6) The amount of this chimera is determined by quantitative PCR, and is the readout of the amount of D-loops extended past this restriction site on the donor in the cell population. R denotes a restriction site in its dsDNA form and (R) a restriction site in its ssDNA form (uncuttable). (B) Experimental system in haploid *S. cerevisiae* used to study initiation of DNA synthesis of an ectopic donor located at *LYS2* on chromosome (Chr.) II upon site-specific DSB induction at *URA3* on Chr. V. The DSB is repairable only by BIR, which in this context is inviable so as to prevent dilution of repair intermediates by fast repair events that would proliferate. The extent of DNA synthesis measured in this assay is at least 405 nt (396 nt up to the *Hin*dIII site, 6 nt of restriction site, and 3 nt of overhang to enable cutting by the enzyme).

entails several hundred kilobases of DNA synthesis, is not temporally resolved by these assays (Jain et al., 2009; Lydeard, Jain, Yamaguchi, & Haber, 2007; Malkova, Naylor, Yamauchi, Ira, & Haber, 2005). This is unexpected given that the conservative DNA synthesis of BIR produces long-lived newly synthesized ssDNA that can be detected by physical means (Saini et al., 2013) and that one study showed a correlation between length of DNA synthesis required for BIR completion and the time of appearance of the dsDNA BIR product (Donnianni & Symington, 2013). These results indicate a delay between DNA synthesis initiation and final BIR product formation. Finally, detection of DNA synthesis initiation by PCR-based assays was found to be dependent on Pol32 (a nonessential Polδ subunit), while DSB repair by gene conversion, which requires DNA synthesis, is not (Jain et al., 2009; Lydeard et al., 2010). It suggests that these methodologies either fail to detect the ssDNA produced upon D-loop extension (DLE) or disproportionately detect the final dsDNA product.

These considerations prompted the development of an alternative and independent methodology for the study of HR-associated DNA synthesis. We developed a quantitative assay based on the proximity ligation principle common to chromosome conformation techniques that overcomes these limitations (Fig. 1A) (Dekker, Rippe, Dekker, & Kleckner, 2002). We also discuss possible refinements of this methodology.

2. METHODS

2.1 Materials

2.1.1 Equipment

- Standard yeast culture equipment (120 rpm shaker and rotator at 30°C, 100- and 250-mL culture flasks, 15- and 50-mL conical tubes).
- Dry bath at 30°C, 37°C, 55°C, and 65°C.
- Water bath at 16°C.
- A quantitative PCR instrument (in our case a Roche LightCycler 96).
- Benchtop centrifuge with 50-mL conical tubes adaptors.
- Microcentrifuge.
- Vortex.

2.1.2 Reagents

- YPD medium: 1% (w/v) of Bacto yeast extract (BD Biosciences), 2% (v/w) Bacto peptone (BD Biosciences), 2% (v/w) D-glucose. Autoclave and store at room temperature (RT).

- YEP-lactate medium: 1% (w/v) of Bacto yeast extract (BD Biosciences), 2% (v/w) Bacto peptone (BD Biosciences), 2% Sodium DL-lactate (60%, Sigma-Aldrich). Autoclave and store at RT.
- Galactose 20% (w/v). Dissolve by stirring upon moderate heating. Filter sterilize and store at RT.
- Spheroplasting buffer: 0.4 *M* sorbitol, 0.4 *M* KCl, 40 m*M* Sodium Phosphate Buffer pH 7.2, 0.5 m*M* $MgCl_2$. Filter sterilize and store at 4°C.
- Zymolyase 100T solution: 2% D glucose, 50 m*M* Tris–HCl pH 7.5, 5 mg/mL Zymolyase 100T (US Biological). Vortex well and store up to 3 months at 4°C.
- Cutsmart buffer 10 ×: 500 m*M* Potassium Acetate, 200 m*M* Tris–Acetate pH 8.0, 100 m*M* Magnesium Acetate, 1 mg/mL BSA. Store at 4°C or −20°C.
- *Hin*dIII-HF 20,000 U/mL (NEB, R3104L).
- Ligation buffer 10 ×: 500 m*M* Tris–HCl pH 8.0, 100 m*M* $MgCl_2$, 100 m*M* DTT, 10 m*M* ATP pH 8.0, 25 μg/mL BSA. Store at −20°C.
- Ligation mix (660 μL per sample): Ligation buffer 1.2 ×, 82.5 μg/mL BSA, 7 μg/mL (Bayou Biolabs), or 240 U/mL (NEB M0202L) DNA T4 ligase. Make fresh right before use.
- 10% Triton X-100.
- 1% and 10% SDS solution.
- 25:24:1 Phenol:chloroform:isoamyl alcohol.
- 3 *M* Sodium Acetate pH 5.2.
- 100% Isopropanol.
- 70% Ethanol.
- TE pH 8.0.
- RNase A 10 mg/mL (Promega, cat. EN0531).
- Proteinase K 10 mg/mL.
- Power SYBR Green PCR Master Mix (Thermo Fisher Scientific, cat. 4367659).
- LightCycler Multiwell plate 96 white (Roche, cat. 04729692001).
    - Hybridization oligonucleotides

      The "hybridization oligonucleotides" (ol2007 and ol2046; Table 1) are used to restore the restriction sites on the ssDNA produced by resection and DNA synthesis (in the 5′ and 3′ unique sequences flanking the region of homology "A," respectively). They must be properly oriented and encompass the restriction site with at least ~80 nt internal to the restriction fragment, in order to remain bound during the enzyme inactivation steps (55°C) and up to the ligation step of the procedure.

Table 1 Primers Used in This Study.

Name	Sequence (5′–3′)	Purpose
Quantitative PCR		
olWDH1760	CAGCGGGCTTGCAGAAGTTG	Reference locus (*ARG4*, Chr. VIII); used for copy number reference and dsDNA loading control in all quantitative PCR experiments
olWDH1761	GGCCAATTAGTTCACCAAGACG	
olWDH2052	ATGTGCCTTCCTACCGCTC	Quantify the intramolecular ligation (i.e., circularization) efficiency of a control 765 bp *Hin*dIII fragment (*YLR050C*; Chr. XII)
olWDH2053	TCAAGCGTGGTTACATTCCTTAC	
olWDH1766	GTTTCAGCTTTCCGCAACAG	Measure DSB formation at the HOcs. DSB induction causes a loss of signal
olWDH1767	GGCGAGGTATTGGATAGTTCC	
olWDH2010	TGCTCGGAGATTACCGAATC	Quantify the *Hin*dIII cutting efficiency on the broken molecule (resected)
olWDH2012	CGAGGCATATTTATGGTGAAGG	
olWDH2009	CACCACTTTGCCATTCAACAC	Quantify the *Hin*dIII cutting efficiency downstream of the donor
olWDH2011	TGCGAGGTTTTCTTGGTCAG	
olWDH2009	CACCACTTTGCCATTCAACAC	Quantify the chimera formed upon circularization of the extended invading molecule (DLE signal). To be normalized on the circularization efficiency
olWDH2010	TGCTCGGAGATTACCGAATC	
HindIII restriction sites restoration		
olWDH2007	TCTGCTCGGAGATTACCGAATCAAAAAAATTTCAAAGAAACCGGAATCAAAAAAAAGAACAAAAAAAAAAAAGATGAATTGAAAAGCTTTATGGACCgac	Hybridize at the *Hin*dIII site in the *URA3* promoter on the resected broken molecule
olWDH2046	AATCTTTGTGAAGCTTCGCAAGTATTCATTTTAGACCCATGGTGGAACCCTAGTGTTGAATGGCAAAGTGGTGATAGAGTTCATAGAATTGGTCAGTAT	Hybridize at the *Hin*dIII site downstream of the A donor on the extended invading molecule

- Quantitative PCR primers

  The qPCR primers must have similar T_m and be positioned so as to generate amplicons of similar size (100–180 bp). The primers used in this study are listed in Table 1 and their purpose described below. Amplification efficiency must be determined for each pair of primers.

2.1.3 S. cerevisiae *Strains*

The genotype of the wild-type (WDHY5511), *rad51Δ* (WDHY5573), and *pol32Δ* (WDHY5574) haploid *S. cerevisiae* strains (W303 *RAD5* background) used in this study is listed in Table 2. The *HO* gene (which encodes the site-specific HO endonuclease) under the control of the *GAL1* promoter has been inserted at the *trp1-1* locus alongside the *hphMX* marker to allow for controlled HO endonuclease expression upon galactose addition in the culture media. The HO-cut site (HOcs) at *MAT* on chromosome (Chr.) III has been mutated (*MAT*-inc) to preclude cutting at *MAT*. The DSB-inducible construct and the donor locus are depicted in Fig. 1B, and their annotated sequences are provided in Dataset S1 in the online version at https://doi.org/10.1016/bs.mie.2017.11.024. The HOcs together with a 2-kb-long sequence "A" (2065 bp of the *LYS2* gene, coordinates 4-2068) and a 327-bp fragment of the PhiX genome flanked by multiple restriction sites replaces the *URA3* gene on Chr. V. The donor consists of the A sequence replacing the *LYS2* gene on Chr. II. These two loci have been extensively used by others to study DSB repair by HR (Inbar & Kupiec, 1999; Mine-Hattab & Rothstein, 2012). They are located in the middle of chromosome arms and represent unconstrained and untethered chromosomal regions (Agmon, Liefshitz, Zimmer, Fabre, & Kupiec, 2013).

Table 2 Relevant Genotype of Haploid *Saccharomyces cerevisiae* Strains Used in This Study

Strain	Relevant Genotype
WDHY5511	*ura3::A-HOcs, lys2::A, trp1::GAL-HO-hphMX, his3D200, can1-100, leu2-3,112, ade2-1, RAD5*
WDHY5573	*ura3::A-HOcs, lys2::A, trp1::GAL-HO-hphMX, his3D200, can1-100, leu2-3,112, ade2-1, RAD5 rad51::LEU2*
WDHY5574	*ura3::A-HOcs, lys2::A, trp1::GAL-HO-hphMX, his3D200, can1-100, leu2-3,112, ade2-1, RAD5 pol32::KanMX*

2.2 The D-Loop Extension Assay

2.2.1 Rationale

The rationale of the DLE assay is depicted in Fig. 1A. It relies on the proximity ligation principle (Dekker et al., 2002) that the collision (and ligation) probability of two physically tethered DNA extremities will remain unaffected by the overall strand concentration. Hence, ligation performed in dilute conditions, which disfavor intermolecular ligation, allows detecting physically linked DNA extremities. In the DLE assay, the copying of a restriction site uniquely present in the 3′ flanking region of the donor upon DLE will cause its physical tethering with the restriction site present on the 5′ flanking region of the broken molecule (Fig. 1B). The two restriction sites ablated in ssDNA can be restored upon hybridization of long oligonucleotides (see below). Detection of the final dsDNA repair product does not require hybridization oligonucleotides. This criterion is used to discriminate the initiation of DNA synthesis from the final product formation. The chimeric molecule formed between the 5′ and the 3′ flanking regions of the donor upon intramolecular ligation can be detected by a short quantitative PCR that does not encompass the region of homology (Fig. 1A).

2.2.2 Procedure

2.2.2.1 Pregrowth, Site-Specific DSB Induction, and Sample Collection

- Inoculate 5 mL of YPD with a fresh colony and grow to saturation overnight at 30°C with shacking.
- Dilute in 80 mL YEP-lactate in a 250-mL flask so that the cells are in exponential phase at an OD_{600} ~0.5 the next morning.
- Harvest a "no DSB" sample prior to galactose addition. Add 2% galactose. At each time point, harvest 5×10^7 cells by centrifugation at 4°C. Wash once in cold TE and store the pellet at −20°C.

2.2.2.2 Yeast Cells Spheroplasting and Restriction Sites Restoration

- Resuspend cells in 1 mL of spheroplasting buffer.
- Add 4 μL of Zymolyase 100T solution (20 μg/mL final concentration Zymolyase 100T) and spheroplast cells 20 min at 30°C. This time may need to be adjusted with each Zymolyase 100T solution.
- Wash cells two times in 1 mL of spheroplasting buffer at 2500 × *g*. Prepare 3 mL of Cutsmart 1 × buffer per sample.
- Wash cells two times in 1 mL of Cutsmart 1 × buffer at 16,000 × *g*. These washing steps eliminate phosphatase contaminants of the

Zymolyase solution and gently lyse cells. Prepare 360 μL of Cutsmart 1.4× with and without hybridization oligonucleotides (olWDH2007 and olWDH2046) at a 3.5 n*M* final concentration for each sample.

- Split the last Cutsmart 1× wash in two tubes. Resuspend one pellet in Cutsmart 1.4× with, and the other in Cutsmart 1.4× without, hybridization oligonucleotides. The final cell concentration is ~7000 cells/μL. When present, hybridization oligonucleotides are in a 3000-fold excess over cells and will restore restriction sites present on resected and newly synthesized ssDNA. In the absence of hybridization oligonucleotides, dsDNA products (i.e., final BIR products) can be detected.
- Store the lysed cells at −80°C or proceed immediately.

2.2.2.3 Restriction Digestion and Ligation

- Thaw cells on ice. Prewarm dry baths at 65°C and 37°C. Prewarm a water bath at 16°C.
- Take 36 μL of resuspended spheroplasts (2.5×10^5 cells) in a new microfuge tube. Solubilize chromatin by adding 4 μL of SDS 1% (0.1% final concentration). Mix by vortex.
- Incubate at 65°C for 15 min. Vortex every 5 min.
- Put on ice. Quench SDS by adding 4.5 μL of Triton X-100 10% (1% final concentration). Mix with pipette.
- Digest DNA with 20–50 U of *Hin*dIII-HF for 1 h at 37°C. Finger-vortex every 20 min. Prewarm a dry bath at 55°C.
- Denature *Hin*dIII with 9 μL of SDS 10% (2% final concentration) at 55°C for 10 min. Vortex.
- Put on ice. Quench SDS with 80 μL of Triton X-100 10%. Mix by pipetting up and down.
- Add 660 μL of ligation mix on ice. The cell concentration is $\sim 3 \times 10^3$ cells/μL in 800 μL. Mix by inversion and incubate at 16°C for 90 min. Mix by inversion every 30 min. Thaw Proteinase K on ice.

2.2.2.4 DNA Purification

- Add 20 μg of Proteinase K and vortex. Digest 30 min at 65°C.
- Put on ice. Transfer in a 2-mL microfuge tube.
- Add an equal volume (i.e., 800 μL) of phenol:chloroform:isoamyl alcohol (25:24:1) and vortex thoroughly for at least 30 s.
- Centrifuge 5 min at 16,000 × *g* and carefully recover ~600 μL of the upper phase in a new 1.5-mL microfuge tube.

- Add 1:10 volume of 3 *M* Sodium Acetate pH 5.2 and 1 volume of isopropanol. Mix by inverting several times.
- Incubate at RT for 30 min.
- Pellet DNA at 16,000 × *g* for 15 min at RT. Discard supernatant and drain well. A small pellet may be visible.
- Wash the pellet in 200 μL ethanol 70%. Carefully remove the supernatant with a pipette. Drain well.
- Dry the pellet at 37°C for 15 min.
- Rehydrate the pellet in 50 μL TE pH 8.0 at RT for 30 min. Vortex.
- Add 20 μg of RNase A. Incubate at 37°C for 30 min. Put on ice.

2.2.2.5 Quantitative PCR

- Set up the 20 μL qPCRs according to the Power SYBR Green PCR Master Mix manufacturer instructions. Use 2 μL of DNA (~10^5 genomes) per reaction. Each reaction is run in duplicate. The different primer pairs to be used and their purpose are shown in Table 1.
- The qPCR cycles are as follow: one initial denaturation step (5 min at 95°C), 50 amplification steps (15 s at 95°C and 30 s at 60°C followed by a fluorescence recording), and a final melting curve (65–97°C with 0.11°C/s increment and continuous fluorescence recording).

2.2.3 Analysis, Normalization, and Controls

Quantitative PCR results were analyzed on the LightCycler 96 Software 1.1 (Roche). With the exception of the DLE PCR (see below), all the amplification results are normalized to the reference *ARG4* locus on Chr. VIII (primers olWDH1760 and olWDH1761), with correction of the amplification efficiency (determined experimentally for each couple of primers). Several control reactions are performed (Fig. 2):

- The efficiency of DSB formation at HOcs is measured with the olWDH1766 and olWDH1767 primers. It is normalized to the uncut levels (i.e., prior to galactose addition). The results obtained by this method are not different from Southern blot quantification (Fig. 3A).
- The efficiency of *Hin*dIII digestion 5′ of the region of homology on the broken molecule with primers olWDH2010 and olWDH2012. It is normalized to an "uncut" control.
- The efficiency of *Hin*dIII digestion 3′ of the region of homology on the donor molecule with primers olWDH2009 and olWDH2011. It also measures primer extension as the copy number increases over the time

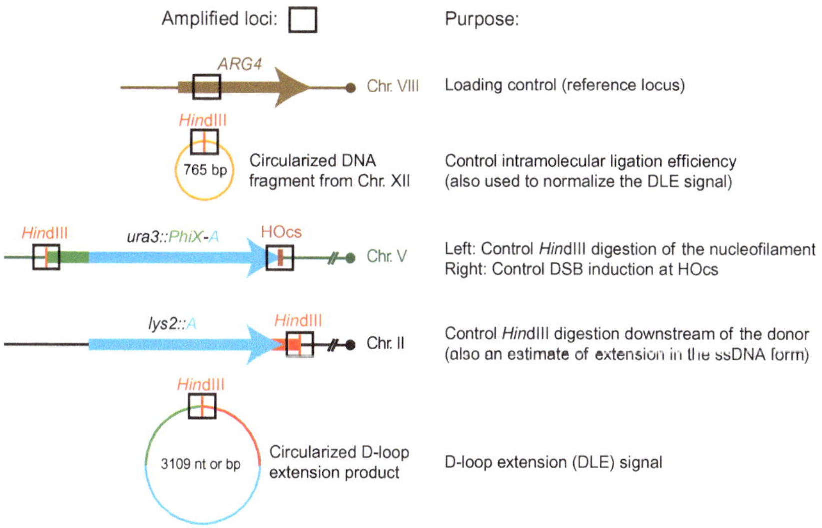

Fig. 2 Quantitative PCRs performed in the DLE assay. Loci for PCR amplification and their purpose are indicated.

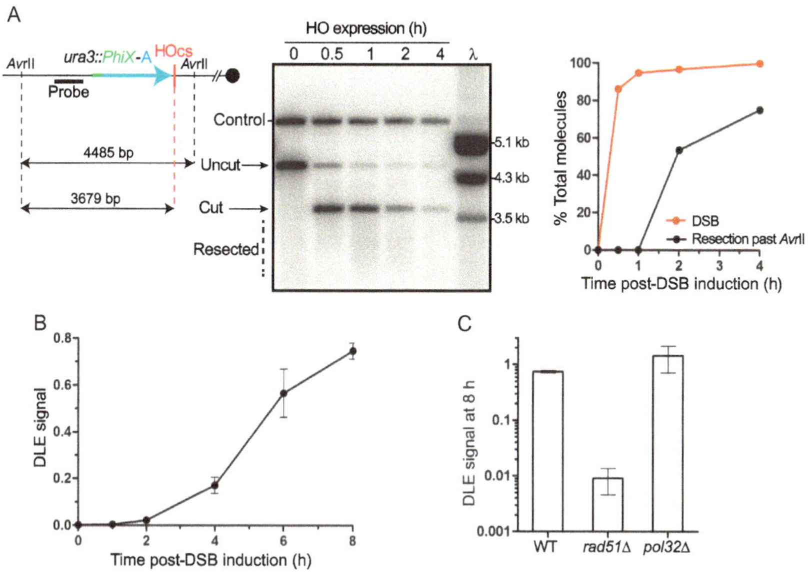

Fig. 3 Kinetics of D-loop extension and genetic requirements. (A) Southern blot analysis of the kinetics of DSB formation at HOcs following HO expression induction and resection at the *Avr*II site located 1.5 kb upstream of the region of homology "A." The loading control corresponds to an additional probe hybridizing the *RAD54* locus. The Southern blot procedure is as described (Piazza, Wright, & Heyer, 2017). (B) Kinetics of D-loop extension in wild-type cells. The DLE signal is normalized to the intramolecular ligation efficiency (Fig. S1A in the online version at https://doi.org/10.1016/bs.mie.2017.11.024). Data represent mean ± SEM of biological triplicates. (C) The DLE signal is Rad51 dependent and Pol32 independent. Intramolecular ligation efficiency is similar in all strains (Fig. S1C in the online version at https://doi.org/10.1016/bs.mie.2017.11.024), and DSB formation at HOcs was verified in the *rad51Δ* mutant (Fig. S1B in the online version at https://doi.org/10.1016/bs.mie.2017.11.024). Data represent mean ± SEM of parallel biological replicates.

course (Fig. S2B in the online version at https://doi.org/10.1016/bs.mie.2017.11.024).

- The intramolecular ligation efficiency of a 765-bp linear dsDNA fragment from Chr. XII (overlapping *YLR050C* and *YLR051C*) with primers olWDH2052 and olWDH2053. In a typical experiment, ~20% of the molecules have been circularized (Fig. S1A in the online version at https://doi.org/10.1016/bs.mie.2017.11.024). The DLE signal is normalized to this signal.

The DLE signal produced by the intramolecular ligation of the extended molecule is obtained with the primers olWDH2009 and olWDH2010. This amplification is normalized to the intramolecular ligation efficiency determined with primers olWDH2052 and olWDH2053.

3. RESULTS AND DISCUSSION

3.1 DLE Kinetics in Wild-Type Cells and Genetics Requirements

The formation of the site-specific DSB at the HOcs is achieved rapidly upon HO expression, with >90% of the molecules cut and resection initiated within an hour after induction (Fig. 3A). The chimeric ligation product expected upon D-loop extension (hereafter referred to as the DLE signal) appears with the expected slower kinetics than the previous steps of DSB formation and resection (Fig. 3B). The first unambiguous amplification is detected 2 h post-HO induction, the amount of which increases sharply (13-fold) between 2 and 6 h (Fig. 3B). At that time the DLE signal reaches approximately half that of the dsDNA control, consistent with the fact that the initial substrate is ssDNA, not dsDNA. The additional 1.4-fold increase observed at 8 h can be attributed to dsDNA repair product formation (see below).

As expected, the DLE signal depends on a functional HR pathway, as it is reduced to background levels in a *rad51Δ* mutant (Fig. 3C). Furthermore, we find that the initiation of DNA synthesis and extension of at least 405 nt (Fig. 1B) is achieved independently of Pol32 (Fig. 3C). Pol32 is the Polδ subunit dispensable for gene conversion but required for the extensive synthesis underlying BIR (Lydeard et al., 2007, 2010). This result suggests that Pol32 promotes BIR not by enabling the initiation of DNA synthesis, but rather by promoting extensive DNA synthesis required for BIR, consistent with its characterized biochemical role as a processivity factor for Polδ (Burgers & Gerik, 1998; Johansson, Garg, & Burgers, 2004).

The PCR-based assay showed a dependence on Pol32 for extension of 45 nt during BIR (Jain et al., 2009; Lydeard et al., 2007). We thus suspect that the amplification resulted not from the ssDNA extension intermediate but from the final BIR product (see also below).

3.2 Restriction Sites Restoration Allows to Distinguish DNA Synthesis Initiation and BIR Product Formation

An anticipated requirement for the DLE assay is the restoration of the restriction sites in the ssDNA produced by the extension of the D-loop (Fig. 4A). This is achieved upon hybridization of long oligonucleotides at these two loci prior to the digestion step (Fig. 4A). Oligonucleotides are designed (i) to encompass the restriction site and (ii) to hybridize a sufficient amount of ssDNA (we used ~80 nt) internal to the restriction fragment to withstand the restriction enzyme denaturation step, so as to enable ligation at the subsequent step. Following induction of DSB formation at HOcs, we monitored the *Hin*dIII digestion of both the upstream (5′) region flanking the A homology on the DSB-inducible Chr. V and the downstream (3′) region flanking the A homology on the donor Chr. II in the presence or absence of their respective hybridization primers ol2007 and ol2046 (Fig. S2 in the online version at https://doi.org/10.1016/bs.mie.2017.11.024A and B, respectively). DNA upstream of the DSB-inducible construct is efficiently digested in its dsDNA form prior and up to 1 h after DSB induction at HOcs, independently of the presence of ol2007 (Fig. S2A in the online version at https://doi.org/10.1016/bs.mie.2017.11.024). After that, restriction digestion depends on ol2007 (Fig. S2A in the online version at https://doi.org/10.1016/bs.mie.2017.11.024), consistent with the resection kinetics observed by Southern blot (Fig. 3A).

The digestion of the *Hin*dIII site downstream of the donor is achieved independently of ol2046 up to 2 h post-DSB induction (Fig. S2B in the online version at https://doi.org/10.1016/bs.mie.2017.11.024). After that, the digestion requires ol2046, consistent with the rise of the DLE signal (Fig. 3B). Although less sensitive than the DLE assay, this decrease in *Hin*dIII digestion in the absence of ol2046 is a secondary readout of DLE, limited to the molecule in the ssDNA form as final dsDNA BIR products are digested. Consequently, the undigested signal plateaus at ~50% of the dsDNA control at 6 and 8 h (Fig. S2B in the online version at https://doi.org/10.1016/bs.mie.2017.11.024).

As expected, the DLE signal up to 6 h depends on both ol2007 and ol2046 (Fig. 4B). At the latest time point ~15% of the DLE signal is independent of the hybridization oligonucleotides (Fig. 4C), indicating that the circularized molecule was dsDNA. The conversion of a fraction of the extended ssDNA into dsDNA satisfyingly explains why the DLE signal keeps increasing from 6 to 8 h past the 50% expected from a normalization on dsDNA, while the undigested donor plateaus (Fig. S3 in the online version at https://doi.org/10.1016/bs.mie.2017.11.024). Hence, the DLE signal obtained in the presence and in the absence of hybridization oligonucleotides enables kinetic analysis of both the initiation of DLE (ssDNA) and the formation of the BIR product (dsDNA). Subtraction of the dsDNA component of the DLE signal (i.e., half of the dsDNA signal) enables a quantitative determination of the molecules that have primed and extended DNA synthesis up to the restriction site in a cell population (Fig. 4D).

These results show that the initiation of DLE and the completion of BIR, which involves asynchronous lagging strand synthesis (Donnianni & Symington, 2013; Saini et al., 2013), are temporally distinct: initiation of DLE occurs mainly between 4 and 6 h post-DSB induction, while conversion to dsDNA occurs in a delayed fashion, after 6 h. These observations and the Pol32-independent nature of DNA synthesis up to 405 nt differ from results obtained with PCR-based methods, which might preferentially detect final dsDNA products (Jain et al., 2009; Jain, Sugawara, Mehta, Ryu, & Haber, 2016; Lydeard et al., 2007; Malkova et al., 2005).

3.3 Possible Refinements of the DLE Assay

The assay is dependent on the availability of compatible restriction sites in the vicinity of the region of homology between the DSB-inducible and the ectopic donor. To overcome this constraint, which dictates the minimum extension length detected at any given locus, genetic engineering may be required. By placing periodically spaced restriction sites, and by using site-specific hybridization oligonucleotides and qPCR primers in independent reactions, such engineering could allow mapping the kinetics of DLE tracts of various length (Fig. 5A). The resolution would only be limited by the hybridization oligonucleotide length (~80 nt), to avoid restriction sites overlap in case a single restriction enzyme is being used.

A second possible development aimed at weeding out dsDNA repair products is to include a restriction site within the region of homology (Fig. 5B). A benefit of such site would be to more straightforwardly allow distinguishing the extending D-loops (ssDNA) from the repair product

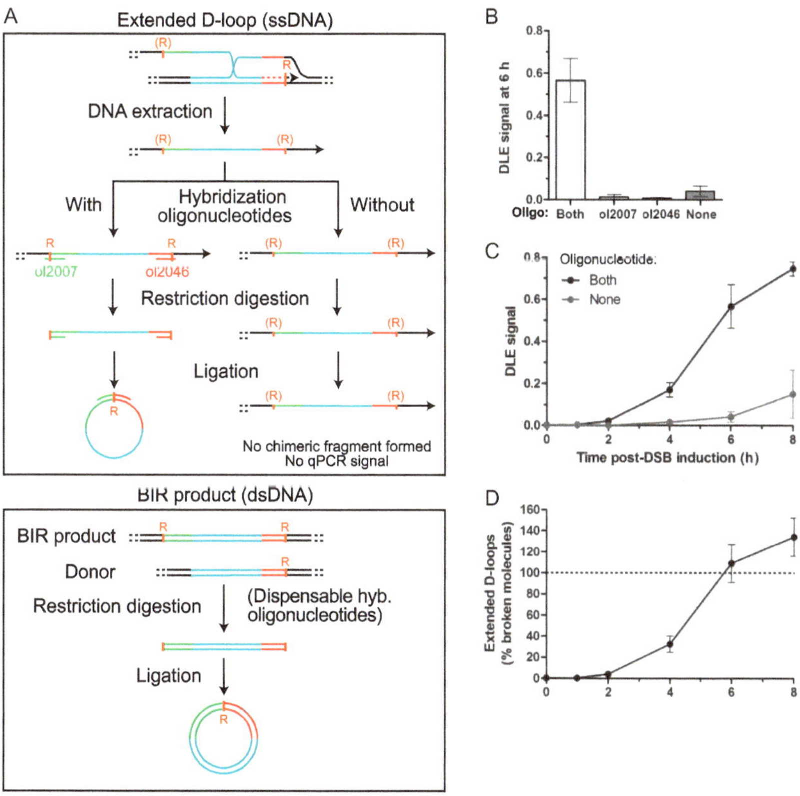

Fig. 4 Restoration of the restriction sites in ssDNA is required for DLE signal. (A) The extended D-loop prior to BIR product formation is in ssDNA form. The restriction digestion step of the DLE assay thus requires restoration of these sites. This is achieved upon annealing of the "hybridization oligonucleotides" ol2007 and ol2046 to the sites located upstream and downstream of the region of homology, respectively. This digestion step enables the subsequent intramolecular ligation reaction which provides the readout of the DLE assay. Digestion of the BIR product (dsDNA) does not require restriction site restoration. Consequently, the intramolecular ligation product is generated independently of the presence of the hybridization oligonucleotides. (B) The DLE signal at 6 h post-DSB induction depends on both ol2007 and ol2046. (C) DLE kinetics in the absence of hybridization oligonucleotides reveal the fraction of DLE signal originating from dsDNA products at 6 and 8 h. (D) Proportion of extended molecules in the cell population deduced from the DLE signal in the presence and absence of hybridization oligonucleotides (C). Data represent mean ± SEM of biological replicates.

(dsDNA) than to compare DLE levels in the presence or absence of the ol2046 oligonucleotide. Indeed, in such a case restriction digestion of dsDNA would prevent the ligation of the two unique flanking regions by physically untethering them while leaving the ssDNA extension product intact (Fig. 5B).

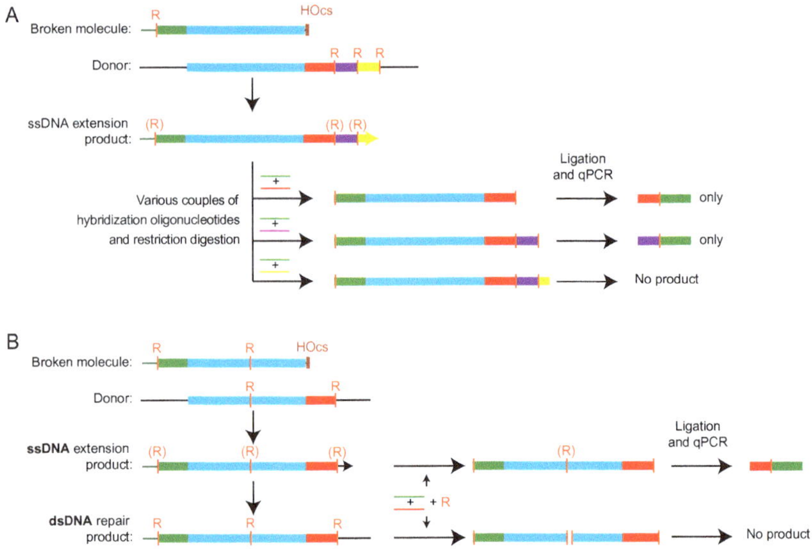

Fig. 5 Possible refinements of the DLE assay. (A) Multiple restriction sites downstream of the donor provided with the appropriate hybridization oligonucleotides would enable determining the kinetics of initiation as well as the extent of DNA synthesis primed from a D-loop. Either the same or compatible restriction enzymes could be used. (B) Restriction digestion of a site present in the homologous region would restrict the analysis to ssDNA extension product by eliminating the dsDNA repair product from the procedure. (A, B) R denotes a restriction site in its dsDNA form and (R) a restriction site in its ssDNA form (uncuttable).

Finally, our knowledge of mitotic DNA synthesis initiation gained with PCR-based assays mostly comes from ectopic DSB repair systems (Lydeard et al., 2007, 2010; Tsaponina & Haber, 2014; White & Haber, 1990). The assay presented here is amenable to the study of DNA synthesis in the context of allelic DNA repair by using restriction site polymorphisms between the DSB-containing homolog and the donor homolog.

4. CONCLUSION

We developed a quantitative proximity ligation-based assay for the study of DNA synthesis during recombinational DSB repair. It provides an alternative and independent approach to semiquantitative PCR-based methods and overcomes certain technical limitations. The versatility of the assay grants further developments to study in greater details the initiation and extension steps of DNA synthesis on different templates and various physiological and mutant contexts.

Supplementary data to this article can be found online at https://doi.org/10.1016/bs.mie.2017.11.024.

ACKNOWLEDGMENTS

We thank members of the Heyer laboratory, especially William Wright, Shanaya Shah, and Paula Cerqueira, for stimulating discussions. We also thank members of the Koszul laboratory, in particular Agnès Thierry and Charlie Cockram, for technical help. We are grateful to the Lambrechts and the Arcangioli labs for the use of their quantitative PCR machines.

FUNDING

A.P. was supported by fellowships from the Fondation ARC pour la Recherche sur le Cancer, the EMBO (ALTF-238-2013), and the Framework Project 7 of the European Union (Marie Curie International Outgoing Fellowship 628355) administered by the Institut Pasteur, France, and received financial support from the Philippe Foundation. This research was supported by MeioRec ANR-13-BSV6-0012-02 to R.K. and NIH grants GM58015 and CA92276 to W.-D.H.

REFERENCES

Agmon, N., Liefshitz, B., Zimmer, C., Fabre, E., & Kupiec, M. (2013). Effect of nuclear architecture on the efficiency of double-strand break repair. *Nature Cell Biology*, *15*, 694–698.

Aylon, Y., Liefshitz, B., Bitan-Banin, G., & Kupiec, M. (2003). Molecular dissection of mitotic recombination in the yeast *Saccharomyces cerevisiae*. *Molecular Biology of the Cell*, *23*, 1403–1417.

Burgers, P. M. J., & Gerik, K. J. (1998). Structure and processivity of two forms of *Saccharomyces cerevisiae* DNA polymerase delta. *The Journal of Biological Chemistry*, *273*, 19756–19762.

Dekker, J., Rippe, K., Dekker, M., & Kleckner, N. (2002). Capturing chromosome conformation. *Science*, *295*, 1306–1311.

Donnianni, R. A., & Symington, L. S. (2013). Break-induced replication occurs by conservative DNA synthesis. *Proceedings of the National Academy of Sciences of the United States of America*, *110*, 13475–13480.

Forget, A. L., & Kowalczykowski, S. C. (2012). Single-molecule imaging of DNA pairing by RecA reveals a three-dimensional homology search. *Nature*, *482*, 423–427.

Heyer, W. D. (2015). Regulation of recombination and genomic maintenance. *Cold Spring Harbor Perspectives in Biology*, *7*, a016501.

Hicks, W. M., Yamaguchi, M., & Haber, J. E. (2011). Real-time analysis of double-strand DNA break repair by homologous recombination. *Proceedings of the National Academy of Sciences of the United States of America*, *108*, 3108–3115.

Inbar, O., & Kupiec, M. (1999). Homology search and choice of homologous partner during mitotic recombination. *Molecular and Cellular Biology*, *19*, 4134–4142.

Inbar, O., Liefshitz, B., Bitan, G., & Kupiec, M. (2000). The relationship between homology length and crossing over during the repair of a broken chromosome. *The Journal of Biological Chemistry*, *275*, 30833–30838.

Jain, S., Sugawara, N., Lydeard, J., Vaze, M., Tanguy Le Gac, N., & Haber, J. E. (2009). A recombination execution checkpoint regulates the choice of homologous recombination pathway during DNA double-strand break repair. *Genes & Development, 23*, 291–303.

Jain, S., Sugawara, N., Mehta, A., Ryu, T., & Haber, J. E. (2016). Sgs1 and Mph1 helicases enforce the recombination execution checkpoint during DNA double-strand break repair in *Saccharomyces cerevisiae*. *Genetics, 203*, 667–675.

Jinks-Robertson, S., Michelitch, M., & Ramcharan, S. (1993). Substrate length requirements for efficient mitotic recombination in *Saccharomyces cerevisiae*. *Molecular and Cellular Biology, 13*, 3937–3950.

Johansson, E., Garg, P., & Burgers, P. M. (2004). The Pol32 subunit of DNA polymerase delta contains separable domains for processive replication and proliferating cell nuclear antigen (PCNA) binding. *The Journal of Biological Chemistry, 279*, 1907–1915.

Lydeard, J. R., Jain, S., Yamaguchi, M., & Haber, J. E. (2007). Break-induced replication and telomerase-independent telomere maintenance require Pol32. *Nature, 448*, 820–823.

Lydeard, J. R., Lipkin-Moore, Z., Sheu, Y. J., Stillman, B., Burgers, P. M., & Haber, J. E. (2010). Break-induced replication requires all essential DNA replication factors except those specific for pre-RC assembly. *Genes & Development, 24*, 1133–1144.

Malkova, A., Naylor, M. L., Yamauchi, M., Ira, G., & Haber, J. E. (2005). RAD51-dependent break-induced replication differs in kinetics and checkpoint responses from RAD51-mediated gene conversion. *Molecular Biology of the Cell, 25*, 933–944.

Mehta, A., Beach, A., & Haber, J. E. (2017). Homology requirements and competition between gene conversion and break-induced replication during double-strand break repair. *Molecular Cell, 65*, 515–526.

Mine-Hattab, J., & Rothstein, R. (2012). Increased chromosome mobility facilitates homology search during recombination. *Nature Cell Biology, 14*, 510–517.

Piazza, A., Wright, W. D., & Heyer, W. D. (2017). Multi-invasions are recombination byproducts that induce chromosomal rearrangements. *Cell, 170*, 760–773.

Saini, N., Ramakrishnan, S., Elango, R., Ayyar, S., Zhang, Y., Deem, A., et al. (2013). Migrating bubble during break-induced replication drives conservative DNA synthesis. *Nature, 502*, 389–393.

Tsaponina, O., & Haber, J. E. (2014). Frequent interchromosomal template switches during gene conversion in *S. cerevisiae*. *Molecular Cell, 55*, 615–625.

White, C. I., & Haber, J. E. (1990). Intermediates of recombination during mating type switching in *Saccharomyces cerevisiae*. *The EMBO Journal, 9*, 663–673.

CHAPTER THREE

Inducing and Detecting Mitotic DNA Synthesis at Difficult-to-Replicate Loci

Lorenza Garribba*^,2^, Wei Wu*^,2^, Özgün Özer*, Rahul Bhowmick*, Ian D. Hickson*^,†^, Ying Liu*^,1^

*Center for Chromosome Stability, University of Copenhagen, Copenhagen, Denmark
†Center for Healthy Aging, University of Copenhagen, Copenhagen, Denmark
[1]Corresponding author: e-mail address: ying@sund.ku.dk

Contents

Abstract

Our conventional understanding of the process of DNA replication is that it occurs in the S-phase of the cell division cycle. However, during investigations into the mechanism by which common fragile sites (CFSs) drive genome instability, we observed that some DNA synthesis was still occurring in early mitosis at these loci. This curious phenomenon of mitotic DNA synthesis (which we now term "MiDAS") appears to be a form of break-induced DNA replication (BIR), a DNA repair process based on homologous recombination that has been characterized in detail only in lower eukaryotes. During MiDAS, it is proposed that parts of the human genome that are not fully replicated when cells enter mitotic prophase complete their replicative cycle at that point. To date, the loci that most depend upon this process are those whose replication can be affected by

[2] These authors contributed equally to this work.

Methods in Enzymology, Volume 601
ISSN 0076-6879
https://doi.org/10.1016/bs.mie.2017.11.025

oncogene-induced DNA replication stress (RS), most notably, CFSs. From our studies, it is clear that the successful completion of MiDAS at CFSs can minimize chromosome missegregation and nondisjunction. Nevertheless, it is still not clear which loci that can undergo MiDAS, whether MiDAS is associated with mutations or genome rearrangements, or whether MiDAS really is a form of BIR. In this review, we describe methods for detecting MiDAS both in prometaphase cells and directly on isolated metaphase chromosomes. In addition, we have included methods for combining MiDAS detection either with immunofluorescence (IF) detection of proteins that are recruited to the MiDAS loci, or with fluorescence in situ hybridization using probes that target specific genomic loci.

ABBREVIATIONS

BSA bovine serum albumin
CDK1 cyclin-dependent kinase 1
DAPI 4′,6-diamidino-2-phenylindole
DNA deoxyribonucleic acid
dTTP deoxythymidine triphosphate
EGTA ethylene glycol-bis(β-aminoethyl ether)-*N*,*N*,*N*′,*N*′-tetraacetic acid
EME1 essential meiotic structure-specific endonuclease 1
G2 gap phase 2
h hour
M mitotic phase
MeOH methanol
MUS81 MUS81 structure-specific endonuclease subunit
PBS phosphate-buffered saline
PFA paraformaldehyde
PIPES piperazine-*N*,*N*′-bis(2-ethanesulfonic acid)
POLD3 DNA polymerase delta 3, accessory subunit 3
rcf relative centrifugal force
rpm revolutions per minute
RS replication stress
RT room temperature

1. INTRODUCTION

The continued proliferation of any cell requires that the genetic material be passed onto the next generation of cells in an undamaged state. In human cells, the genetic information is stored in 23 pairs of chromosomes (22 pairs of autosomes and 1 pair of sex chromosomes). The replication and segregation of these 46 chromosomes are highly regulated during the cell division cycle. This cycle comprises interphase, when the cells grow in size and the chromosomes are replicated, and M-phase when the chromosomes and cytoplasm are segregated equally to the two daughter cells. Mitosis is subdivided into five subphases: prophase, prometaphase, metaphase, anaphase, and telophase. Until recently, the conventional view of the cell cycle was that DNA is fully replicated during interphase otherwise entry into mitosis would be prohibited. However, this view has been challenged recently by the finding that some regions of the human genome seem only to be fully replicated in the early stages of mitosis. These "tardy" regions are not randomly distributed in the genome, but instead concentrate at loci that are known as common fragile sites (CFSs) (Minocherhomji et al., 2015). They are referred to as being "fragile" because they lie within chromosomal loci that have a propensity to appear as a gap or break on otherwise fully condensed metaphase chromosomes. This phenomenon is usually termed CFS expression.

CFSs are recognized as a driver of genome instability in human cells and are hotspots for deletions or translocations in cancers (Richards, 2001). The generally accepted model for their expression is that the CFS locus shows a delay in chromatin condensation in early mitosis caused by the persistence of a region of underreplicated DNA (El Achkar, Gerbault-Seureau, Muleris, Dutrillaux, & Debatisse, 2005). This phenomenon is rare in cultured normal cells, but can be exacerbated by the exposure of the cells to agents that partially inhibit DNA synthesis during S-phase (such as aphidicolin; APH) (Glover, Berger, Coyle, & Echo, 1984). One reason why cancer cells seem susceptible to CFS instability is because of the widespread occurrence of oncogene activation in cancer cells, which is a known driver of replication stress (RS) (Hills & Diffley, 2014; Lecona & Fernandez-Capetillo, 2014; Ren et al., 2017).

The mitotic DNA synthesis occurring at CFSs (now termed "MiDAS") (Bhowmick, Minocherhomji, & Hickson, 2016) was discovered, while we were trying to accurately determine when CFS replication is completed during the cell cycle in the presence of RS. It had been shown previously that replication was still occurring after the completion of S-phase (Bergoglio et al., 2013; Pedersen, Kruse, Nilsson, Oestergaard, & Lisby, 2015). In our previous studies, the dTTP analogue 5-ethynyl-2′-deoxyuridine (EdU) was utilized to mark sites where DNA replication was still occurring. Because labeling with EdU takes 20–30 min, any EdU found in mitotic cells was assumed to have been incorporated into the genome in G2. The use of the CDK1 inhibitor, RO3306 (Vassilev, 2006), to arrest cells at the G2/M boundary allowed us to separate G2 from M-phase, and therefore to add EdU only after cells had entered the prophase of mitosis. Under these circumstances, we could detect EdU on metaphase chromosomes and demonstrate that DNA synthesis was still occurring in early mitosis. Importantly, we could prove that the sites of MiDAS correspond to broken CFS loci by combining EdU detection with fluorescent in situ hybridization (FISH) (Minocherhomji et al., 2015). Moreover, we observed that the entry into prophase triggers the recruitment of DNA structure-specific endonuclease, MUS81–EME1, to CFSs, and that the nuclease action of MUS81–EME1 promotes POLD3-dependent DNA synthesis at CFSs. Although not characterized in detail, our results suggest that MiDAS is a DNA repair-based "salvage" pathway based on a form of BIR that helps cells to complete DNA synthesis at difficult-to-replication loci-like CFSs when they encountered RS. If MiDAS fails to occur, then chromosome missegregation and nondisjunction ensues, particularly in cancer cells with intrinsically high levels of chromosome instability (CIN + ve) (Minocherhomji et al., 2015).

This chapter describes the protocols that we have developed to detect DNA synthesis in mitosis in either prometaphase cells or directly on fixed metaphase chromosomes, using the "Click-iT EdU" reaction. We have also included two additional protocols for (i) detection of specific proteins that are recruited to the sites of ongoing MiDAS sites, by combining EdU detection with IF and (ii) identification of genomic regions where MiDAS is occurring, by combining EdU detection with FISH. We hope that these methods will be helpful for the further characterization of MiDAS and for the characterization of how MiDAS impacts on genome stability of human cells.

2. MATERIALS

2.1 Cell Lines

U2OS (an adherent cancer cell line) and GM09237 (a lymphocyte cell line growing in suspension). Levels of MiDAS vary greatly between cell types, with cancer cells such as HeLa and U2OS showing a high level, and normal fibroblasts and lymphocytes showing a lower level.

2.2 Reagents

(1) Standard cell culture reagents
(2) RO3306 (Merck; stock concentration, 10 m*M*)
(3) EdU (ThermoFisher Scientific; stock concentration, 40 m*M*)
(4) APH (Sigma-Aldrich; stock concentration, 1 m*M*)
(5) Click-iT EdU Alexa Fluor 594 Imaging Kit, Click-iT EdU Alexa Fluor 488 Imaging Kit green (ThermoFisher Scientific)

(6) Vectashield mounting medium with DAPI (Vector Laboratories, Inc.)
(7) Standard immunofluorescence (IF) analysis-related reagents (optional, for detection of EdU and protein of interests)
(8) Standard FISH analysis-related reagents (optional, for detection of EdU and genomic loci of interests)

2.3 Equipment/Tools

(1) Standard cell culture facility
(2) Standard tools for performing IF (or FISH) analysis (optional, for detection of EdU and proteins or genomic loci of interests)
(3) Standard slides and cover slips
(4) Fluorescence microscope (we use an Olympus BX63 microscope)
(5) Poly-L-lysine adhesion slides (ThermoFisher Scientific)
(6) Cytospin4 centrifuge (ThermoFisher Scientific). Only required for analysis of isolated metaphase chromosomes

3. METHOD

The method is largely divided into two parts: see Sections 3.1 and 3.2. In each part, the experimental steps are grouped into four categories: (i) to induce RS and arrest cells in G2 phase; (ii) to label cells with EdU in prometaphase or metaphase; (iii) to harvest and fix cells in prometaphase or metaphase for various analysis (i.e., EdU detection, EdU combined with IF, or EdU combined with FISH); and (iv) the detection of the various signals. The experimental flow and the representative images from each type of analysis are illustrated in Figs. 1 and 2.

3.1 To Analyze MiDAS in Prometaphase

To induce RS and arrest cells in the G2 phase

- 3.1.1: (day 1, seed cells): Exponentially growing cells are seeded into one T75 flask at 30% confluency with 10 mL of cell culture medium.
- 3.1.2: (day 2, treatment with or without APH and RO3306): To enrich prometaphase cells that might be undergoing MiDAS following APH treatment, the following can be performed: change the medium from step 3.1.1 and add 2–4 μL APH into 10 mL new medium (the final APH concentration can be 0.2–0.4 μ*M*) for 16 h and add 7 μL of RO3306 (final concentration: 7 μ*M*) in the last 6–8 h of APH treatment (see Note 1, 2).

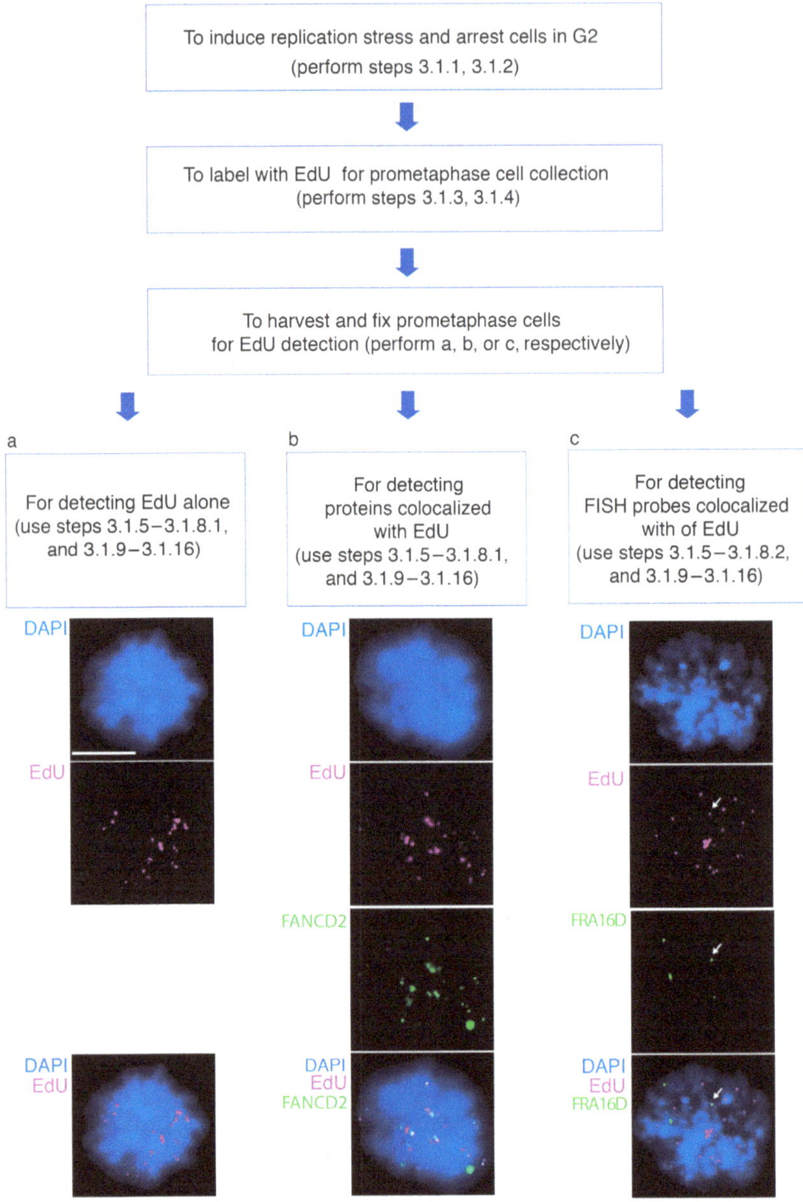

Fig. 1 A summary of the experimental procedures for detecting MiDAS in prometaphase cells, the proteins recruited to sites of MiDAS, and the genomic regions where MiDAS occurs. Representative images obtained from each type of analysis in U2OS cells are shown below the flow chart. The colocalization of EdU with a FISH probe is denoted by the *white arrow*. Scale bar, 5 μm.

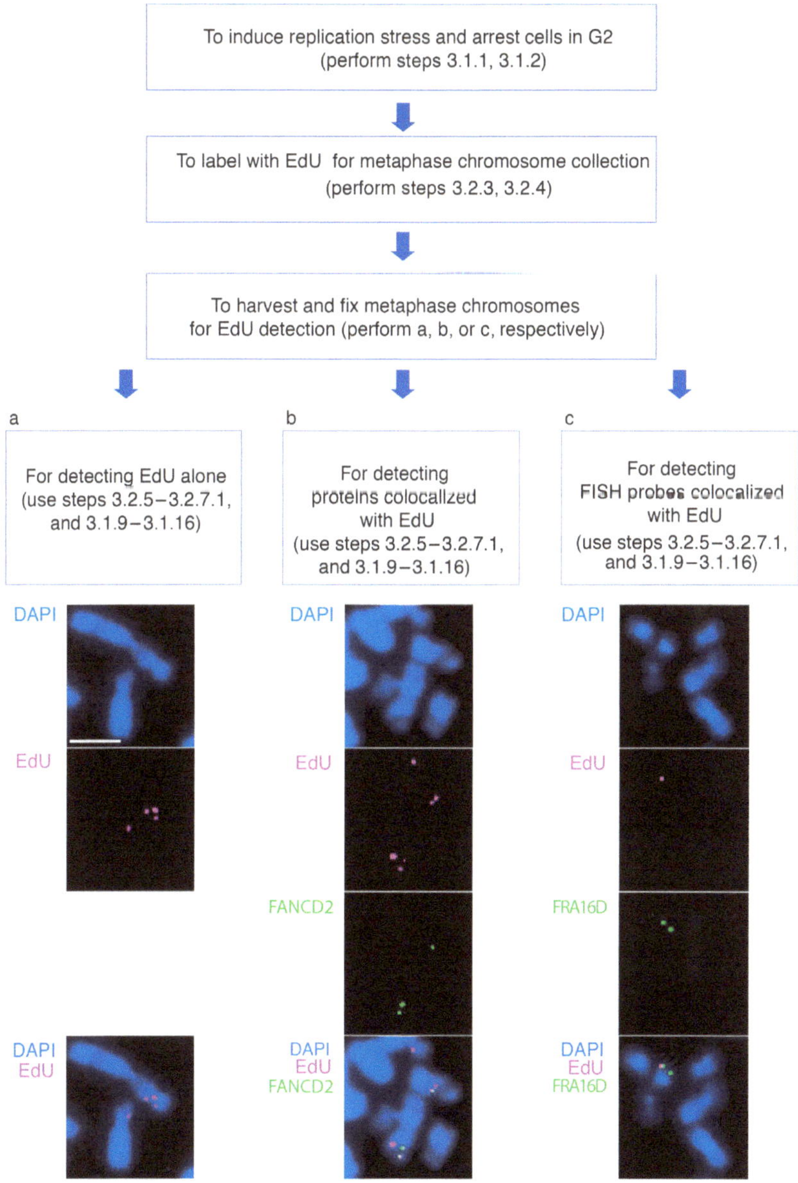

Fig. 2 A summary of the experimental procedures for detecting MiDAS on metaphase chromosomes, the proteins recruited to sites of MiDAS, and the genomic regions where MiDAS occurs. Representative images obtained from each analysis in U2OS cells (or GM09237 cells when EdU is combined with FISH analysis) are shown below the flow chart. Scale bar, 5 μm.

To label cells with EdU (day 3)

- 3.1.3: Prepare fresh medium with EdU (final concentration: 20 μ*M*) (see Note 3) and keep it warm at 37°C.
- 3.1.4: Discard the old medium from the flask with synchronized cells and wash them with prewarmed (37°C) PBS three times (1–5 min each wash) (see Note 4, 5). After the last wash, add prewarmed medium that contains EdU to the flask and put the flask back into the incubator for 20–30 min (see Note 6).

To harvest cells in prometaphase and fix cells for EdU detection (day 3)

- 3.1.5: Collect the mitotic cells in a 15-mL Falcon tube using the "shake-off" method (see Note 7). For cells growing in suspension, this step is omitted.
- 3.1.6: Spin down the cells at 300 rcf for 5 min at room temperature (RT).
- 3.1.7: Discard the supernatant and add a small amount of fresh medium (250 μL per slide to be analyzed) to resuspend cells very gently by pipetting up and down with 1 mL stripettes (see Note 8). Seed the 250 μL cell/medium mixture onto poly-L-lysine adhesion slides. Let the cells settle down on the slides for 2–5 min at RT.
- 3.1.8: Drain the medium from the slide by tilting it onto its side and dabbing it carefully on a piece of tissue paper (see Note 9). At this step, cells can be fixed in two different ways. The choice depends on whether EdU detection will be performed alone or combined with either IF or FISH. If EdU detection is to be performed in combination with FISH, the steps involved in FISH should be performed prior to EdU detection.
  - 3.1.8.1: (if EdU detection alone, or in combination with IF, will be performed): add 100 μL PTMEF buffer (4% paraformaldehyde (PFA), 200 m*M* PIPES pH 6.8, 200 m*M* $MgCl_2$, 10 m*M* EGTA pH 8.0, 0.2% Triton X-100; to be prepared no earlier than 30 min before use) per slide to fix cells for 10 min at RT in a fume hood. Then, remove the fixative by tilting the slide onto a piece of tissue paper and wash the slide three times with PBS. If EdU detection is combined with IF, IF steps can be performed following this step. When IF steps are completed, store the slide in PBS in a 4°C fridge until the EdU detection step is performed.
  - 3.1.8.2: (if EdU detection and FISH will be performed): immerse the slides in a Coplin jar containing MeOH–acetic acid (3:1 ratio) in a fume hood and fix them for 2 min at RT. Take the slides out and let them dry (it is preferable to leave step this overnight). The FISH analysis can be performed following this stage. When FISH steps are

completed, refix the slides in 4% PFA in PBS for 4 min at RT in a fume hood. Then, remove the fixative by tilting the slide onto a piece of tissue paper and wash the slide three times in PBS. Proceed to the EdU detection step.

To detect EdU (day 3)

- 3.1.9: Block the slides in 3% BSA in PBS for 30 min at RT (necessary when EdU is detected alone or to be combined with FISH).
- 3.1.10: Permeabilize the cells in 0.5% Triton X-100 in PBS for 20 min at RT (not necessary when EdU is to be combined with IF).
- 3.1.11: Perform EdU detection according to the instructions provided by the supplier of the kit, with a modification in the composition of the buffer, as shown below:

Reaction Components

1 × Click-iT® reaction buffer	430 μL
$CuSO_4$ (Component E)	20 μL
Alexa Fluor® azide	1.2 μL
10 × Click-iT® EdU buffer additive	50 μL
Total volume	500 μL

Note: Add the ingredients in the order listed in the table.

The reaction mix is prepared freshly and then applied to the slide. A coverslip is placed on the top of the reaction/cell mix to prevent evaporation.

- 3.1.12: Incubate the slides with 50 μL Click-iT reaction mix (for each spot). Each reaction/cell mix should be loosely covered with a coverslip for between 45 min and 1 h at RT in the dark (see Note 10).
- 3.1.13: Gently remove the coverslips and wash slides with 3% BSA/0.5% Triton in PBS three times (10–20 min each time) (see Note 11).

To mount the slides and perform microscopic analysis (day 3)

- 3.1.14: Rinse the slides in a Coplin jar with water and then remove the water as much as possible.
- 3.1.15: Mount the slides with Vectashield Mounting Medium containing DAPI and place a cover slip on top. Seal the slides with nail varnish and let them dry.
- 3.1.16: Observe the slides with a fluorescence microscope.

3.2 To Analyze EdU on Metaphase Chromosome Spreads

To induce RS and arrest cells at G2 phase

- Follow the procedure as described in steps 3.1.1 and 3.1.2.

To label cells with EdU

- 3.2.3: The same as that for step 3.1.3.
- 3.2.4: Discard the medium from the APH and RO3306-treated cells, and wash them with prewarmed (37°C) PBS three times for 5 min each. After the last wash, add prewarmed medium containing 20 μ*M* EdU and 0.1 μg/mL Karyomax (Colcemid), and incubate for 1 h at 37°C.

To harvest metaphase cells and fix metaphase spreads for EdU detection (day 3)

- 3.2.5, 3.2.6: The same as that described in steps 3.1.5 and 3.1.6.
- 3.2.7: At this step, cells can be fixed in two different ways. The choice depends on whether EdU detection will be performed alone or alternatively combined with either IF or FISH to detect specific regions of the genome. If EdU detection is to be performed in combination with FISH, the various steps involved in FISH detection should be performed prior to the EdU detection.
  - 3.2.7.1: (if EdU is to be detected alone, or in combination with IF): resuspend the pellet in an appropriate amount (depending on the amount of cells, see Note 8) of hypotonic solution (75 m*M* KCl) and incubate for 20 min at 37°C. Centrifuge 100 μL of cell suspension per coated cytospin slide in a Cytospin4 at 1100 rpm for 2 min (see Note 12). Check the cell density under a microscope and air-dry the slides until they are at least 90% dried (no visible liquid should be seen on the slide). Fix cells as described in step 3.1.8.1 and leave them in cold PBS until IF steps (in case of IF plus EdU detection) or EdU detection is to be performed.
  - 3.2.7.2: (if EdU and FISH will be performed): resuspend the cells in 5–10 mL hypotonic solution (75 m*M* KCl) and incubate for 20 min at 37°C. Harvest the cells by centrifugation for 10 min at 300 rcf. Fix them by adding 2 mL MeOH–CH3COOH (3:1 ratio) drop by drop while vortexing (start very slowly to avoid the formation of cell clumps). Add extra fixative in order to have 5 mL in each 15-mL tube and then leave the suspensions at RT for 30 min (to make slides on the same day). Alternatively, store the tubes at −20°C for future analysis. Drop the cells onto standard microscopy slides (see Note 13). Let the slides dry overnight at RT (see Note 14) and then proceed with the FISH steps (in case of combination FISH plus EdU detection).

At the end of FISH protocol, refix the slides in 4% PFA in PBS for 4 min at RT in a fume hood. Remove the fixative by tilting the slide onto a piece of tissue paper and wash the slide three times in PBS. Proceed with EdU detection.

To detect EdU (or in combination with IF or FISH) metaphase chromosomes

- Follow the procedure as described in steps 3.1.9 to 3.1.16.

4. NOTES

1. The RO3306 concentration for reversibly arresting each cell type needs to be optimized due to the inherent differences in the cell cycle of each cell line, and the fact that RO3306 produced by different companies can have a different potency according to our analyses. In addition, dormant replication origins could be activated when RO3306 is added (via its inhibition of CDK1) (Ge & Blow, 2010), which means that RS caused by APH might be partially counteracted. Therefore, it is not advisable to add RO3306 alongside APH for 16 h to analyze MiDAS induced by APH.
2. MiDAS can also be detected in asynchronous cells treated with APH. This can be achieved by seeding cells onto coverslips in suitable plates (with approx. 35% confluency) on day 1, incubating cells on day 2 with APH (0.4 μ*M*) for 16 h and, on day 3, incubating cells with EdU (10–20 μ*M*) in fresh new medium (without APH) for 30 min, and then fixing cells for EdU detection.
3. We found that incubation with 20 μ*M* EdU gave a stronger EdU signal than did 10 μ*M* EdU.
4. After treatment with RO3306, the total wash time to remove the drug should be optimized very carefully for each cell line. This is because MiDAS commences very quickly after release into prophase, and each cell line recovers from an RO3306 arrest with different kinetics. For example, for U2OS cells, the total wash time should be 5 min, while for lymphocytes this can be up to 15 min. The indication for a proper release is the number of EdU foci. Following APH treatment, each U2OS prometaphase cell usually has more than 10 EdU foci.
5. It is important to maintain the cells at 37°C as much as possible. Therefore, the flask of cells should be put back into the incubator during the washing steps, where possible.

6. We found that 25–30 min after RO3306 release is a suitable time point for collecting prometaphase U2OS cells.
7. The strength and length of time applied should be adjusted according to the mitotic status of the cells. The general rule is to shake off mitotic cells without shaking off interphase cells, or causing mitotic cells excessive stress or physical damage. Therefore, it is recommended not to shake the flask too vigorously when collecting mitotic cells. Instead, try to collect as many as possible of the healthy mitotic cells by gentle tapping of the flask.
8. Starting from a T75 cm flask, you should have enough cells to make two slides, and therefore you should resuspend your pellet in 500 μL fresh medium and then seed 250 μL onto each slide. The number of mitotic cells that you can harvest from one flask can vary from time to time. Therefore, it is important to get an idea of the number of mitotic cells at the end of RO3306 treatment by looking at the cells under the microscope. In general, when seeding cells onto a poly-L-lysine adhesion slide, we drop 250 μL per slide.
9. From here on, every step should be performed very gently to avoid washing off the cells.
10. Air bubbles should be avoided in the Click-iT reaction and cell mix.
11. This is a crucial step to remove background.
12. The rotor speed depends on the cell type and should be optimized. For lymphocytes, for example, we use 850 rpm.
13. To make metaphase chromosome spreads: first, you should keep normal slides without coating in a Coplin jar with distilled water and store the Coplin jar at 4°C, preferably overnight so that the slides become cold. Also, keep a tray with wet paper towels in the fridge. When you are preparing spreads, centrifuge the cell suspensions at 300 rcf for 10 min and resuspend the pellets in a proper amount of fresh fixative such that the suspension becomes slightly milky. It is always recommended to make a test slide and check cell density under a microscope. Once the cells are resuspended properly, perform the standard "dropping" of cells to make metaphase chromosome spreads. Again, it is important to do a few test runs to get the spreading of the chromosomes to occur in such a way that they are not too sparse or, on the other hand, too dense and lying on top of one another.
14. The slides can also be dried for a few days at RT, or 65°C for 10 min, or 42°C overnight.

5. CONCLUSIONS

Since its discovery in 2015 as a putative "salvage" pathway for protecting vulnerable regions of the human genome from potentially being missegregated to the daughter cells in mitosis, MiDAS has emerged as a very useful marker denoting those chromosomal sites where RS has been encountered during S-phase. When combined with other methods such as FISH or IF, the detection of MiDAS is a powerful tool not only to define specific loci susceptible to RS, but also to discover which proteins are recruited to loci undergoing MiDAS. Although MiDAS analysis has widespread applicability, it is particularly relevant to studies on cancer cells possessing intrinsic RS driven by oncogene activation. In this regard, a goal of our current studies is to identify a protein involved in MiDAS that might serve as a biomarker for indicating the degree of RS in tumor biopsy specimens, with the ultimate aim of improving the clinical management of cancer patients (Ren et al., 2017).

ACKNOWLEDGMENTS

We thank all members in Liu and Hickson groups for helpful discussions, M. Clausen and T.A. Wass for technical assistance, and Ms. S. Wang for arranging the figures. Work in the authors' laboratory is supported by the European Union (Horizon 2020, Grant 665233; L.G., Ö.Ö., I.D.H., and Y.L.), Chinese Scholarship Council (W.W.), Independent Research Fund Denmark (R.B.), The Nordea Foundation (I.D.H.), and Danish National Research Foundation (DNRF115; I.D.H. and Y.L.).

REFERENCES

Bergoglio, V., Boyer, A. S., Walsh, E., Naim, V., Legube, G., Lee, M. Y., et al. (2013). DNA synthesis by Pol eta promotes fragile site stability by preventing under-replicated DNA in mitosis. *The Journal of Cell Biology*, *201*, 395–408.

Bhowmick, R., Minocherhomji, S., & Hickson, I. D. (2016). RAD52 facilitates mitotic DNA synthesis following replication stress. *Molecular Cell*, *64*, 1117–1126.

El Achkar, E., Gerbault-Seureau, M., Muleris, M., Dutrillaux, B., & Debatisse, M. (2005). Premature condensation induces breaks at the interface of early and late replicating chromosome bands bearing common fragile sites. *Proceedings of the National Academy of Sciences of the United States of America*, *102*, 18069–18074.

Ge, X. Q., & Blow, J. J. (2010). Chk1 inhibits replication factory activation but allows dormant origin firing in existing factories. *The Journal of Cell Biology*, *191*, 1285–1297.

Glover, T. W., Berger, C., Coyle, J., & Echo, B. (1984). DNA polymerase-alpha inhibition by Aphidicolin induces gaps and breaks at common fragile sites in human-chromosomes. *Human Genetics*, *67*, 136–142.

Hills, S. A., & Diffley, J. F. (2014). DNA replication and oncogene-induced replicative stress. *Current Biology*, *24*, R435–R444.

Lecona, E., & Fernandez-Capetillo, O. (2014). Replication stress and cancer: It takes two to tango. *Experimental Cell Research*, *329*, 26–34.

Minocherhomji, S., Ying, S., Bjerregaard, V. A., Bursomanno, S., Aleliunaite, A., Wu, W., et al. (2015). Replication stress activates DNA repair synthesis in mitosis. *Nature*, *528*, 286–290.

Pedersen, R. T., Kruse, T., Nilsson, J., Oestergaard, V. H., & Lisby, M. (2015). TopBP1 is required at mitosis to reduce transmission of DNA damage to G1 daughter cells. *The Journal of Cell Biology*, *210*, 565–582.

Ren, L., Chen, L., Wu, W., Garribba, L., Tian, H., Liu, Z., et al. (2017). Potential biomarkers of DNA replication stress in cancer. *Oncotarget*, *8*, 36996–37008.

Richards, R. I. (2001). Fragile and unstable chromosomes in cancer: Causes and consequences. *Trends in Genetics*, *17*, 339–345.

Vassilev, L. T. (2006). Cell cycle synchronization at the G2/M phase border by reversible inhibition of CDK1. *Cell Cycle*, *5*, 2555–2556.

CHAPTER FOUR

An Approach to Detect and Study DNA Double-Strand Break Repair by Transcript RNA Using a Spliced-Antisense RNA Template

Havva Keskin, Francesca Storici[1]
Georgia Institute of Technology, Atlanta, GA, United States
[1]Corresponding author: e-mail address: storici@gatech.edu

Contents

Abstract

A double-strand break (DSB) is one of the most dangerous DNA lesion, and its repair is crucial for genome stability. Homologous recombination is considered the safest way to repair a DNA DSB and requires an identical or nearly identical DNA template, such as a sister chromatid or a homologous chromosome for accurate repair. Can transcript RNA serve as donor template for DSB repair? Here, we describe an approach that we developed to detect and study DNA repair by transcript RNA. Key features of the method are: (i) use of antisense (noncoding) RNA as template for DSB repair by RNA, (ii) use of intron splicing to distinguish the sequence of the RNA template from that of the DNA that generates the RNA template, and (iii) use of a *trans* and *cis* system to study how RNA repairs a DSB in homologous but distant DNA or in its own DNA, respectively. This chapter provides details on how to use a spliced-antisense RNA template to detect and study DSB repair by RNA in *trans* or *cis* in yeast cells. Our approach for detection of DSB repair by RNA in cells can be applied to cell types other than yeast, such as bacteria, mammalian cells, or other eukaryotic cells.

Methods in Enzymology, Volume 601
ISSN 0076-6879
https://doi.org/10.1016/bs.mie.2017.11.026

1. INTRODUCTION

Chromosomal DNA is highly at risk of being damaged as a result of metabolic processes or environmental factors, and repair of damaged DNA is critical for maintaining genome stability. A double-strand break (DSB) is a highly dangerous DNA lesion. A DSB can be generated by endogenous and exogenous sources, and its repair is crucial for genome stability (Bhattacharjee & Nandi, 2016). If a DSB is not repaired properly, it could cause mutations and genome rearrangements (Symington & Gautier, 2011). There are two major types of DSB repair mechanisms: non-homologous end-joining (NHEJ) and homologous recombination (HR) (Liu & Huang, 2014; San Filippo, Sung, & Klein, 2008). Broken DNA ends are religated together often exploiting microhomologies in NHEJ, and the DSB is repaired without having a homologous DNA template. NHEJ is mostly an error-prone process due to the possibility of introducing small insertions or deletions (Bhattacharjee & Nandi, 2016; Ma et al., 2005). HR utilizes a homologous template such as a sister chromatid or homologous chromosome to repair the DSB; thus, DSB repair by HR is much more accurate than repair via NHEJ (Bhattacharjee & Nandi, 2016).

It has been shown that RNA can be converted into complementary DNA, cDNA, via reverse transcription and can transfer genetic information to DNA in retroviruses, retrotransposons, and telomere synthesis (Autexier & Lue, 2006; Baltimore, 1985). Early studies showed that the reverse transcriptase (RT) enzyme from the transposon of yeast (Ty) can reverse transcribe not only Ty RNA but also a generic RNA to produce cDNA, and these cDNAs can recombine with homologous DNA, integrate into DNA (Curcio & Garfinkel, 1991; Derr & Strathern, 1993; Derr, Strathern, & Garfinkel, 1991), or be captured at the site of a DSB via NHEJ (Moore & Haber, 1996; Teng, Kim, & Gabriel, 1996).

Our group recently demonstrated that also transcript RNA can be a direct template for DNA DSB repair in budding yeast (Keskin et al., 2014). We found that in the absence of functional RNases H1 and H2, which are enzymes that cleave RNA in an RNA/DNA hybrid (Cerritelli & Crouch, 2009), an RNA transcript can directly repair a DSB induced in the same DNA locus generating the repairing transcript (in *cis*), or in a homologous but different locus from the one generating the repairing transcript (in *trans*). Moreover, we showed that DSB repair by RNA in *cis* is more efficient than in *trans*. These data suggest that DSB repair might be due to the proximity of the transcript to the target broken DNA (Keskin et al., 2014). To distinguish DSB repair by RNA from repair by cDNA, we deleted the *SPT3* gene, which has an important role in the transcription of the RT from the yeast retrotransposons (Tys), and thus in formation of cDNA (Boeke, Styles, & Fink, 1986; Keskin et al., 2014; Teng et al., 1996).

Our method to detect DSB repair by RNA using a spliced-antisense RNA template has the following key features. First, the template for DSB repair is a noncoding antisense RNA, which cannot be translated into a functional protein, thus its transcription does not lead to a phenotype. The antisense RNA has the correct sequence of marker gene that is disrupted by an artificial intron. Second, because the artificial intron can only be spliced from the antisense and not from the sense RNA of the marker, only the antisense RNA can be devoid of the intron and serve as donor to restore the functional sequence of the marker gene. After splicing, the antisense RNA is different from the same DNA sequence that was used to generate the antisense RNA. Thus, if the DNA that generates the antisense RNA is used as template for DSB repair, the marker will remain nonfunctional, still disrupted by the intron sequence. Furthermore, we developed two systems to detect DSB repair by RNA in *trans* or in *cis*. In the *trans* system, the spliced-antisense RNA serves as template to repair a DSB in a homologous but distant locus. In the *cis* system, the spliced-antisense RNA serves as template to repair a DSB in its own DNA gene.

The role of RNA in DNA break repair needs to be further investigated. The approach and systems described here to detect DSB repair by transcript RNA are useful to provide further important insights to better understanding properties and functions of the newly discovered HR mechanism between RNA and DNA molecules in eukaryotic cells. In this chapter, we describe the methodology for the *trans* and *cis* systems for RNA-templated DSB repair in yeast *Saccharomyces cerevisiae* using a spliced-antisense RNA as template for DSB repair. Moreover, similar systems can be developed to detect DNA DSB repair in different cell types.

2. REPAIR SYSTEMS TO DETECT RNA-TEMPLATED REPAIR

We developed two experimental systems to detect DSB repair by RNA in yeast *S. cerevisiae* by using the following background strain FRO-767, which is the haploid strain (*leu2::HOcs, mata*Δ*::hisG, ho*Δ*,* *hmlΔ::ADE1, hmrΔADE1, ade1, leu2-3,112, lys5, trp1::hisG, ura3-52, ade3::GAL::HO*): the *trans* and the *cis* systems (Keskin et al., 2014).

2.1 The *trans* System

The features of the *trans* system are:

1. The *HIS3* reporter gene is used in this system, and yeast cells are auxotrophic for histidine (His^- cells), thus they cannot grow on His^- medium.
2. On chromosome XV, the endogenous *HIS3* gene is disrupted by the HO endonuclease cutting site. Because the HO cutting site is large (124 bp), the likelihood to have His^+ cells by end joining is very low ($<10^6$).
3. The *HO* endonuclease gene is at the *ade3* locus, and it is expressed by the galactose-inducible promoter (*pGAL1*).
4. On chromosome III, another copy of the *HIS3* gene is inserted in the middle of the *LEU2* gene and is disrupted by an artificial intron (AI) in the antisense orientation that was previously developed (Curcio & Garfinkel, 1991; Dombroski et al., 1994).
5. This nonfunctional *his3* gene on chromosome III has the endogenous *HIS3* promoter in the sense orientation and the galactose-inducible promoter (*pGAL1*) to transcribe the antisense RNA of *his3*.
6. The system on chromosome III was inserted at the *LEU2* locus. The upstream Ty2 sequence was deleted to avoid possible transcription of *his3* from Ty2.
7. Moreover, translation of the repairing transcript does not produce a functional His3 protein. For this reason, the antisense of *HIS3*, which is not translated into functional His3 protein, is used as template for repair of the DSB induced on chromosome III.
8. The *trans* system is designed to test the repair capacity of the spliced *his3* antisense transcript from chromosome III to repair the broken *his3* allele on chromosome XV, in which the DSB is occurring in the middle of the *his3* gene by HO endonuclease cleavage (Fig. 1).
9. After the artificial intron is spliced out from the *his3* antisense transcript, the antisense transcript has perfect *HIS3* sequence to be used as a template to repair the broken *his3* region on chromosome XV.
10. The broken ends of the *his3* gene in the *trans* system can be repaired to functional *HIS3* by recombination with a homologous template after induction of the HO DSB. This template could be a cDNA of *his3*

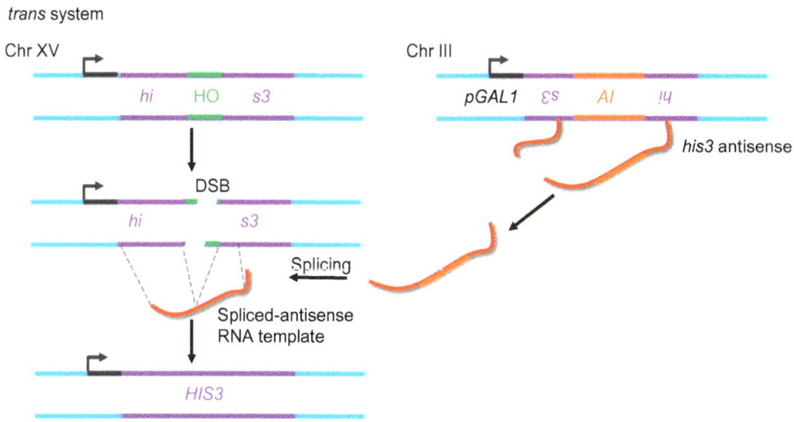

Fig. 1 Scheme of *trans* system to detect RNA-templated DSB repair. *AI*, artificial intron (*orange*); *Chr III*, chromosome III; *Chr XV*, chromosome XV; *HOcs*, homothallic switching endonuclease (*green*) cutting site; *pGAL1*, galactose-inducible promoter, *right turn black arrow*, promoter of *HIS3* or *pGAL1*.

or the antisense transcript RNA of *his3*. Because Ty transposons are very abundant in yeast, Ty RT could reverse transcribe the *his3* antisense transcript into cDNA. To distinguish the repair by *his3* transcript RNA from repair by its cDNA, we deleted the *SPT3* gene. Inactivation of *SPT3* minimizes Ty RT activity in yeast cells.

2.2 The *cis* System

The features of the *cis* system are:

1. The *HIS3* reporter gene is used in this system, and yeast cells are auxotrophic for histidine (His$^-$ cells), thus they cannot grow on His$^-$ medium.
2. The *HIS3* gene is disrupted by an AI in the antisense orientation that was previously developed (Curcio & Garfinkel, 1991; Dombroski et al., 1994).
3. The *his3* allele disrupted by the AI was inserted on chromosome III at the *LEU2* locus. The upstream Ty2 sequence was deleted to avoid possible transcription of *his3* from Ty2.
4. This nonfunctional *his3* gene on chromosome III has the endogenous *HIS3* promoter in the sense orientation and the galactose-inducible promoter (*pGAL1*) to transcribe the antisense RNA of *his3*.
5. The HO endonuclease cutting site was inserted in the middle of the AI which is located inside the *his3* gene on chromosome III. Because the

HO cutting site is large (124 bp), the likelihood to have His$^+$ cells by end joining is very low $(<10^6)$.

6. Moreover, translation of the repairing transcript does not produce a functional His3 protein. For this reason, the antisense of *HIS3*, which is not translated into functional His3 protein, is used as template for repair.
7. The *cis* system is designed to test the repair capacity of the *his3* antisense transcript to repair a DSB occurring in its own DNA, in which the HO endonuclease cutting site is inserted in the middle of the AI (chromosome III) (Fig. 2).
8. After the artificial intron is spliced out from the *his3* antisense transcript, the antisense transcript has perfect *HIS3* sequence to be used as a template to repair the broken *his3* gene on chromosome III.
9. The repairing transcript and the HO gene are regulated by the galactose-inducible promoter (*pGAL1*), and yeast cells are auxotrophic for histidine (His$^-$ cells), thus they cannot grow on His$^-$ media.
10. The broken ends of the *his3* gene in the *cis* system can be repaired to functional *HIS3* by recombination with a homologous template after induction of the HO DSB. This template could be a cDNA of *his3* or the spliced-antisense RNA transcript of *his3*. Because Ty transposons are very abundant in yeast, Ty RT could reverse transcribe the *his3* antisense transcript into cDNA. To distinguish the repair by *his3* transcript RNA from repair by its cDNA, we deleted the *SPT3* gene. Inactivation of *SPT3* minimizes Ty RT activity in yeast cells.

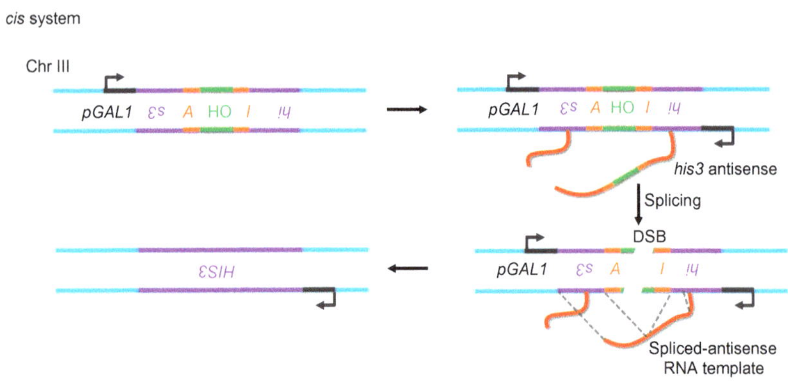

Fig. 2 Scheme of *cis* system to detect RNA-templated DSB repair. *AI*, artificial intron (*orange*); *Chr III*, chromosome III; *HOcs*, homothallic switching endonuclease (*green*) cutting site; *pGAL1*, galactose-inducible promoter, *right turn black arrow*, endogenous *HIS3* promoter or *pGAL1*.

2.3 Note

In the *trans* system, HO DSB repair can occur via recombination with the DNA sequence that carries the *his3* gene disrupted by AI. Similarly in the *cis* system, if the DSB occurs after DNA replication when both sister chromatids are present and only one sister chromatid has the DSB, the intact *his3* sequence of one chromatid could serve as donor to repair the broken *his3* sequence on the other chromatid. The presence of the AI serves to discriminate DSB repair by the spliced-antisense RNA template from DSB repair by the *his3* DNA region disrupted by AI that generates the antisense RNA transcript. After the AI is spliced out from the antisense RNA transcript, spliced-antisense RNA template can repair the DSB and restore the sequence of the functional *HIS3* gene. Differently, if there is repair by *his3* DNA disrupted by AI, the *his3* gene will remain nonfunctional, and cells be His^-.

3. ASSAYS TO DETECT RNA-TEMPLATED DSB REPAIR IN *CIS* AND *TRANS* SYSTEMS

Here we describe two assays to detect DSB repair by RNA using a spliced-antisense RNA template in the *trans* and *cis* systems. The plating assay is an easy and visual assay based on replica-plating. The fluctuation assay is a quantitative assay to determine the frequency of His^+ colonies after DSB induction in *his3* and repair by the spliced-antisense RNA template.

3.1 Patching Assay

3.1.1 Materials

- Autoclave
- Petri dishes (plates)
- YPD solid medium: Combine 10 g of yeast extract (1%), 20 g peptone (2%), 20 g dextrose (2%), and 20 g (2%) agar. Add water to 1 L, autoclave and then pour into Petri dishes.
- YPGal solid medium: Combine 10 g of yeast extract (1%), 20 g peptone (2%), and 20 g (2%) agar, add water to 900 mL, and autoclave. Dissolve 20 g galactose in 100 mL sterile water and filter sterilize. Cool the media to 55–60°C and slowly add 100 mL of galactose, then pour into Petri dishes
- His^- solid medium: Combine 1.7 g yeast nitrogen base (0.17%), 5 g ammonium sulfate (0.5%), 20 g dextrose (2%), 20 g agar (2%), and amino

acids (L-tryptophan, L-arginine hydrochloride, L-methionine, L-tyrosine, L-leucine, L-lysine hydrochloride, L-phenylalanine, L-glutamic acid, L-aspartic acid, L-valine, L-uracil, L-threonine, L-serine, adenine sulfate). Add water to 1 L, autoclave, and then pour into Petri dishes.

- Burner
- Yeast strains of the *cis* and/or *trans* system
- Sterile toothpicks
- Replica-plating tool
- Sterile velveteen squares for replica-plating
- 30°C incubator

3.1.2 Procedure

1. Prepare all the media beforehand.
2. Yeast cells are patched when burner is on and grown on YPD (rich medium) for 1 day.
3. YPD plates are replica-plated on YPGal medium.
4. Incubate for 2 days in a 30-C incubator.
5. YPGal plates are replica-plated on His$^-$ medium and incubated for 3 days.
6. Photograph His$^-$ plates.

3.1.2.1 Note

If there is a significant cell background on His$^-$ plates after replica-plating, the His$^-$ plates can be replica-plating on new His$^-$ plates to eliminate the background (Fig. 3).

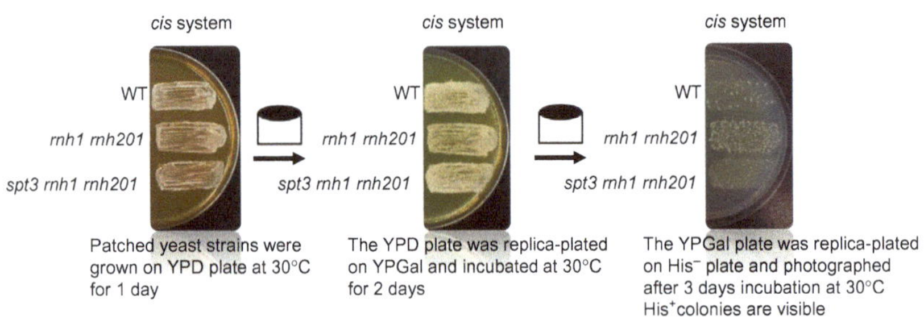

Fig. 3 Scheme of patching assay. In this example, the yeast strains with different backgrounds of the *cis* system were used for patching assay and photographed.

3.2 Fluctuation Assay

3.2.1 Materials

- Autoclave
- Burner
- Petri dishes (plates)
- Lactic acid
- YPLac liquid medium. Combine 10 g yeast extract (1%), 20 g peptone (2%), and 27 mL lactic acid (2.7%). Add water to 1 L, adjust pH to 5.5 with lactic acid, and then autoclave
- YPGal solid medium
- His^- solid medium
- 150-mL sterile Erlenmeyer flasks
- Sterile toothpicks
- Yeast strains of *cis* and/or *trans* system
- 50-mL falcon tubes
- Centrifuge (with a rotor of 15 cm radius used for 50- and 15-mL tubes)
- Microcentrifuge (with a rotor of 7 cm radius used for 1.5-mL tubes)
- Micropipettes
- Sterile tips
- Vortex
- Hemocytometer
- 30°C shaking incubator
- Sterile RNases H free water
- Glass beads
- Replica-plating tool
- Sterile velveteen squares (for replica-plating)
- Sonicator
- 30°C incubator

3.2.2 Procedure

This procedure is an example for testing chosen strain in the *cis* or *trans* system to determine the frequency of His^+ colonies as result of RNA-templated DSB repair. At least two isolates should be inoculated from each chosen strain sample into two different cultures.

1. Prepare all the media beforehand.
2. Yeast cells are inoculated by a sterile toothpick in a 150-mL sterile Erlenmeyer flask with 50 mL lactose medium (YPLac) when burner

is on. The amount of yeast cells taken for the inoculations is around 5 mm^2 (size of a flat top of toothpick). Yeast cells are incubated overnight by shaking at 30°C.

3. Transfer each cell culture into a 50-mL falcon tube and centrifuge at 1500 × *g* (3000 rpm) for 2 min.
4. Discard the supernatant.
5. Transfer each pellet after centrifugation into a 1.5-mL sterile microcentrifuge tube.
6. Perform 10^3 dilution by using sterile RNase free water for counting the cells.
7. Add 10 μL of each diluted sample into a hemocytometer to count the cells.
8. Calculate the number of cells per 1 μL.
9. Prepare YPGal plates by adding 8–12 sterile glass beads.
10. Plate 10^7 cells on three identical YPGal plates per each sample for replica-plating on His^-. Plate 10^4 or 10^5 cells on two identical YPGal plates per sample to calculate the survival.
11. Incubate YPGal plates at 30°C for 2 days.
12. After 2-day incubation, replica-plate YPGal plates on His^- medium.
13. Incubate His^- plates at 30°C for 3 days.
14. Count and record the number of His^+ colonies.
15. Calculate His^+ frequency by this formula: (# of colonies on His^+ medium/# of colonies on YPGal medium (survival plate) * dilution rate) * 10^7 viable cells.
16. Calculate survival by this formula: (# of colonies on YPGal medium/# of cells plated on YPGal medium) * 100.
17. Conduct statistical analysis of data and pairwise comparison by using two-tailed Mann–Whitney *U*-test (Heyer, Ehmsen, & Liu, 2010).

3.2.2.1 Notes

1. During cells counting, if cells are clumpy, they can be separated by using a sonicator.
2. Different strains genotypes can affect the growth of the cells such as some recombination mutants. In this case, plate 10^8 cells instead of 10^7 and plate 10^5 cells instead of 10^4 on YPGal plates (Fig. 4).

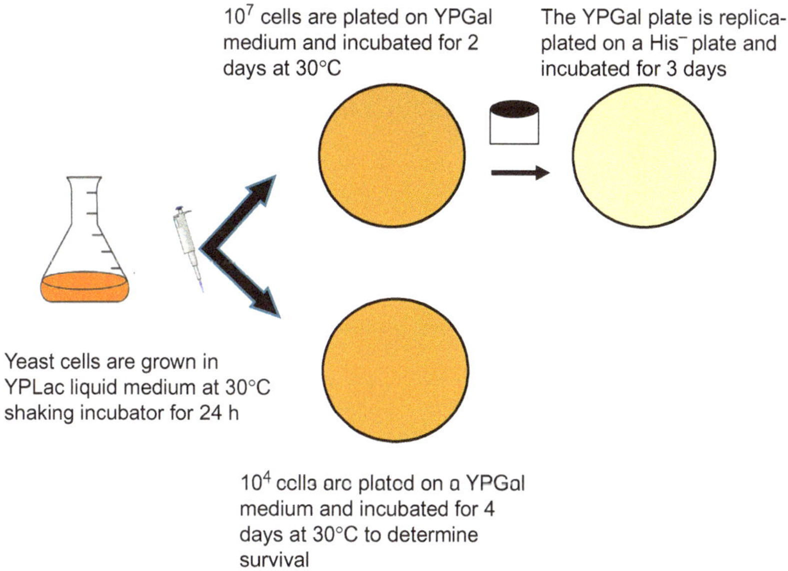

Fig. 4 Scheme of fluctuation assay. Visual representation of the steps of the fluctuation assay.

REFERENCES

Autexier, C., & Lue, N. F. (2006). The structure and function of telomerase reverse transcriptase. *Annual Review of Biochemistry*, *75*, 493–517.

Baltimore, D. (1985). Retroviruses and retrotransposons: The role of reverse transcription in shaping the eukaryotic genome. *Cell*, *40*(3), 481–482.

Bhattacharjee, S., & Nandi, S. (2016). Choices have consequences: The nexus between DNA repair pathways and genomic instability in cancer. *Clinical and Translational Medicine*, *5*(1), 45.

Boeke, J. D., Styles, C. A., & Fink, G. R. (1986). Saccharomyces cerevisiae SPT3 gene is required for transposition and transpositional recombination of chromosomal Ty elements. *Molecular and Cellular Biology*, *6*(11), 3575–3581.

Cerritelli, S. M., & Crouch, R. J. (2009). Ribonuclease H: The enzymes in eukaryotes. *The FEBS Journal*, *276*(6), 1494–1505.

Curcio, M. J., & Garfinkel, D. J. (1991). Single-step selection for Ty1 element retrotransposition. *Proceedings of the National Academy of Sciences of the United States of America*, *88*(3), 936–940.

Derr, L. K., & Strathern, J. N. (1993). A role for reverse transcripts in gene conversion. *Nature*, *361*(6408), 170–173.

Derr, L. K., Strathern, J. N., & Garfinkel, D. J. (1991). RNA-mediated recombination in S. cerevisiae. *Cell*, *67*(2), 355–364.

Dombroski, B. A., et al. (1994). An in vivo assay for the reverse transcriptase of human retrotransposon L1 in Saccharomyces cerevisiae. *Molecular and Cellular Biology*, *14*(7), 4485–4492.

Heyer, W. D., Ehmsen, K. T., & Liu, J. (2010). Regulation of homologous recombination in eukaryotes. *Annual Review of Genetics*, *44*, 113–139.

Keskin, H., et al. (2014). Transcript-RNA-templated DNA recombination and repair. *Nature*, *515*(7527), 436–439.

Liu, T., & Huang, J. (2014). Quality control of homologous recombination. *Cellular and Molecular Life Sciences*, *71*(19), 3779–3797.

Ma, Y., et al. (2005). Repair of double-strand DNA breaks by the human nonhomologous DNA end joining pathway: The iterative processing model. *Cell Cycle*, *4*(9), 1193–1200.

Moore, J. K., & Haber, J. E. (1996). Capture of retrotransposon DNA at the sites of chromosomal double-strand breaks. *Nature*, *383*(6601), 644–646.

San Filippo, J., Sung, P., & Klein, H. (2008). Mechanism of eukaryotic homologous recombination. *Annual Review of Biochemistry*, *77*, 229–257.

Symington, L. S., & Gautier, J. (2011). Double-strand break end resection and repair pathway choice. *Annual Review of Genetics*, *45*, 247–271.

Teng, S. C., Kim, B., & Gabriel, A. (1996). Retrotransposon reverse-transcriptase-mediated repair of chromosomal breaks. *Nature*, *383*(6601), 641–644.

CHAPTER FIVE

Assaying Repair at DNA Nicks

Luther Davis[1], Yinbo Zhang, Nancy Maizels
University of Washington, Seattle, WA, United States
[1]Corresponding author: e-mail address: lwd@uw.edu

Contents

Abstract

Nicks are the most common form of DNA damage, but they have only recently been shown to initiate damage that requires repair. Analysis of the pathways of nick repair in human cells has benefited from the development of enzymes that target nicks to specific sites in the genome and of reporters that enable rapid analysis of homology-directed repair and mutagenic end joining. Nicks undergo efficient repair by single-stranded oligonucleotide donors complementary to either the nicked or intact DNA strand, via pathways that are normally suppressed by RAD51. Here we discuss the details of reporter assays that take advantage of the convenience and sensitivity of flow cytometry to analyze pathways of repair at targeted DNA nicks. These assays are readily carried out in 96-well format cell culture plates, enabling mechanistic questions to be addressed by determining the contributions of specific factors by depletion and/or ectopic expression.

Methods in Enzymology, Volume 601
ISSN 0076-6879
https://doi.org/10.1016/bs.mie.2017.12.001

1. INTRODUCTION

Nicks are the most common form of DNA damage but were long thought to undergo immediate ligation that prevented them from initiating recombination or mutagenesis. The demonstration that nicks initiate efficient homology-directed repair (HDR) and mutagenic end-joining (mutEJ) was met with some surprise, as many believed that nicks could only initiate recombination if they escaped ligation and were converted to DSBs upon replication. Nonetheless, there were clear biological precedents for nick-initiated HDR. The mechanism of immunoglobulin gene diversification in all fowl depends on regulated gene conversion that is initiated by nicks that result from processing sites that have undergone C to U deamination by AID (Maizels, 2005). Moreover, nicks targeted by the phage f1 gene II protein (Strathern, Weinstock, Higgins, & McGill, 1991) or by derivatives of the RAG proteins (Lee, Neiditch, Salus, & Roth, 2004) had been shown to initiate HDR. However, those systems did not provide experimental traction for systematic studies of nick repair.

The availability of enzymes that reliably target nicks to specific sites has enabled detailed analyses of the pathways of nick repair. The first "nickase" was generated by mutation of one of two active sites in the monomeric *S. cerevisiae* meganuclease, I-SceI (Niu, Tenney, Li, & Gimble, 2008). Nick repair became of particular interest when a nickase derivative of the I-AniI meganuclease was subsequently shown to initiate HDR in a reporter in the human genome at a frequency only one-third below the frequency of DSB-initiated HDR at the same site (McConnell Smith et al., 2009). Nick-initiated HDR is advantageous for genome engineering and gene therapy, as nicks initiate far less mutEJ than DSBs, whether targeted by nickases derived from meganucleases (Davis & Maizels, 2011), zinc finger nucleases (Ramirez et al., 2012; Wang et al., 2012), or CRISPR/Cas9 (Cong et al., 2013; Ran et al., 2013; Trevino & Zhang, 2014). Direct comparison of repair outcomes at targeted nicks and DSBs has revealed that even though HDR occurs by different pathways at nicks and DSBs, comparable efficiency can be achieved if repair is routed along the optimum pathway.

The structure of the exogenous repair donor determines the pathway of HDR (Fig. 1). At both nicks and DSBs, double-stranded plasmid DNA donors support HDR by the well-characterized canonical pathway in which RAD51 forms filaments on the exposed 3′ end of the target DNA strand to enable invasion of the donor duplex (Heyer, Ehmsen, & Liu, 2010; Jasin &

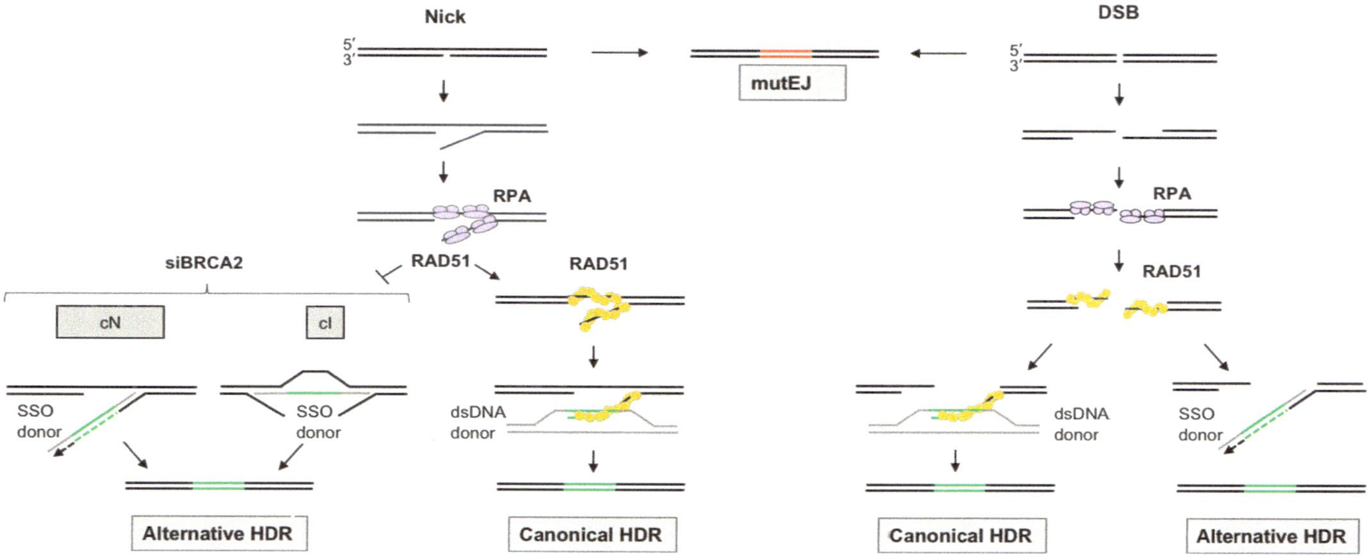

Fig. 1 Repair of targeted nicks and DSBs by exogenous donors. Targeted nicks (*left*) undergo repair by HDR or mutEJ. HDR depends upon RPA, and the pathway is determined by donor structure. HDR supported by SSO donors is inhibited by RAD51, which appears to promote reannealing of complementary strands to make them inaccessible to SSOs. SSO donors support very efficient alternative HDR at nicks under conditions which prevent RAD51 loading onto DNA (e.g., siBRCA2, shown) or inhibit RAD51 activity. Two pathways support alternative HDR at nicks by SSO donors, with pathway use determined by whether the SSO is complementary to the nicked (cN) or intact (cI) strand of the target. dsDNA donors can also support canonical HDR at nicks, in a pathway that requires RAD51, as it does at DSBs. Targeted DSBs (*right*) can undergo repair by HDR or mutEJ, with NHEJ as the major mutEJ pathway. DSBs are immediately resected to generated free 3′ ends, eliminating the possibility of reannealing. RPA binds the free ends and is replaced by RAD51, which promotes invasion of duplex DNA donors and canonical HDR. SSO donors can also support efficient alternative HDR at DSBs, via a pathway that is independent of RAD51. Donors, *gray*; heterologous region, *green*; mutated region, *red*; newly synthesized DNA, *dotted line*.

Rothstein, 2013; Symington & Gautier, 2011). Canonical HDR requires RAD51 as well also factors that load RAD51 on DNA, including BRCA2, PALB2, and DSS1, and frequencies of HDR supported by dsDNA donors are reduced upon depletion or inactivation of these factors.

Single-stranded oligonucleotide (SSO) donors have attracted considerable recent interest as they are short-lived and do not readily integrate into the genome, making them especially useful for genome engineering and gene therapy applications. SSOs are also cheap to synthesize and easy to use. HDR by SSO donors proceeds via alternative HDR pathways distinct from canonical HDR (Fig. 1). HDR by SSO donors is suppressed by RAD51/BRCA2 at nicks and independent of RAD51/BRCA2 at DSBs (Davis & Maizels, 2014, 2016). An explanation for this difference is that binding by RAD51 may prevent annealing to an exogenous SSO donor by promoting reannealing of the two strands of the target. In contrast DSBs are resected immediately after cleavage, eliminating the complementary strand and exposing single-stranded 3′ ends for annealing to an exogenous donor.

Frequencies of HDR at nicks by SSO donors can be increased by a variety of treatments that prevent RAD51 loading onto DNA (Davis & Maizels, 2014, 2016), including depletion of RAD51 itself by siRAD51 treatment; inhibition of RAD51 activity by expression of the dominant-negative mutant, $RAD51^{K133R}$; depletion of factors necessary to load RAD51 onto DNA (siBRCA2, siPALB2, siDSS1); or expression of the BRC3 peptide, which interferes with the BRCA2/RAD51 interaction necessary to load RAD51 on DNA. These treatments stimulate HDR at nicks by SSO donors to achieve frequencies comparable to HDR at DSBs by dsDNA donors.

2. CHROMOSOMAL REPORTERS FOR ANALYSIS OF REPAIR BY FLOW CYTOMETRY

2.1 Chromosomal Reporters to Quantify Repair by Exogenous Donors

To analyze repair at nicks, we have used chromosomal recombination reporters designed to report on repair of a target gene supported by DNA donors supplied exogenously. Interest in exogenous donors reflects their utility in genome editing in contexts in which no endogenous donor is available. The use of exogenous donors was central to the establishing that donor structure can determine HDR pathway, and that distinct pathways promote HDR by donors complementary to the nicked or intact strand (Fig. 1).

The recombination reporters are fluorescent proteins which generate a signal only following repair. Repair events are quantified by flow cytometry, a convenient and very sensitive method to quantify repair, reproducibly detecting events that occur with frequencies as low as 10^{-4}. The use of flow cytometry to quantify recombination in mammalian cells was pioneered by Jasin and collaborators, who developed the DR-GFP reporter construct in which a defective GFP gene and its homologous repair cassette are carried on the same DNA molecule and separated by about 3kb (Pierce, Johnson, Thompson, & Jasin, 1999).

The assay procedures described below have been designed to achieve HDR frequencies of 1%–5% and reliably report on modulations in assay conditions that stimulate or inhibit HDR or mutEJ. Transduction of cells with RNPs containing both CRISPR gRNAs and Cas9 nuclease makes it possible to achieve very high (≥50%) editing frequencies (Kim, Kim, Cho, Kim, & Kim, 2014; Liao et al., 2015) and to edit almost any target gene in almost any primary or transformed cell type. Very high frequencies of editing are critical for experiments that use repair as a tool to engineer a specific gene or a specific cell type, but not for studies of mechanism which require a dynamic assay range capable of detecting both increased and decreased repair frequencies.

2.2 Traffic Light Reporter Construct

The "Traffic Light" (TL) green/red reporter (Certo et al., 2011) enables both HDR and mutEJ to be measured at nicks and DSBs targeted to the same site.

2.2.1 Key Genetic Elements of the Reporter

The TL reporter (Certo et al., 2011) carries a defective GFP gene joined to an mCherry gene by a linker that encodes the T2A self-cleaving peptide (Szymczak et al., 2004), and an upstream promoter that drives expression of a transcript that includes the GFP, T2A, and mCherry cassettes (Fig. 2A). Translation utilizes a single ATG start codon just upstream of the GFP gene. Cells expressing this transcript are GFP^-: they do not synthesize functional GFP because the GFP gene is interrupted by a short (38bp) insertion which bears two in-frame stop codons, 121 and 127nt downstream of the ATG start codon (Fig. 2B). They are also $mCherry^-$: they do not synthesize functional mCherry because the mCherry gene is not in the correct reading frame for translation. Nicks or DSBs targeted to (or very near) the 38bp insertion in GFP can initiate either HDR or mutEJ.

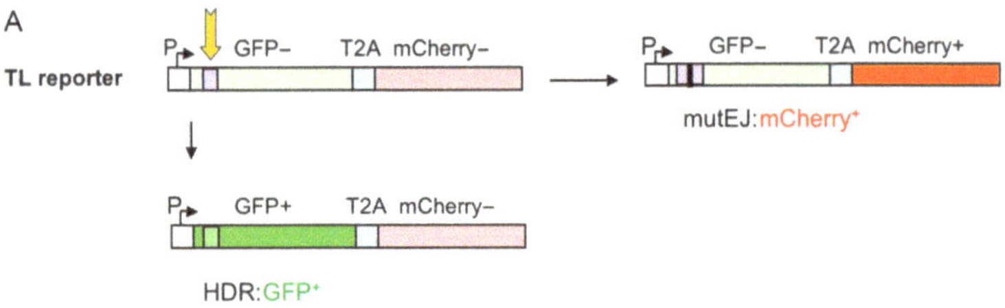

B

GFP insertion

5′-atggtgagcaagggcgagggggctgttcaccggggtggtgcccatcctggtcgagcTGGACGGCGACG
TAAACGGCCACAAGTTCAGCGTGTCCGGCctcgaccgtgaggaggtttctctgtaaagctaagagctCAC
CTACGGCAAGCTGACCCTGAAGTTCATCTGCACCACCGgcaagctgcccgtgccctggcccaccctcgt-3′

C

Correction cassettes

TOP: 5′-TGGACGGCGACGTAAACGGCCACAAGTTCAGCGTGTCCGGCgagggcgagggcgatgcCAC
CTACGGCAAGCTGACCCTGAAGTTCATCTGCACCACCG-3′

BOT: 5′-CGGTGGTGCAGATGAACTTCAGGGTCAGCTTGCCGTAGGTGgcatcgccctcaccctcGCC
GGACACGCTGAACTTGTGGCCGTTTACGTCGCCGTCCA-3′

BOT A1H3: 5′-CGGTGGTGCAGATGAACTTCAGGGTAAGCTTGCCGTAGGTGgcatcgccctcaccctcGCC
GGACACGCTGAATTTGTGGCCGTTTACGTCGCCGTCCA-3′

Fig. 2 Traffic Light (TL) reporter assay for HDR and mutEJ. (A) The TL reporter (Certo et al., 2011) carries a GFP gene rendered defective by a 38 bp insertion (*purple*) joined by a linker that encodes the T2A self-cleaving peptide (Szymczak et al., 2004) to an mCherry gene. Repair is initiated by nicks or DSBs targeted to the insertion (*arrow*). An upstream SFFV promoter (P) drives expression of a transcript that includes the GFP, T2A, and mCherry cassettes. Translation initiates at the ATG start codon of the GFP gene (+1 reading frame). There are two in-frame stop codons within the 38 bp insertion in GFP that prevent its expression. HDR that replaces the 38 bp insertion with a 17 bp correction cassette (*light green*) will correct the defective GFP gene (*dark green*), resulting in GFP^+ cells. The mCherry gene encodes functional protein when read in the +2 reading frame, but is out of frame when translation initiates at the ATG at +1. mutEJ that shifts mCherry into the correct +2 reading frame (*red*) will result in $mCherry^+$ cells. (B) Sequence of the region of the TL reporter including the ATG start codon of the GFP gene (*underlined*); the 38 bp insertion in GFP (*lowercase purple font*) containing two TAA stop codons in frame with the initiating ATG at positions 121 and 127 (*underlined*); and out-of-frame stop codons at positions 68 and 105 (TAA and TGA, respectively; *red*). Regions homologous with 99 nt SSOs used for gene correction are shown in *uppercase*; other regions in *lowercase*. (C) Sequences of 99 nt SSOs used for gene correction. Each carries a 17 nt region (*green lowercase font*) that when swapped for the 38 bp insertion in the TL reporter (*purple lowercase font* in (B)) will correct the defective GFP gene and enable GFP expression. Homology arms, *uppercase*. TOP: In this SSO, the regions that flank the correction cassette are identical to the nontranscribed (coding) strand and complementary to the transcribed (noncoding) strand in the reporter. This SSO will support HDR by the cI pathway if the nontranscribed strand is nicked, and by the cN pathway if the transcribed strand is nicked. BOT: Complement of TOP. In this SSO, the regions that flank the correction cassette are identical to the transcribed strand and complementary to the nontranscribed strand in the reporter. BOT-A1H3: BOT modified by two single base changes (*underlined*) to carry *Apo*I (RAATTY) and *Hin*dIII (AAGCTT) sites at indicated positions.

2.2.2 Scoring HDR

GFP gene correction depends upon provision of an exogenous donor bearing a correction cassette flanked by regions homologous to the target (e.g., Fig. 2C). HDR that swaps the correction cassette for the insertion in the GFP gene will generate a sequence that encodes a functional GFP protein in the +1 reading frame resulting in GFP^+ cells. The requirement for replacement of the 38 nt insertion ensures that GFP^+ cells are not the result of reversion mutations, and eliminates the risk of secondary nuclease cutting of the repaired reporter that might occur if only single-nucleotide changes resulted in GFP^+ cells. Table 1 presents typical frequencies for HDR with this donor. HDR frequencies are significantly greater (3.5- to 13-fold) if only a 1 nt insertion is required (Davis & Maizels, 2016).

2.2.3 Scoring mutEJ

Insertions or deletions (indels) that shift mCherry into the same reading frame as the ATG start codon will enable translation of functional mCherry and result in $mCherry^+$ cells. To enable mCherry expression, the indel must occur within a restricted 17 nt zone of the 38 bp insertion in the GFP gene, downstream of the stop codon in the +2 reading frame (position 105) but upstream of the two stop codons in the +1 reading frame (positions 121 and 127). An indel upstream of the stop codon at position 105 that places mCherry in frame would also place that stop codon in frame, preventing

Table 1 Typical HDR and mutEJ Frequencies

	Treatment	Donor	HDR (%)	mutEJ (%)
Nick	—	SSO	0.05–0.2	0.01–0.1
	siBRCA2	SSO	2–5	0.05–0.5
	—	dsDNA plasmid	0.2–0.5	0.01–0.1
	siBRCA2	dsDNA plasmid	0.010.1	0.3–0.8
DSB	—	SSO	2–4	1–3
	siBRCA2	SSO	2–4	1–3
	—	dsDNA plasmid	1–3	1–3
	siBRCA2	dsDNA plasmid	0.2–0.5	1–2

Frequencies were determined by scoring the fraction of GFP^+ or $mCherry^+$ cells among BFP^+ cells in 293T TL reporter cells. As described in the text, the frequency of mutEJ is determined by a frame-shift assay, based on the fraction of $mCherry^+$ cells, and the actual frequency of mutEJ is predicted to be at threefold higher than the value shown.

translation of mCherry. Similarly, an indel downstream of the stop codon at position 121 would not permit translation to bypass this stop. Only indels into only one of the three possible reading frames result in translation of mCherry, so the frequency of mCherry$^+$ cells is predicted to be only one-third the frequency of mutEJ events. We typically report frequencies as the fraction of BFP$^+$ cells (e.g., Table 1).

2.3 Donors for HDR

Single-stranded DNA donors may be provided by transfection with SSOs from 75 to 120 nt in length, which carry the short (17 bp) correcting sequence flanked by 20–50 bp regions homologous to the target. Duplex DNA donors may be provided by transfection with plasmids which carry the correcting sequence flanked by 200–500 bp regions homologous to the target.

3. CELL LINES AND POPULATIONS HARBORING CHROMOSOMAL REPORTERS

We have used two human cell lines for analysis of recombination: HEK293T and HT1080. HEK293T cells are widely used as a model for recombination, as they grow robustly (16–24 h doubling time) and are readily transfectable. HEK293T cells were derived by SV40 transfection of the human embryonic kidney cell line, HEK293. Expressed T antigen supports the replication of plasmids containing the SV40 origin in these cells and renders p53 and Rb nonfunctional. These cells are deficient in mismatch repair due to silencing of MLH1, and they do not respond to normal cell cycle checkpoint controls (Cejka et al., 2003; Trojan et al., 2002).

HT1080 cells derive from a human fibrosarcoma and have a diploid karyotype, functional p53 and Rb, and they respond to cell cycle checkpoint controls. They are less readily transfectable than 293T cells.

3.1 Generating Cell Lines and Populations Harboring Chromosomal Reporters

Reporter constructs are delivered to cells by transduction using a lentiviral vector, which integrates throughout the genome. Recombination frequencies can be influenced by local chromatin structure and direction and timing of DNA replication, and integration at random sites avoid these potential pitfalls of assays of constructs integrated at a single site. Transduction is carried out at relatively low multiplicity of infection so that cells will carry more a single reporter construct. Following selection of cells bearing the reporter, early passage cell populations or clonal lines are stored and tested for recombination activity.

3.1.1 Lentivirus Production and Transfection

1. Day 0: Seed 1 × 10^5 HEK293T cells in 0.5 mL DMEM media per well in a 24-well plate. If large amounts of lentiviral vector is desired, this can be readily scaled up to greater cell numbers in larger plates (e.g., 3 × 10^6 HEK293T cells in 10 mL in a 10-cm plate).
2. Day 1: Transfect cells with the reporter pTL reporter lentiviral construct (Addgene 31480) and the pMD2.G (Addgene 12259, VSV *env* plasmid) and psPAX2 (Addgene 12260, packaging protein) plasmids in a ratio of 4:1:2 μg pTL:pMD2G:psPAX2 per well. Use Lipofectamine LTX transfection reagent (Life Technologies) following manufacturer's instructions. The pTL reporter lentiviral construct bears a puromycin-resistance marker for selection of integrants.
3. Day 2: Aspirate medium and replace with 0.5 mL prewarmed DMEM.
4. Day 3: Collect supernatant, which will contain lentiviral particles that can transduce the reporter. Supernatant containing virus can be stored at −80°C indefinitely. There is no need to titer, since the optimum amount for transduction will be determined experimentally.

3.1.2 Generating Cell Lines or Populations Harboring the Virus

1. Day 0: Plate cells of interest in 24-well plates (2 × 10^5 cells in 500 μL).
2. Day 1: Infect cells with virus by adding 2.5 μg polybrene (Santa Cruz Biotechnology) to each well, followed by viral supernatant (0, 0.5, 1, 2, 5, and 10 μL—a total of six wells).
3. Day 2: Transfer cells from each well to a 10-cm plate (in 10 mL DMEM), culture 24 h.
4. Day 3: Add puromycin (Sigma) to a final concentration of 0.5 μg/mL by adding one-tenth volume of medium containing 5 μg/mL puromycin.
5. Days 10–13: Determine the number of puromycin-resistant colonies on the 10-cm plates at each dilution of virus (step 2). There should be no colonies on the plate that did not receive virus, and colony frequency should increase with increasing viral supernatant. Plates that are nearly confluent or with very high colony counts should be discarded to reduce the likelihood of multiple reporter insertions per cell, as this may affect repair frequencies and jeopardize the ability to make useful comparisons between different lines or populations.

6. Generate a population of cells harboring the reporter by pooling puromycin-resistant cells from 1 to 3 plates with a total of 100–500 colonies, then expanding to confluence on two 10-cm plates with 10 mL DMEM supplemented with puromycin.
7. Alternatively, generate clonal lines by using a cloning disc to transfer colonies from the 10-cm plate with the fewest colonies (<100) to wells of a 24-well plate containing 1 mL medium supplemented with puromycin. Establishing clones from the plate with fewest colonies minimizes the likelihood of multiple insertions and facilitates establishing lines that originate from single colonies. Expand colonies on one 10-cm plate to confluency.

3.2 Storage and Quality Control of Early Passage Cells

1. Days 14–21: Early passage cells are recovered for long-term storage in a liquid nitrogen-cooled freezer. To store cell populations, recover cells from two 10-cm plates and freeze 10–12 vials (2×10^6 cells/vial). To store clonal lines, recover cells from a single plate and freeze a fraction of cells in 2–3 vials (2×10^6 cells/vial). Expand remaining cells to perform quality control.
2. Verify HDR and mutEJ activity, as described later. If using clonal lines, it is useful to compare HDR frequencies in four to six lines before focusing on one or two for further experiments, as lines may vary as much as 10-fold in HDR frequency, and some lines may carry the reporter but exhibit no HDR.
3. Isolate genomic DNA from 2×10^6 cells and confirm the presence of full-length reporter by PCR. Reporter copy number may be estimated by comparison with a control single-copy gene, taking into account that diploid cells will carry twice as many copies of a single-copy endogenous gene as of a single-copy reporter.
4. Expand remainder of cells for HDR assays and freezing of additional vials.
5. After expansion, puromycin selection may not be required to maintain the chromosomal reporter. This should be determined empirically by comparison of repair frequencies in populations cultured with and without puromycin. If using pooled populations of transfectants, note that the composition of pools may change if puromycin is discontinued as different insertion events within the pool may become transcriptionally inactive at different rates.

4. ASSAYING REPAIR FREQUENCIES

4.1 Cell Culture, siRNA Depletion, and Ectopic Protein Expression

The procedure is described in detail for HEK293T TL cell lines or populations in 96-well format. The procedure can be adapted for other cell types by modifying transfection conditions and grow out times. It can be scaled up if larger samples are necessary.

4.1.1 Cell Culture and Fixation

1. Day 0: Seed cells at approximately 8×10^3 cells per well in 100 µL medium in a 96-well cell culture plate. Cultures proliferate to a density of approximately 1–2 $\times 10^5$ cells per well in the standard 5-day culture period, so each well of a 96-well plate can contain as many as 2×10^5 cells. For each experimental condition, assays are typically carried out in triplicate. In that case three identical wells are set up for each experimental condition and each well is analyzed separately by flow cytometry.
2. Day 2: Transfect cells with Cas9 or Cas9^{D10A} (or other variant) expression plasmids and single-stranded or duplex DNA donors for HDR, in mixes of 20 µL Optimem per well, to which has been added: 30 ng of Cas9 or Cas9^{D10A} expression plasmid, 15 ng of guide RNA (gRNA) expression plasmid, 30 ng of pCVL SFFV d14GFP dsDNA plasmid donor (approximately 0.08 pmol), or 2.5 pmol single-stranded synthetic deoxyoligonucleotide donor and 0.24 µL Lipofectamine LTX. Culture for 3 days.
3. Day 5: Collect samples for analysis by flow cytometry. Aspirate the growth medium and wash the cells once in 50–100 µL PBS. Detach the cells by adding 35 µL 0.05% trypsin per well for 1–2 min.
4. Fix cells by adding 70 µL of 3% formaldehyde in PBS, for a final concentration of 2% formaldehyde
5. Analyze samples by flow cytometry (below).

4.1.2 Assays Involving siRNA Depletion

1. Day 0: Seed cells at approximately 4×10^3 cells per well in 100 µL medium in a 96-well cell culture plate. Reduced growth or viability due to depletion of some factors will reduce cell proliferation. In that case, assays may be carried out in a 24-well plate, by scaling up all

components. Alternatively, it can be convenient to use 96-well plates and pool cells from pairs of wells cultured under identical conditions just prior to analysis by flow cytometry. In anticipation of this, six identical wells are set up for each experimental condition, and cells in individual wells are fixed and then pooled prior to analysis.

2. Day 1: Transfect cells with siRNAs in 10.5 μL total volume per well, containing 0.125 μL RNAiMAX, 0.5 μL of 0.625 μ*M* siRNA, and 9.875 μL of OptiMEM (Life Technologies).
3. Day 2: Transfect cells with Cas9 or $Cas9^{D10A}$ (or other variant) expression plasmids and single-stranded or duplex DNA donors for HDR, in mixes of 20 μL OptiMEM per well, to which has been added: 30 ng of Cas9 or $Cas9^{D10A}$ expression plasmid, 15 ng of gRNA expression plasmid, 30 ng of pCVL SFFV d14GFP dsDNA plasmid donor (approximately 0.08 pmol), or 2.5 pmol single-stranded synthetic deoxyoligonucleotide donor and 0.24 μL Lipofectamine LTX. Culture for 3 days.
4. Day 5: Collect samples for analysis by flow cytometry. Aspirate medium and wash cells as above, detach the cells by adding 35 μL 0.05% trypsin per well for 1–2 min. If pairs of wells are to be pooled, combine trypsinized cells from two identical wells, fix cells by addition of 35 μL of 6% formaldehyde in PBS, for a final concentration of 2% formaldehyde.

4.1.3 Assays Involving Ectopic Protein Expression

1. Day 0: Seed cells as for siRNA depletion assays, culture for 2 days.
2. Day 2: Transfect cells as above, but include in the transfection mix 15 ng of the construct expressing the gene of interest. The amount can be adjusted to increase or decrease the level of expression, with the caveat that high concentrations of plasmids can interfere with transfection efficiency. Samples with no vector and empty vector should be analyzed as controls for the effects of expression.
3. Day 5: Collect and fix cells, as above.

4.2 Monitoring siRNA Depletion or Ectopic Protein Expression

- Western blots are frequently used to monitor protein overexpression or siRNA depletion. It may be necessary to scale up cultures to produce sufficient material for a Western blot. Depending on the potency of the detection antibody, sufficient material for a single lane of the blot may be obtained by culturing an inoculum of 2.5×10^4 cells in 500 μL in a 24-well plate, or further scale up (10^6 cells per 2 mL in a 6-well plate) may be necessary.

- As an alternative to Western blotting, it can be convenient to monitor siRNA depletion or ectopic protein expression by flow cytometry of stained cells. This requires fewer cells than a Western blot and provides a rapid and quantitative measure of individual cells and the population. Antibodies that give specific signals on Western blots do not necessarily do so when used for cell staining, and this must be verified for each antibody. If siRNA efficiently depletes the transcript of interest, verification is readily carried out by comparing the signal of the encoded protein from cells treated with the specific siRNA to cells treated with a nonspecific control siRNA (e.g., siNT2, ID# 4390847 ThermoFisher Scientific).

4.3 Identification of Transfectants and Determination of Transfection Frequency

- Typically, we have coexpressed Cas9 or Cas9^{D10A} and BFP, using expression constructs in which the two genes are separated by a T2A "self-cleaving" peptide (Szymczak et al., 2004). The accuracy of using the coexpressed BFP signal as a proxy for Cas9 or Cas9^{D10A} expression depends on a nearly 1:1 correspondence of expression of Cas9 or Cas9^{D10A} and BFP linked by the T2A peptide, and similar folding kinetics and stability of Cas9 or Cas9^{D10A} and BFP. Cells expressing BFP can first be scored at 16 h posttransfection, and their frequency increases more than 10-fold by 40 h then slowly reaches a peak at 72 h (Fig. 3). However, the BFP signal is transient, while the GFP and mCherry signals reflect permanent changes resulting from HDR or mutEJ, so the latter signals persist while the BFP signal levels off.
- Transfection frequency can also be determined by transfection in parallel of cells with constructs that express BFP or other fluorescent proteins. This allows HDR frequency to be calculated by dividing the frequency of GFP$^+$ cells by the frequency of BFP$^+$ cells. However, parallel measurements are not inherently as accurate as internal controls, and use of the T2A fusion enables transfected cells to be sorted as a separate population for further analysis.

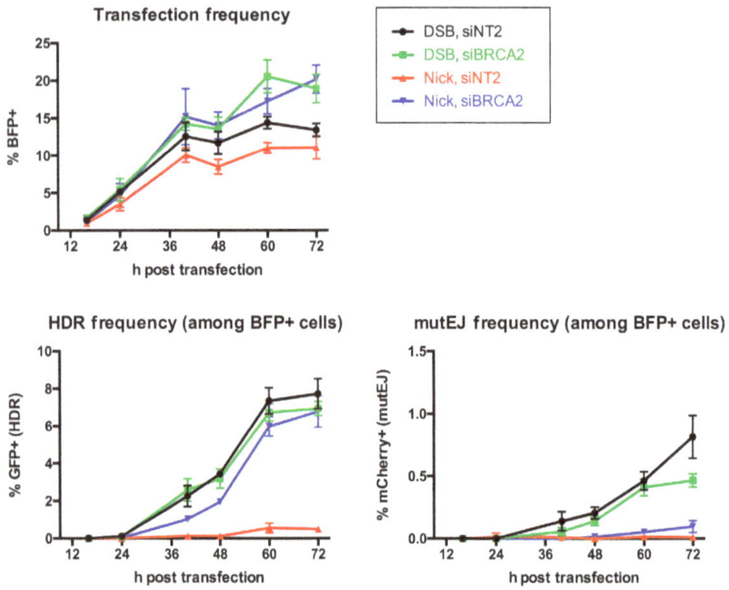

Fig. 3 Time course of appearance of BFP$^+$, GFP$^+$, and mCherry$^+$ TL cells. *Above*, representative example of frequency of BFP$^+$ cells 0–72 h after transfection of Cas9-BFP (DSBs) and Cas9^{D10A}-BFP (nicks) into 293T TL cells provided with a guide RNA expression plasmid and SSO donor for HDR. *Below*, frequency of GFP$^+$ cells (HDR) and mCherry$^+$ cells (mutEJ) among BFP$^+$ transfectants.

4.4 Timing of Cell Harvest and Analysis

- Control experiments determining frequencies of HDR at nicks and DSBs supported by SSO donors showed that analysis of HDR and mutEJ frequencies 5 days after initial plating of cells (i.e., 3 days after transfection with the nuclease expression construct) provides a convenient snapshot that is informative for mechanistic studies and highly reproducible. Fig. 3 shows a representative example of repair in 293T TL cells transfected with SSO donors for HDR. The GFP and mCherry signals are not detectable before 24 h posttransfection. The GFP signal increases steadily thereafter and levels off at about 60 h, possibly reflecting limited persistence of the donors. The mCherry signal continues to increase modestly through 72 h.
- Termination of the culture at 5 days after initial seeding (3 days posttransfection) prevents overgrowth of the culture that would require that wells be split prior to analysis, minimizing variation due to cell handling.

- Caveats: Recombination by different pathways may proceed at different paces. For example, as evident in Fig. 3, the end point of the frequency of HDR by SSOs in cells treated with siBRCA2 is comparable at nicks and DSBs, but at early time points the signal for HDR at a DSB is as much as twice as high as the signal for HDR at a nick.
- Harvested cells do not need to be analyzed immediately. Once fixed they can be safely stored at 4°C for at least 7 days.

4.5 Flow Cytometry

Flow cytometry is carried out on an instrument able to detect GFP (excitation 488 nm, emission 507 nm), mCherry (excitation 587 nm, emission 610 nm), and mTagBFP (referred to as BFP herein; excitation 402 nm, emission 457 nm). We use an LSR II flow cytometer (Becton Dickinson, Franklin Lakes, NJ), and the procedure described below is based on this equipment. Other flow cytometers with similar capabilities may also be used, but the procedure may require adaptation to a different instrument or software.

4.5.1 Cell Analysis and Signal Recording

1. Using BD FACSDiva flow software, set parameters to detect the linear area, height and width of the forward scatter (FSC) and side scatter (SSC) signals. For the fluorescent channels (GFP, mCherry, and BFP) the log area of the signal is collected.
2. To use the 96-well plate reader (BD High Throughput Sampler, HTS), add plate experiment and choose the wells that will be sampled, flow rate, and sample volume. We typically use a flow rate of 1.0 μL/s and sample 70 μL of the 105 μL sample to avoid air being pumped into the system as a well empties.
3. In experimental layout, set events to record as 200,000 total cells. This will cause recording to terminate when and if the cell number reaches 200,000 independent of the total number of cells in the sample. This provides a ceiling high enough for recording low-frequency events with good statistical significance but confines the run to a reasonable duration when analyzing cells at a typical flow rate of 3000 cells/s.
4. On the global worksheet, roughly draft the gating plots and use them as a monitor for the flow process. Occasionally, an air bubble may be trapped in the line, causing distortion of the data, or clumped cells may block the line. In such cases, abort the reading and troubleshoot as results will not be reliable.

4.5.2 Data Analysis Using FlowJo (Tree Star, Ashland, OR)

Fig. 4 shows an example of successive gating, as described below:

1. Draw the first gate in FSC-A vs SSC-A to include all cells (removes debris).
2. Among cells, draw the second gate in SSC-W vs SSC-H for single cells (eliminates doublets).
3. Fluorescent gates (BFP, GFP, and mCherry) should be based on comparison of experimental samples and nontransfected controls. These gates are always compromises. If they are drawn far enough from the bulk negative population to exclude all control cells from the gate, then some positive cells in the experimental population will not be counted, and if they are drawn too close to the negative population, some negative cells will be misscored as positive.
4. Among single cells, the third gate selects BFP^+ cells.
5. GFP^+ and $mCherry^+$ among all single cells (from second gate).
6. GFP^+ and $mCherry^+$ among BFP^+ cells (from third gate).

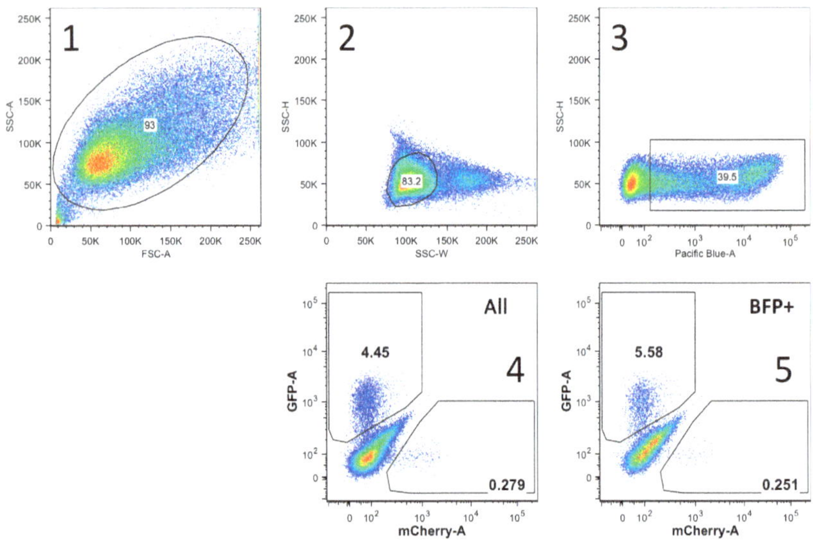

Fig. 4 Example of gating for BFP^+, GFP^+, and $mCherry^+$ 293T TL cells. Representative example of gating in an experiment analyzing HDR and mutEJ initiated by nicking the TL reporter. 1. First a gate is established for cells, eliminating dead cells and debris. 2. The second gate eliminates doublets and defines single cells. 3. The third gate defines BFP^+ cells among single cells. This gate is based on comparison with a nontransfected control. 4. Gates are defined for GFP^+ (*y*-axis) and $mCherry^+$ (*x*-axis) cells among all single cells defined by the second gate. 5. Gates identify GFP^+ (*y*-axis) and $mCherry^+$ (*x*-axis) cells among BFP^+ cells, defined by the third gate.

7. Export the gated data as a table.
8. Perform statistical analysis (two-tailed *t*-test) in either Microsoft Excel or Prism.

4.6 PCR Assays

The TL reporter can be used to assay for sequence changes that cooccur with the either an HDR or mutEJ event by PCR of the region of interest after sorting either the GFP^{+} or $mCherry^{+}$ cells. For example, to determine the extent to which HDR with a defined donor causes coconversion of sites 5′ and 3′ of the selected repair event (Davis & Maizels, 2016), we have used a 99 nt SSO donor bearing silent single-nucleotide changes that create *Apo*I and *Hin*dIII restriction fragment polymorphisms (Fig. 2C). Transfections are carried out in 6-well plates, and 3 days posttransfection cells are transferred to 10 cm plates and cultured for 5–7 days prior to sorting live cells on a Becton Dickinson Aria II flow cytometer. GFP^{+} cells are sorted into one well per experimental condition and cultured for 4–6 days. Genomic DNA is prepared (Qiagen), and the 574 bp region spanning the conversion site is then PCR amplified using primers 5′-CCAAGGACCTGAAATGACC-3′ and 5′-GTCCTCCTTGAAGTCGATGC-3′ with Taq DNA polymerase in ThermoPol buffer (NEB, Ipswich, MA). *Apo*I and *Hin*dIII digestions are performed directly in the ThermoPol buffer and DNA fragments resolved on a 1.5% agarose gel. Frequencies of marker conversion are determined by quantifying the fraction of cleaved vs total DNA on a BioRad Imaging system.

ACKNOWLEDGMENTS

This research was supported by NIH R01 CA183967 and NIH R21 CA190675.

We thank Donna Prunkard for assistance with flow cytometry.

REFERENCES

Cejka, P., Stojic, L., Mojas, N., Russell, A. M., Heinimann, K., Cannavo, E., et al. (2003). Methylation-induced G(2)/M arrest requires a full complement of the mismatch repair protein hMLH1. *The EMBO Journal*, *22*(9), 2245–2254. https://doi.org/10.1093/emboj/cdg216.

Certo, M. T., Ryu, B. Y., Annis, J. E., Garibov, M., Jarjour, J., Rawlings, D. J., et al. (2011). Tracking genome engineering outcome at individual DNA breakpoints. *Nature Methods*, *8*(8), 671–676. https://doi.org/10.1038/nmeth.1648.

Cong, L., Ran, F. A., Cox, D., Lin, S., Barretto, R., Habib, N., et al. (2013). Multiplex genome engineering using CRISPR/Cas systems. *Science*, *339*(6121), 819–823. https://doi.org/10.1126/science.1231143.

Davis, L., & Maizels, N. (2011). G4 DNA: At risk in the genome. *EMBO Journal*, *30*(19), 3878–3879. https://doi.org/10.1038/emboj.2011.342.

Davis, L., & Maizels, N. (2014). Homology-directed repair of DNA nicks via pathways distinct from canonical double-strand break repair. *Proceedings of the National Academy of Sciences of the United States of America*, *111*, E924–932.

Davis, L., & Maizels, N. (2016). Two distinct pathways support gene correction by single-stranded donors at DNA nicks. *Cell Reports*, *17*(7), 1872–1881. https://doi.org/10.1016/j.celrep.2016.10.049.

Heyer, W. D., Ehmsen, K. T., & Liu, J. (2010). Regulation of homologous recombination in eukaryotes. *Annual Review of Genetics*, *44*, 113–139. https://doi.org/10.1146/annurev-genet-051710-150955.

Jasin, M., & Rothstein, R. (2013). Repair of strand breaks by homologous recombination. *Cold Spring Harbor Perspectives in Biology*, *5*(11), a012740, https://doi.org/10.1101/cshperspect.a012740.

Kim, S., Kim, D., Cho, S. W., Kim, J., & Kim, J. S. (2014). Highly efficient RNA-guided genome editing in human cells via delivery of purified Cas9 ribonucleoproteins. *Genome Research*, *24*(6), 1012–1019. https://doi.org/10.1101/gr.171322.113.

Lee, G. S., Neiditch, M. B., Salus, S. S., & Roth, D. B. (2004). RAG proteins shepherd double-strand breaks to a specific pathway, suppressing error-prone repair, but RAG nicking initiates homologous recombination. *Cell*, *117*(2), 171–184. https://doi.org/10.1016/S0092-8674(04)00301-0.

Liao, W. Q., Qi, Y. L., Wang, L., Dong, X. M., Xu, T., Ding, C. D., et al. (2015). Recql5 protects against lipopolysaccharide/D-galactosamine-induced liver injury in mice. *World Journal of Gastroenterology*, *21*(36), 10375–10384. https://doi.org/10.3748/wjg.v21.i36.10375.

Maizels, N. (2005). Immunoglobulin gene diversification. *Annual Review of Genetics*, *39*, 23–46. https://doi.org/10.1146/annurev.genet.39.073003.110544.

McConnell Smith, A., Takeuchi, R., Pellenz, S., Davis, L., Maizels, N., Monnat, R. J., Jr., et al. (2009). Generation of a nicking enzyme that stimulates site-specific gene conversion from the I-AniI LAGLIDADG homing endonuclease. *Proceedings of the National Academy of Sciences of the United States of America*, *106*(13), 5099–5104. https://doi.org/10.1073/pnas.0810588106.

Niu, Y., Tenney, K., Li, H., & Gimble, F. S. (2008). Engineering variants of the I-SceI homing endonuclease with strand-specific and site-specific DNA-nicking activity. *Journal of Molecular Biology*, *382*(1), 188–202. https://doi.org/10.1016/j.jmb.2008.07.010.

Pierce, A. J., Johnson, R. D., Thompson, L. H., & Jasin, M. (1999). XRCC3 promotes homology-directed repair of DNA damage in mammalian cells. *Genes & Development*, *13*(20), 2633–2638.

Ramirez, C. L., Certo, M. T., Mussolino, C., Goodwin, M. J., Cradick, T. J., McCaffrey, A. P., et al. (2012). Engineered zinc finger nickases induce homology-directed repair with reduced mutagenic effects. *Nucleic Acids Research*, *40*(12), 5560–5568. https://doi.org/10.1093/nar/gks179.

Ran, F. A., Hsu, P. D., Lin, C. Y., Gootenberg, J. S., Konermann, S., Trevino, A. E., et al. (2013). Double nicking by RNA-guided CRISPR Cas9 for enhanced genome editing specificity. *Cell*, *154*(6), 1380–1389. https://doi.org/10.1016/j.cell.2013.08.021.

Strathern, J. N., Weinstock, K. G., Higgins, D. R., & McGill, C. B. (1991). A novel recombinator in yeast based on gene II protein from bacteriophage f1. *Genetics*, *127*(1), 61–73.

Symington, L. S., & Gautier, J. (2011). Double-strand break end resection and repair pathway choice. *Annual Review of Genetics*, *45*, 247–271. https://doi.org/10.1146/annurev-genet-110410-132435.

Szymczak, A. L., Workman, C. J., Wang, Y., Vignali, K. M., Dilioglou, S., Vanin, E. F., et al. (2004). Correction of multi-gene deficiency in vivo using a single 'self-cleaving' 2A peptide-based retroviral vector. *Nature Biotechnology*, *22*(5), 589–594. https://doi.org/10.1038/nbt957.

Trevino, A. E., & Zhang, F. (2014). Genome editing using Cas9 nickases. *Methods in Enzymology*, *546*, 161–174. https://doi.org/10.1016/B978-0-12-801185-0.00008-8.

Trojan, J., Zeuzem, S., Randolph, A., Hemmerle, C., Brieger, A., Raedle, J., et al. (2002). Functional analysis of hMLH1 variants and HNPCC-related mutations using a human expression system. *Gastroenterology*, *122*(1), 211–219.

Wang, J., Friedman, G., Doyon, Y., Wang, N. S., Li, C. J., Miller, J. C., et al. (2012). Targeted gene addition to a predetermined site in the human genome using a ZFN-based nicking enzyme. *Genome Research*, *22*(7), 1316–1326. https://doi.org/10.1101/gr.122879.111.

CHAPTER SIX

Recovery of Alternative End-Joining Repair Products From *Drosophila* Embryos

Terrence Hanscom, Varandt Y. Khodaverdian, Mitch McVey[1]
Tufts University, Medford, MA, United States
[1]Corresponding author: e-mail address: mitch.mcvey@tufts.edu

Contents

Abstract

In this chapter, we describe a method for the recovery and analysis of alternative end-joining (alt-EJ) DNA double-strand break repair junctions following I-*Sce*I cutting in *Drosophila melanogaster* embryos. Alt-EJ can be defined as a set of Ku70/80 and DNA ligase 4-independent end-joining processes that are typically mutagenic, producing deletions, insertions, and chromosomal rearrangements more frequently than higher-fidelity repair pathways such as classical nonhomologous end joining or homologous recombination. Alt-EJ has been observed to be upregulated in HR-deficient tumors and is essential for the survival and proliferation of these cells. Alt-EJ shares many initial

Methods in Enzymology, Volume 601
ISSN 0076-6879
https://doi.org/10.1016/bs.mie.2017.11.027

processing steps with homologous recombination, specifically end resection; therefore, studying alt-EJ repair junctions can provide useful insight into aborted HR repair. Here, we describe the injection of plasmid constructs with specific cut sites into *Drosophila* embryos and the subsequent recovery of alt-EJ repair products. We also describe different analytical approaches using this system and how amplicon sequencing can be used to provide mechanistic information about alt-EJ.

ABBREVIATIONS

alt-EJ	alternative end joining
c-NHEJ	classical nonhomologous end joining
CRISPR	clustered regularly interspaced short palindromic repeats
ddH$_2$O	distilled, deionized water
DSB	double-strand break
PCR	polymerase chain reaction
SDS	sodium dodecyl sulfate

1. INTRODUCTION

DNA end-joining mechanisms are a set of double-strand break (DSB) repair processes that can be used throughout the cell cycle, preventing improper recombination and loss of heterozygosity (Ceccaldi, Rondinelli, & D'Andrea, 2016; Lieber, 2010; Lieber, Gu, Lu, Shimazaki, & Tsai, 2010; Shrivastav, De Haro, & Nickoloff, 2008). End-joining repair can be separated into two main classes. The first, known as classical nonhomologous end joining (c-NHEJ), occurs when the Ku70/80 heterodimer binds DNA ends with high affinity, ends are juxtaposed and processed, and the DNA ligase 4/XRCC4/XLF complex completes the reaction by ligating the DNA backbone (Chiruvella, Liang, & Wilson, 2013; Lieber, 2010). Because it does not require the presence of a homologous template, c-NHEJ is the dominant DNA repair pathway outside of S and G2 phases. Ku and ligase 4-independent end joining, also referred to as alternative end joining (alt-EJ), is kinetically slower than c-NHEJ, occurs throughout the cell cycle, and can happen even when c-NHEJ is available (Chang, Pannunzio, Adachi, & Lieber, 2017). Alt-EJ repair is generally imprecise, leading to larger insertions and deletions than are typically observed with c-NHEJ repair (Frit, Barboule, Yuan, Gomez, & Calsou, 2014; Rodgers & McVey, 2016).

One well-characterized alt-EJ mechanism, known as microhomology-mediated end joining, utilizes 5′–3′ resection to expose short, generally 1–20 nucleotide, single-stranded microhomologies on both ends of the break (Seol, Shim, & Lee, 2017; Sfeir & Symington, 2015). This dependence on resection highlights one of the aspects common between alt-EJ and homologous recombination (Truong et al., 2013). Annealing of the complementary bases in these microhomologies, followed by flap trimming, fill-in synthesis, and ligation, leads to deletions of varying sizes. Microhomology-bearing junctions are commonly found at sites of chromosomal translocations, implicating alt-EJ in the formation of genomic rearrangements and mutations associated with cancer (Soni, Siemann, Pantelias, & Iliakis, 2015; Wood & Doublie, 2016).

Drosophila melanogaster has proven to be an excellent model in which to study alt-EJ. The *Drosophila* genome lacks many c-NHEJ proteins which are found in various other organisms, including DNA-PKcs, Artemis, and the X family polymerases mu and lambda (Sekelsky, 2017). Furthermore, flies lacking the c-NHEJ component DNA ligase 4 are able to carry out alt-EJ with high efficiency (McVey, Radut, & Sekelsky, 2004). Recently, we have utilized *Drosophila* to investigate the precise mechanisms of alt-EJ and how sequence context may influence the spectrum of alt-EJ repair products (Khodaverdian et al., 2017; Yu & McVey, 2010). Here, we describe our recently developed, rapid, high-throughput methodology for the recovery, and analysis of alt-EJ products in a variety of sequence contexts.

1.1 A Comparison of Methods for Recovering Alt-EJ Repair Products

Researchers studying DSB repair utilize various methods to induce site-specific DSBs, including endonucleases such as I-*Sce*I and I-*Ppo*I, zinc finger nucleases, transcription activator-like effector nucleases, and clustered regularly interspaced short palindromic repeats (CRISPR) systems (Gaj, Gersbach, & Barbas, 2013). Site-specific DSBs offer advantages over ionizing radiation and chemical agents that cause multiple nonspecific DSBs, with the biggest advantage being the ability to easily discern sequence changes that occur during repair. We have previously worked with an I-*Sce*I recognition sequence integrated via *P*-element transposition on the second chromosome of *D. melanogaster* (Rong & Golic, 2003). By using standard genetic crosses to introduce the I-*Sce*I endonuclease into male flies with the recognition site, multiple independent DSB repair events can be recovered from these flies' germ lines and sequenced. However, there are several drawbacks to using the chromosomal I-*Sce*I site. One is the amount of time that the process requires; induction of DSBs and recovery requires multiple crosses, each of which takes approximately 2 weeks. Another issue is that repair product identification requires a number of downstream steps, including recovery of genomic DNA, polymerase chain reaction (PCR) to amplify the DNA surrounding the break, and either direct sequencing of the PCR product or cloning of the product into a suitable vector for Sanger sequencing.

Because of the time and labor required with the chromosome-based system, we have recently developed a more efficient, plasmid-based injection system. Advantages of this system include significantly reduced time to isolate and sequence repair products and the ability to apply next-generation amplicon sequencing to sequence hundreds of thousands of repair junctions. In addition, early-stage *Drosophila* embryos contain high levels of maternally deposited DNA repair enzymes, including proteins such as DNA polymerase theta and DNA ligase III that have been implicated in alt-EJ (Beagan & McVey, 2016; Paul et al., 2013; Sharma et al., 2015). By utilizing mutant embryos that lack DNA ligase 4, a protein essential for c-NHEJ, we can specifically study alt-EJ repair. Importantly, we have found that alt-EJ repair events recovered from plasmids injected into embryos are similar in type and frequency to those recovered from a chromosomal-based system, suggesting that alt-EJ operates similarly in both systems (Khodaverdian et al., 2017; Yu & McVey, 2010).

2. PREPARING AND INJECTING PLASMIDS INTO EMBRYOS

2.1 Overview of the Process

Plasmids containing an I-*Sce*I recognition site, referred to here as I-*Sce*I repair constructs, are injected into *Drosophila* embryos expressing the I-*Sce*I endonuclease. After a 4-h incubation, during which time the plasmids are cut and repaired, plasmid DNA is extracted and subjected to individual Sanger or high-throughput amplicon sequencing. The available strategies for injection and recovery are broadly outlined in Fig. 1. One strategy involves transformation of the recovered plasmids into bacteria, which results in colonies containing independent repair events. DNA from these colonies can be extracted and subjected to in vitro I-*Sce*I digest to identify inaccurate repair products. These are then Sanger sequenced, and the repair junctions are analyzed. Alternatively, the recovered plasmids can be pooled

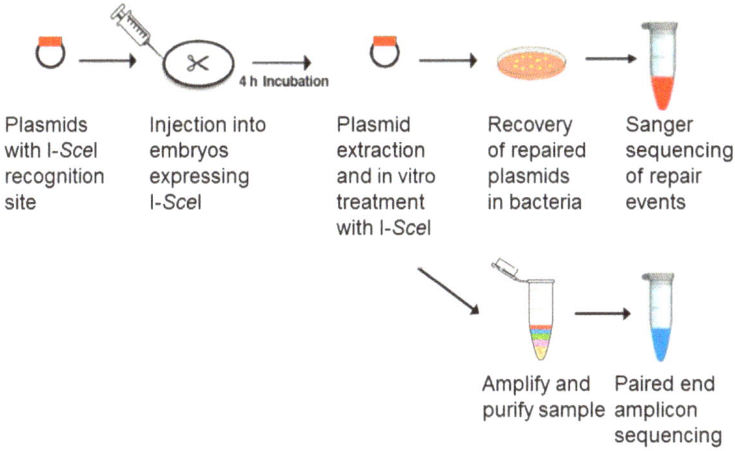

Fig. 1 Injection, recovery, and characterization of alt-EJ repair events. Transgenic embryos expressing I-*Sce*I are injected with plasmids containing the endonuclease recognition site. Following cutting and repair, inaccurately repaired plasmids are preferentially recovered following in vitro treatment with I-*Sce*I, transformed into bacteria, and Sanger sequenced. Alternatively, repair junction sequences are amplified by PCR and deep sequenced.

and the DNA sequence flanking the I-*Sce*I site can be amplified via PCR. After appropriate indexes are added, the libraries are sequenced using next-generation amplicon sequencing. The use of amplicon sequencing produces several million reads per run, of which hundreds of thousands are inaccurate alt-EJ repair products.

2.2 Reagent Preparation

2.2.1 Plasmid Preparation

1. To generate plasmids containing an I-*Sce*I recognition site, amplify the site from a suitable template using PCR. We used Taq polymerase to amplify the I-*Sce*I site and flanking sequence from the *Iw*7 chromosomal locus (Rong & Golic, 2003) and used TA cloning to insert the PCR fragment into the pGem-T Easy vector, but any sequence with an I-*Sce*I recognition site is suitable.
2. Transform the plasmids into a suitable *Escherichia coli* strain, plate on antibiotic-containing medium, and incubate at 37°C overnight.
3. Inoculate 50 mL of LB broth with a single colony, incubate overnight, and purify the DNA using a Macherey-Nagel NucleoBond Xtra Midi kit or other suitable method for obtaining high-purity DNA.

4. Prior to injection, subject a sample of the purified plasmid to sequencing to validate the presence of the I-*Sce*I site.
5. If changes to the DNA sequence flanking the I-*Sce*I site are desired, these can be engineered using site-directed mutagenesis.

2.2.2 Injection Mix Preparation

1. Make a 0.1-*M* sodium phosphate (pH 6.8) solution using 0.1 *M* NaH_2PO_4 (monobasic) and 0.1 *M* Na_2HPO_4 (dibasic). Make a 0.1-*M* potassium chloride solution. Combine these to create a 10-mL stock solution of 1 m*M* sodium phosphate and 50 m*M* potassium chloride. Filter sterilize using a 0.22-μm filter.
2. Bring plasmids to a standard concentration of 500 ng/μL within the injection buffer to create the final injection mix.
3. Store 10 μL aliquots of the injection mix at −20°C. Thaw immediately before using.

2.3 Embryo Collection

The success of this protocol depends upon the injection of I-*Sce*I-bearing plasmids into early-stage embryos (<2 h old) that have not yet cellularized. To do this, females must be primed so that they are healthy and laying large quantities of eggs that can be collected in a short period of time. Egg collection conditions must be optimized to facilitate rapid recovery and processing of embryos.

Additional resources describing effective methods for collecting, preparing, and injecting embryos with plasmid DNA can be found in Kiehart, Crawford, and Montague (2007) and O'Brochta and Atkinson (2004).

2.3.1 Procedure

1. Obtain a stock of flies expressing the I-*Sce*I endonuclease. We use a stock expressing I-*Sce*I under control of a ubiquitin promoter with genotype: w^{1118}; *P{Ubiq::I-SceI}*, *Sp/CyO*, *P{Ubiq::I-SceI}* (Preston, Flores, & Engels, 2006). To recover inaccurate repair events in a c-NHEJ deficient genetic background, an isogenic stock with the $lig4^{169a}$ mutation can be used (McVey et al., 2004). Stocks are maintained in culture bottles containing cornmeal agar in a 25°C incubator on a 12-h light–dark cycle.
2. To prepare grape agar, combine 400 mL distilled, deionized water (ddH_2O), 90 mL frozen grape juice concentrate, 11 g agar, 29 g dextrose, 14.5 g sucrose, and 1.25 mL 10 N sodium hydroxide in a 1-L Pyrex bottle.

3. Microwave until the agar is melted. Monitor carefully as it boils over easily. Set aside to cool.
4. In a 50-mL conical tube, create acid solution by mixing 20.9 mL propionic acid, 2.08 mL phosphoric acid, and ddH_2O to bring the volume to 50 mL.
5. Add 5.6 mL of acid solution to the cooling bottle of agar when it reaches a temperature of 65°C. Store remainder of the acid mix at room temperature. Mix the agar well and pour into 55 mm plastic Petri plates, filling about half full. Plates will last several months stored in plastic bags at 4°C.
6. To create yeast paste, cover the bottom of a 100-mL beaker with active yeast pellets. Add ddH_2O intermittently, 1 mL at a time, and stir until a thick paste is formed. Store excess yeast paste at 4°C.
7. Prepare embryo collection cages (Fig. 2A) by fitting the grape agar plate within the lip of a 100-mL plastic beaker. Secure the plate by wrapping a rubber band around the collection cage.

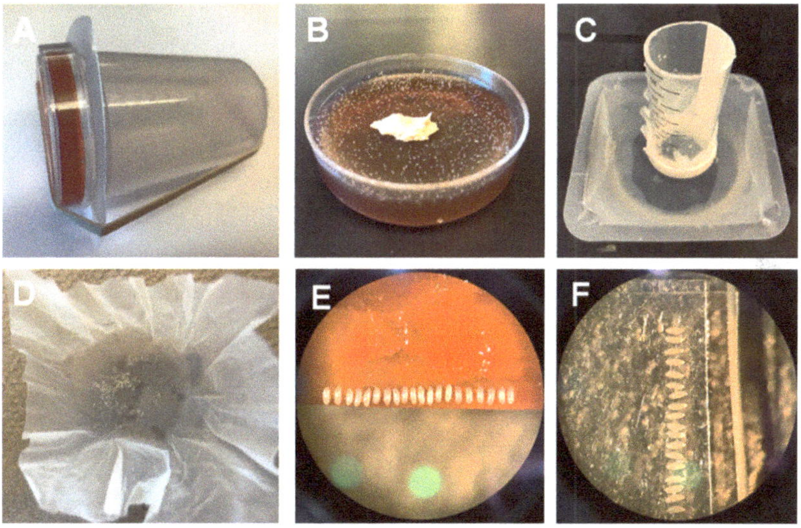

Fig. 2 Apparatus for embryo collection and injection. (A) Embryo collection cage, with rubber band securing a grape agar plate to a tricorner plastic beaker. (B) Grape agar collection plate with yeast paste in the center. A collection of 100 well-fed females will lay hundreds of embryos on the plate in less than an hour. (C) Nylon mesh secured to conical tube cut in half, for dechorionation and washing of embryos. (D) Nylon mesh with embryos after dechorionation and washing is complete. (E) Embryos aligned in a row on a grape agar slab, ready for transfer to a coverslip. (F) Embryos after transfer to double-stick tape attached to a glass coverslip, covered in halocarbon oil.

8. Collect embryos according to the following schedule:

   Day 1: Sort ~100–200 newly eclosed flies (a mixture of females and males) into a bottle/large vial containing standard cornmeal- or molasses-based fly food and a few active yeast pellets on the food surface. The yeast promotes egg laying.

   Day 2: Make fresh yeast paste. Warm a grape plate to 25°C, blot dry, and apply a ~1-cm diameter dab of yeast paste to the center of the plate. Anesthetize the bottle of flies with CO_2 and move them into the embryo collection cage. Cover the cage with the grape plate and wrap tight with a rubber band to secure the plate (Fig. 2A). Lay the cage on its side at 25°C and incubate overnight.

   Day 3: In the morning remove the old grape agar plate and swap with a new plate warmed to 25°C with fresh yeast paste in the center (Fig. 2B). This is best accomplished by tapping the cage on a benchtop, grape plate up, to knock the flies down to the bottom while you change the plates. Alternatively, flies can be anesthetized using CO_2, although this typically decreases egg laying for a period of time. Wrap the plastic beaker of the cage with tin foil, as females generally lay better in the dark. Place the cage on its side at 25°C and incubate for 1 h. It is important to keep the cage on its side so that the flies have enough surface area to walk, can easily access the grape plate (especially important if they cannot fly), and do not get stuck in the yeast paste. Repeat the plate changing process as needed for additional collections. Grape plates can be reused by rinsing them and blotting them dry immediately after rinsing. Flies will generally lay well in a cage for up to 5 days. When finished with the cage, freeze it overnight and dispose of flies.

Note: If the females are not laying sufficient eggs, ensure that the grape plates are warmed to 25°C before placing in the cages, are dry, and the yeast paste is fresh. Additionally, certain fly stocks do not lay as well as others. If using a mutant with low fecundity, add more flies to the cage. Disturbing the flies too frequently is detrimental to the egg-laying process and should be minimized.

2.4 Dechorionation of Embryos

2.4.1 Materials Needed

Squirt bottles
Soft-bristled paint brush
50% Bleach

10 × Embryo Wash Stock: (7% NaCl, 0.7% Triton X-100; dilute to 1 × working concentration)
Razor blade
50 mL Conical tube with the bottom half removed, and a 1-cm hole cut in the center of the cap
Genesee Scientific Mesh, Nitex Nylon 50 μ*M* screen, Item # 57–106 (cut into 4 × 4 cm pieces)
Glass slides
Glass coverslips (1 in. square)
Paper towels
Plastic weighing dishes
3M double-sided tape

2.4.2 Procedure

1. Perform these manipulations at a sink, preferably with a ddH_2O water faucet. Stretch the mesh screen over the top of the conical tube and screw the cap over the top of it. The screen should be stretched tight at the cap opening. Place the conical tube cap-down in the plastic weighing dish.
2. Using a spray bottle filled with ddH_2O, gently dispense water onto the grape plate, and using the bristles of the paint brush, separate the embryos from the grape agar. Break up the yeast paste to recover any embryos that were laid within. While the embryos are floating in the water, carefully pour them into the conical tube and onto the screen. Repeat as necessary until all embryos are in the conical tube (Fig. 2C).
3. Use the ddH_2O water from the faucet to gently wash the embryos in the conical tube, allowing all yeast paste and residual grape agar to flow through the screen until only embryos remain.
4. Submerge and agitate embryos for 1 min in a shallow container (a weighing dish works well) containing bleach diluted to 50% concentration in water (Fig. 2C). Make sure all embryos are covered. If embryos get stuck on side walls, use a spray bottle of 50% bleach solution to get them back onto the screen. After 1 min, remove the embryos from the bleach and rinse them for at least 2 min using ddH_2O from the faucet. The amount of time necessary to remove all of the chorion will vary depending on the strength and age of the bleach and should be adjusted accordingly.
5. Wash the embryos with embryo wash solution by submerging the bottom of the conical in a weighing dish containing 1 × embryo wash.

Agitate for 4 min. If embryos stick to the sides of the conical tube, use a squirt bottle of embryo wash to get them back onto the screen. Rinse the embryo wash from the conical using a squirt bottle containing ddH_2O for at least 2 min. Using the squirt bottle of ddH_2O, gently spray the walls of the conical tube to collect the embryos in the center of the screen. Place aside (Fig. 2D).

2.5 Positioning Embryos for Injection

1. Gently remove the grape agar from the plate and flip upside down on a paper towel. Using a razor blade, cut out a flat, rectangular piece of agar and place on a microscope slide. Pat dry with a paper towel. If any water remains, the surface tension on the agar slab will make it extremely difficult to position the embryos in the next step.
2. With embryos in the center of the screen, pick up the conical tube and grip the excess screen against the outside of the tube. Slowly, unscrew the cap, ensuring that the screen remains pressed against the tube and does not move. Gently blot off any residual water on a paper towel, remove the conical tube, and place the screen down on a dry paper towel. With dry gloves, grip the edges of the screen, gently flip over, and press the embryos onto the grape agar until all embryos are stuck to the agar. This may require several gentle taps.
3. Embryos should be approximately 1.5 h old at this point and should remain uncellularized throughout the injection process. Younger embryos are opaque, bright white, and firm while older, unusable embryos are more translucent. When lining up embryos for the injection process, do not select embryos older than 2 h or those that have a mushy-looking consistency, as they have been overbleached and will rupture when injected.
4. Place the slab of grape agar under a microscope and align at least 60 embryos with a metal-tipped probe. Line them up, side by side, with approximately one-half of an embryo's width between each (Fig. 2E). If you are right-handed, sort the embryos from left to right on the grape agar. If left-handed, sort from right to left. This will prevent disturbing the embryos once they are aligned on the agar.
5. Cut a 1 cm piece of double-sided tape lengthwise down the middle and apply it to the edge of a glass coverslip, cut-side out. The cut side of the tape has a clean edge that is better suited for embryo adhesion and injection.

6. Remove the top adhesive cover of the tape and press the taped coverslip down on the aligned embryos, lightly at first and then adding slightly more pressure. Gradually raise the coverslip off of the grape agar slab (Fig. 2F). Do not slide the tape off the edge as this can cause the embryos to fall off the tape.
7. Place the coverslip in a desiccating chamber for 1–2 min to dry the embryos. This prevents rupturing during the subsequent injections.

2.6 Microinjection of Embryos

2.6.1 Materials Needed

Borosilicate glass capillary tubes pulled to make injection needles
Halocarbon oil 700 (Sigma Aldrich)
Anhydrous calcium sulfate desiccate (Drierite)
Injection microscope (Zeiss injection scope and Parker-Hannifin Picospritzer II)
2 μL Pipettor with long gel loading tips (like Eppendorf GELoader)
Plasmid injection mix (prepared as described earlier)
1% Sodium dodecyl sulfate (SDS) solution in a 0.7% sodium chloride buffer

2.6.2 Preparing Needles

1. Pulling of microinjection needles is best done using an instrument such as the Sutter Instrument Co. Model P-87 Flaming/Brown Micropipette Puller. For this instrument, our settings are as follows: heat: 848; pull: 125; velocity: 70; time: 250 ms; pressure: 200.
2. Pulled injection needles do not have an opening and must be cut at the ends with a razor blade. Use the razor blade to cut off just the tip of the needle at a 45-degree angle. The needle bore should be extremely narrow, wide enough to aspirate the droplet of injection mix, but small enough that the embryos do not rupture when punctured. This will require some trial and error, so it helps to pull more needles than necessary for a single round of injections.
3. Load 1–2 μL of injection mix into several needles using a 2-μL pipettor with a long pipette tip. Extra filled needles can be placed on rolled pieces of clay or colored lab tape in a humid Petri dish at 4°C.

2.6.3 Injecting the Embryos

The injection procedure can be adapted according to the particular injection apparatus that is used. We use a Picospritzer II injector connected to pressurized nitrogen gas, with a foot pedal that releases air in pulses of 25 ms when depressed.

1. Cover embryos with a thin coat of halocarbon oil 700, placing an additional drop on the distal end of the coverslip to be used for air aspiration and test injections to adjust the size of the injection mix droplet. Put a drop of halocarbon oil on a glass slide and position the coverslip on top for stability.
2. Place the loaded needle into the aspirator apparatus and the slide onto the microscope stage. Focus the microscope on the halocarbon oil drop. Using the Picospritzer, dispense any air in the tip of the needle into the oil droplet and adjust the pulse time so that the volume of plasmid mix being injected is approximately 1/20th the size of the embryo (you should be able to see the liquid when it is injected into the embryo).
3. Prior to injection, examine each embryo individually to determine its approximate age. Young embryos should have a clear outer edge and a brown oval center (Fig. 3). Embryos with well-defined, cellularized contours at their periphery are too old.
4. Align the exterior edge of the embryo in the same plane as the injection needle. Puncture the tip of the embryo with the needle and insert the needle so that its tip is in the middle of the embryo. Inject the plasmid mix and immediately remove the needle from the embryo. Some small leakage of cytoplasm may occur, but if leaking continues more than 1 s,

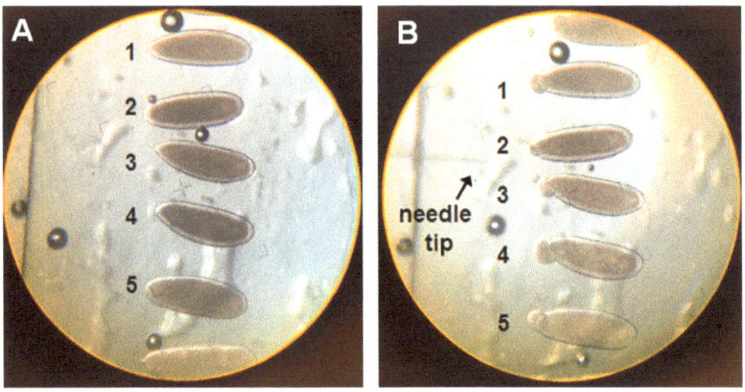

Fig. 3 Injection of embryos. (A) Embryos prior to injection. Embryos 1, 2, and 4 are approximately 2 h old but have not yet cellularized. Embryos 3 and 5 are younger and are surrounded by a translucent ring smaller than that observed in older embryos. (B) Embryos after injection. Some cytoplasmic leakage can be observed at the injection site. This is normal but should be minimized as much as possible.

the needle may be too large. Note: While embryos do not have to survive to hatching, as is required for transgenic procedures (O'Connor & Chia, 1993), their structural integrity should remain intact until plasmid isolation several hours later.

5. As injections proceed down the row of embryos, the stage may need to be adjusted so that both the needle and the embryo edge are focused in the same plane. This avoids such issues as needle breakage, pushing the embryos off the tape, or aspirating injection mix outside of the embryos.
6. Should the needle become clogged, move the needle to the distal droplet of halocarbon oil and disperse the clog with rapid aspirations or by increasing the pulse time. If the needle cannot be cleared, then begin again with a new needle.
7. Maintenance tip: Regularly clean the stage of the injection microscope for smooth movements of the slide, as halocarbon oil can cause sticking. Ensure all gaskets in the injector are sealing correctly, especially those that make a direct seal with the needle. If air is escaping from the needle input junction during injection bursts, droplet size will be inconsistent. Further, a loss of pressure can cause a vacuum in the needle which can result in reuptake of the injection mix and embryo contents.
8. After the injections, place the coverslip of embryos into a small covered container (e.g., Petri dish) and incubate at 25°C for 4 h to allow for cutting of the construct and subsequent repair. Incubation time can be increased or decreased according to preference. Following the incubation, DNA can immediately be extracted or the embryos can be stored at −20°C for up to 2 weeks. The halocarbon oil will freeze over the embryos.
9. Following the injections, the needles can be saved if they are free of clogs and handled carefully. To do this, remove the needle from the injection apparatus and place in a protected container at −20°C.

3. PLASMID RECOVERY AND REPAIR PRODUCT SEQUENCING

Here, we describe the recovery and purification of the repaired plasmid constructs. Following plasmid DNA isolation, the repair constructs can either be transformed into bacteria for Sanger sequencing (protocol 3.3) or amplified for next-generation amplicon sequencing (protocol 3.4).

3.1 Recovery of Injected Embryos

1. If starting with frozen embryos, remove the coverslips from the freezer. The halocarbon oil will melt.
2. Using a dissecting microscope, use a razor blade to carefully scrape excess halocarbon oil off of the double-sided tape without disturbing the embryos or the adhesive.
3. Using a bulb, gently squirt a 1% SDS solution over the embryos to remove any remaining halocarbon oil. The embryos should remain stuck to the adhesive, but stop rinsing if they start to fall off.
4. Wash away the SDS by gently agitating the coverslip in a Petri dish containing ddH_2O.
5. In one smooth motion, scrape a clean razor blade down the row of embryos as if running a spatula underneath them. Avoid scraping up the adhesive as much as possible. The embryos will collect in a pile on the edge of the razor blade.

3.2 Plasmid DNA Recovery by Alkaline Lysis

3.2.1 Reagents

P1 solution: 50 m*M* Tris–HCl, pH 7.5; 10 m*M* EDTA
P2 solution: 0.2 N NaOH, 1% SDS
P3 solution: 1.32 *M* KOAc, pH 4.8. Should be kept at 4°C
Plastic pestles (for homogenization)
95% and 70% Ethanol

3.2.2 Procedure

1. Use a plastic pestle to transfer the embryos from the razor blade to a 1.5-mL tube containing 25 μL of P1 solution.
2. Using the pestle, homogenize the embryos so that any that did not rupture while being scraped up are fully disrupted. Ensure the no embryos remain on the pestle. Vortex thoroughly to promote further lysis.
3. Add 25 μL P2 solution and shake. Mixture should clear immediately, indicating lysis, but can stand up to 5 min for best yields. Incubation for longer than 5 min can cause contamination with genomic DNA.
4. Add 25 μL of cold P3 solution and shake. A white precipitate containing cell debris and proteins should form. Spin in a microcentrifuge for 7 min at maximum speed.
5. Transfer the supernatant to a new tube with 200 μL 95% ethanol and mix by inverting several times. Incubate on ice for at least 10 min.

6. Centrifuge for 5 min at maximum speed. Discard the supernatant.
7. Wash pellet with 200 μL of 70% ethanol to remove excess salts. Centrifuge for 2 min at maximum speed.
8. Discard the ethanol and dry the pellet at room temperature for 10 min or until no ethanol remains in the tube.
9. Resuspend in 20 μL of ddH_2O water. Incubate at 37°C for 30 min with occasional gentle vortexing. Store at 4°C.
10. Optional step: We have found that the percentage of plasmids that are cut and inaccurately repaired is low (on the order of 10%), compared to the percentage of plasmids that are repaired perfectly or not cut. To mitigate this, recovered plasmid samples can be treated with I-*Sce*I endonuclease to reduce the number of uncut or perfectly repaired plasmids and to prevent them from being included in the downstream analysis. Incubate the recovered plasmid DNA, 2 μL of I-*Sce*I buffer, and 1 μL of I-*Sce*I restriction enzyme (NEB) in a 20-μL total reaction volume at 37°C for 1 h. Inactivate the I-*Sce*I by incubating 65°C for 20 min.

3.3 Bacterial Transformation and Sanger Sequencing

1. Use chemically competent *E. coli* XL10 or other suitable bacterial strain. Allow the frozen cells to thaw on ice for 10–15 min.
2. Add 5 μL of DNA directly to the cells. To promote high transformation efficiency, do not mix by pipetting. Cells must remain on ice at all times.
3. Incubate for 30 min on ice.
4. In a heat block, incubate at 42°C for 30–45 s.
5. Place the cells back on ice for 2 min.
6. Add 200 μL of sterile Luria broth and incubate at 37°C for 20 min.
7. Pipette 200 μL of each sample on antibiotic-containing plates that have been prewarmed to 37°C and incubate at 37°C overnight. Use the injection mix as a positive control for transformation efficiency.
8. Each colony that forms will contain an independent repair event. Inoculate a portion of each bacterial colony in 2 mL of Luria broth plus the appropriate antibiotic. Grow overnight at 37°C.
9. Transfer the culture into a 1.5-mL Eppendorf tube. Centrifuge at maximum speed for 10 s to pellet the cells. Discard the supernatant and proceed with a standard alkaline lysis miniprep protocol (described earlier, but increasing the volume of solutions fourfold).

10. Perform Sanger sequencing on the plasmid DNA using sequencing primers that anneal to plasmid sequence about 200 bp on either side of the I-*Sce*I site.
11. Alternative method: To eliminate the need to grow each colony in liquid culture, colony PCR can be performed using cells obtained from each independent colony. The PCR products can then be purified and sequenced using the Sanger method.

3.4 Next-Generation Amplicon Sequencing

3.4.1 Materials Needed

Agencourt AMPure beads (Protocol 000601v024)
Primers for the region flanking the I-*Sce*I recognition site
High-fidelity DNA polymerase (Q5, New England Biolabs)
Nextera sequencing index kit
Fragment analyzer or other equipment to analyze PCR product size
Illumina MiSeq Reagent Kit v3 600-cycle (MS-102-3003)

3.4.2 Amplicon Sequencing Approach

1. Following DNA extraction and enrichment of the inaccurately repaired plasmids, use PCR to amplify ~200 bp of DNA flanking the I-*Sce*I recognition site. To add variation to the amplicon PCR pool and promote efficient sequencing on the Illumina MiSeq platform, we use a set of pooled primers containing adapter sequences and 1–4 nucleotide random sequences at the 5′ end. Use a high-fidelity polymerase such as Q5 (New England Biolabs) and limit the amplification to 15–20 cycles to promote diversity of inaccurately repaired junctions and to prevent overrepresentation of the most common repair constructs.
2. Perform an AMPure bead purification to remove all PCR products less than 100 bp.
3. Perform a second, eight-cycle PCR with Nextera sequencing indexes (Illumina) for amplicon sequencing.
4. Perform a secondary AMPure bead purification step to remove all products less than 100 bp.
5. Use a fragment analyzer or other diagnostic tool to assess the consistency of fragment sizes. Ideally, 99%–100% of the samples should match the desired fragment size ±5 bp.
6. Sequence the amplicons using a paired-end sequencing kit. We use the Illumina MiSeq platform, which we find typically generates 8–10 million

paired-end reads per sequencing run. PhiX genomic DNA should be spiked into the sequencing reactions at concentrations of 5%, or higher if a pooled primer set is not used in step 1.

4. DATA ANALYSIS

Data analysis using either the Sanger or amplicon sequencing can be used to address a number of questions (Fig. 4):

(1) Did the repair process remove DNA (a deletion), add new DNA (an insertion), or both (an indel)?

(2) Did the repair process involve annealing at microhomologous sequences prior to synthesis and ligation?

(3) If there are inserted sequences, were these fully or partially templated from sequences flanking the I-*Sce*I cut site?

(4) Could the repair process utilize secondary-structure-forming sequencing as primers for synthesis of new DNA?

To conduct our sequence analysis, we typically utilize a number of open-source and commercially available sequence analysis tools, together with custom-designed programs in Python and R. Individual sequences from Sanger sequencing can be aligned to the original reference sequence using BLAST or various proprietary software tools. Below, we briefly describe our workflow for analyzing amplicon sequencing data.

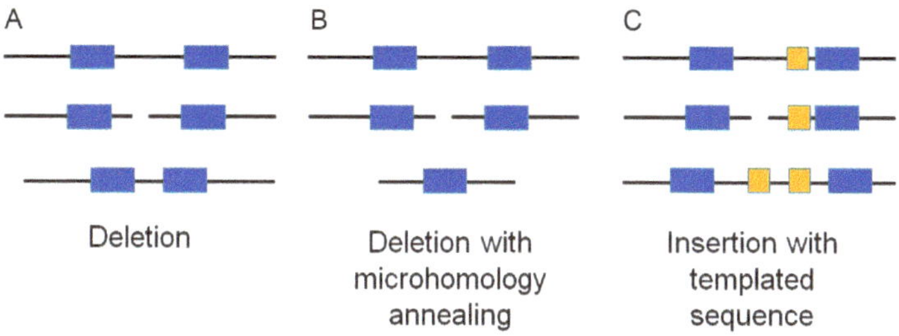

Fig. 4 Types of end-joining repair. (A) Deletion of sequences flanking the I-*Sce*I site. This can occur via both classical and alternative end joining. (B) Deletion of sequences via microhomology annealing (microhomologies are represented by *dark blue boxes*). This type of junction is frequently associated with alt-EJ. (C) Insertion of sequences that are templated from DNA flanking the I-*Sce*I cut site (represented by *light orange boxes*). This type of junction typically results from alt-EJ repair.

4.1 Analyzing Reads From Amplicon Sequencing

1. Raw reads from paired-end amplicon sequencing runs should be trimmed to the amplicon primer sequence. This includes trimming the random primer sequences at the ends of the amplicon. Junctions lacking 10 bp of reference sequence at both the 5′ and 3′ ends of each amplicon should be removed as PCR artifacts as they fail to span both sides of the break.
2. Overlapping read pairs can then be merged into single consensus reads as FASTQ files, using a program such as CLC Bio Genomics Workbench.
3. The reads can be mapped to the reference sequence corresponding to the original repair construct using a program such as Geneious and exported as SAM files for analysis.
4. The structure of repair junctions can be determined by using the CIGAR string to identify sequences matching the 5′ and 3′ ends of the amplicon, deleted sequences relative to the I-*Sce*I cut site, and inserted sequences. More detailed information about potential analytical parameters is described in chapter "High-throughput analysis of DNA break-induced chromosome rearrangements by amplicon sequencing" by Brown et al., in this volume.
5. After removing reads that match the reference sequence exactly (and represent either uncut or perfectly repaired plasmids), the percentage of reads corresponding to each junction can be calculated by dividing the number of reads per junction by the total number of inaccurate reads in the sequencing run. This provides a rough estimate of the overall representation of different repair junctions in each sample. We have found good correlation between repair junction frequency using Sanger and amplicon deep sequencing methods (Khodaverdian et al., 2017).

5. SUMMARY AND CONCLUSION

Alt-EJ is now recognized as a bona fide DSB repair mechanism that operates both in the presence and absence of classical end-joining repair. Elucidation of the mechanisms of alt-EJ will provide insight into the etiology of deletions, templated insertions, and chromosome translocations that are frequently associated with cancer and other human diseases. Here, we have described a process for the rapid identification and recovery of large numbers of alt-EJ repair products, using a high-throughput in vivo embryo injection system in *Drosophila*. While we have focused on the repair of double-stranded breaks induced by the I-*Sce*I endonuclease, the protocol can easily be adapted for other types of DSBs, including those induced by CRISPR–Cas9, which produces blunt-ended cuts that may promote a different spectrum of alt-EJ repair products.

ACKNOWLEDGMENTS

We would like to thank Amy Yu and Juan Castaneda for their help in developing and documenting these protocols.

REFERENCES

Beagan, K., & McVey, M. (2016). Linking DNA polymerase theta structure and function in health and disease. *Cellular and Molecular Life Sciences*, *73*(3), 603–615. https://doi.org/10.1007/s00018-015-2078-9.

Ceccaldi, R., Rondinelli, B., & D'Andrea, A. D. (2016). Repair pathway choices and consequences at the double-strand break. *Trends in Cell Biology*, *26*(1), 52–64. https://doi.org/10.1016/j.tcb.2015.07.009.

Chang, H. H. Y., Pannunzio, N. R., Adachi, N., & Lieber, M. R. (2017). Non-homologous DNA end joining and alternative pathways to double-strand break repair. *Nature Reviews. Molecular Cell Biology*, *18*(8), 495–506. https://doi.org/10.1038/nrm.2017.48.

Chiruvella, K. K., Liang, Z., & Wilson, T. E. (2013). Repair of double-strand breaks by end joining. *Cold Spring Harbor Perspectives in Biology*, *5*(5), a012757. https://doi.org/10.1101/cshperspect.a012757.

Frit, P., Barboule, N., Yuan, Y., Gomez, D., & Calsou, P. (2014). Alternative end-joining pathway(s): Bricolage at DNA breaks. *DNA Repair (Amst)*, *17*, 81–97. https://doi.org/10.1016/j.dnarep.2014.02.007.

Gaj, T., Gersbach, C. A., & Barbas, C. F., 3rd. (2013). ZFN, TALEN, and CRISPR/Cas-based methods for genome engineering. *Trends in Biotechnology*, *31*(7), 397–405. https://doi.org/10.1016/j.tibtech.2013.04.004.

Khodaverdian, V. Y., Hanscom, T., Yu, A. M., Yu, T., Mak, V., Brown, A. J., et al. (2017). Secondary structure forming sequences drive SD-MMEJ repair of DNA double-strand breaks. *Nucleic Acids Research*, *22*, 12848–12861. https://doi:10.1093/nar/gkx1056.

Kiehart, D. P., Crawford, J. M., & Montague, R. A. (2007). Collection, dechorionation, and preparation of Drosophila embryos for quantitative microinjection. *CSH Protocols*, *2007*. pdb prot4717 https://doi.org/10.1101/pdb.prot4717.

Lieber, M. R. (2010). The mechanism of double-strand DNA break repair by the nonhomologous DNA end-joining pathway. *Annual Review of Biochemistry*, *79*, 181–211. https://doi.org/10.1146/annurev.biochem.052308.093131.

Lieber, M. R., Gu, J., Lu, H., Shimazaki, N., & Tsai, A. G. (2010). Nonhomologous DNA end joining (NHEJ) and chromosomal translocations in humans. *Sub-Cellular Biochemistry*, *50*, 279–296. https://doi.org/10.1007/978-90-481-3471-7_14.

McVey, M., Radut, D., & Sekelsky, J. J. (2004). End-joining repair of double-strand breaks in Drosophila melanogaster is largely DNA ligase IV independent. *Genetics*, *168*(4), 2067–2076. https://doi.org/10.1534/genetics.104.033902.

O'Brochta, D. A., & Atkinson, P. W. (2004). Transformation systems in insects. *Methods in Molecular Biology*, *260*, 227–254. https://doi.org/10.1385/1-59259-755-6:227.

O'Connor, M., & Chia, W. (1993). P element-mediated germ-line transformation of Drosophila. *Methods in Molecular Biology*, *18*, 75–85. https://doi.org/10.1385/0-89603-245-0:75.

Paul, K., Wang, M., Mladenov, E., Bencsik-Theilen, A., Bednar, T., Wu, W., et al. (2013). DNA ligases I and III cooperate in alternative non-homologous end-joining in vertebrates. *PLoS One*, *8*(3), e59505. https://doi.org/10.1371/journal.pone.0059505.

Preston, C. R., Flores, C. C., & Engels, W. R. (2006). Differential usage of alternative pathways of double-strand break repair in Drosophila. *Genetics*, *172*(2), 1055–1068. https://doi.org/10.1534/genetics.105.050138.

Rodgers, K., & McVey, M. (2016). Error-prone repair of DNA double-strand breaks. *Journal of Cellular Physiology*, *231*(1), 15–24. https://doi.org/10.1002/jcp.25053.

Rong, Y. S., & Golic, K. G. (2003). The homologous chromosome is an effective template for the repair of mitotic DNA double-strand breaks in Drosophila. *Genetics*, *165*(4), 1831–1842.

Sekelsky, J. (2017). DNA repair in Drosophila: Mutagens, models, and missing genes. *Genetics*, *205*(2), 471–490. https://doi.org/10.1534/genetics.116.186759.

Seol, J. H., Shim, E. Y., & Lee, S. E. (2017). Microhomology-mediated end joining: Good, bad and ugly. *Mutation Research*, *17*, 30041–30046. https://doi.org/10.1016/j.mrfmmm.2017.07.002.

Sfeir, A., & Symington, L. S. (2015). Microhomology-mediated end joining: A back-up survival mechanism or dedicated pathway? *Trends in Biochemical Sciences*, *40*(11), 701–714. https://doi.org/10.1016/j.tibs.2015.08.006.

Sharma, S., Javadekar, S. M., Pandey, M., Srivastava, M., Kumari, R., & Raghavan, S. C. (2015). Homology and enzymatic requirements of microhomology-dependent alternative end joining. *Cell Death & Disease*, *6*, e1697.https://doi.org/10.1038/cddis.2015.58.

Shrivastav, M., De Haro, L. P., & Nickoloff, J. A. (2008). Regulation of DNA double-strand break repair pathway choice. *Cell Research*, *18*(1), 134–147. https://doi.org/10.1038/cr.2007.111.

Soni, A., Siemann, M., Pantelias, G. E., & Iliakis, G. (2015). Marked contribution of alternative end-joining to chromosome-translocation-formation by stochastically induced DNA double-strand-breaks in G2-phase human cells. *Mutation Research, Genetic Toxicology and Environmental Mutagenesis*, *793*, 2–8. https://doi.org/10.1016/j.mrgentox.2015.07.002.

Truong, L. N., Li, Y., Shi, L. Z., Hwang, P. Y., He, J., Wang, H., et al. (2013). Microhomology-mediated end joining and homologous recombination share the initial end resection step to repair DNA double-strand breaks in mammalian cells. *Proceedings of the National Academy of Sciences of the United States of America*, *110*(19), 7720–7725. https://doi.org/10.1073/pnas.1213431110.

Wood, R. D., & Doublie, S. (2016). DNA polymerase theta (POLQ), double-strand break repair, and cancer. *DNA Repair (Amst)*, *44*, 22–32. https://doi.org/10.1016/j.dnarep.2016.05.003.

Yu, A. M., & McVey, M. (2010). Synthesis-dependent microhomology-mediated end joining accounts for multiple types of repair junctions. *Nucleic Acids Research*, *38*(17), 5706–5717. https://doi.org/10.1093/nar/gkq379.

CHAPTER SEVEN

High-Throughput Analysis of DNA Break-Induced Chromosome Rearrangements by Amplicon Sequencing

Alexander J. Brown*, Aneesa T. Al-Soodani*[,2], Miles Saul[†], Stephanie Her[‡], Juan C. Garcia[§], Dale A. Ramsden[§], Chengtao Her*, Steven A. Roberts*[,¶,1]

*School of Molecular Biosciences, College of Veterinary Medicine, Washington State University, Pullman, WA, United States
[†]University of Washington, Seattle, WA, United States
[‡]Dartmouth College, Hanover, NH, United States
[§]Lineberger Comprehensive Cancer Center, Curriculum in Genetics and Molecular Biology, University of North Carolina at Chapel Hill, Chapel Hill, NC, United States
[¶]Center for Reproductive Biology, College of Veterinary Medicine, Washington State University, Pullman, WA, United States
[1]Corresponding author: e-mail address: sroberts@vetmed.wsu.edu

Contents

[2] Current address: Department of Chemistry, University of Utah, Salt Lake City, UT 84112, United States.

Methods in Enzymology, Volume 601
ISSN 0076-6879
https://doi.org/10.1016/bs.mie.2017.11.028

Abstract

The mechanistic understanding of how DNA double-strand breaks (DSB) are repaired is rapidly advancing in part due to the advent of inducible site-specific break model systems as well as the employment of next-generation sequencing (NGS) technologies to sequence repair junctions at high depth. Unfortunately, the sheer volume of data produced by these methods makes it difficult to analyze the structure of repair junctions manually or with other general-purpose software. Here, we describe methods to produce amplicon libraries of DSB repair junctions for sequencing, to map the sequencing reads, and then to use a robust, custom python script, Hi-FiBR, to analyze the sequence structure of mapped reads. The Hi-FiBR analysis processes large data sets quickly and provides information such as number and type of repair events, size of deletion, size of insertion and inserted sequence, microhomology usage, and whether mismatches are due to sequencing error or biological effect. The analysis also corrects for common alignment errors generated by sequencing read mapping tools, allowing high-throughput analysis of DSB break repair fidelity to be accurately conducted regardless of which suite of NGS analysis software is available.

ABBREVIATIONS

BWA Burrows–Wheeler Aligner
CIGAR Compact Idiosyncratic Gapped Alignment Report
C-NHEJ classical nonhomologous end joining
DSB double-strand break
Hi-FiBR high-throughput fidelity of break repair
MMEJ microhomology-mediated end joining
NGS next-generation sequencing

1. INTRODUCTION

DNA double-strand breaks (DSBs) are an extremely deleterious form of DNA damage that, if not repaired accurately, can result in cell death (Bennett, Lewis, Baldwin, & Resnick, 1993; Jeggo, 1990; Resnick & Martin, 1976) or carcinogenic mutations (Aparicio, Baer, & Gautier, 2014; Hromas, Williamson, Lee, & Nickoloff, 2016). Historically, DSBs have been thought to be repaired via homologous recombination (HR) or classical nonhomologous end joining (C-NHEJ) (Ciccia & Elledge, 2010). HR is recognized as error-free since it uses homologous template DNA to direct the repair process (Nickoloff, 2017). In contrast, C-NHEJ involves local processing of the DNA sequence surrounding the break until the DNA ends are compatible for direct ligation (Lieber, 2010). As a consequence, C-NHEJ is frequently error-prone, resulting in insertions, deletions, and base substitutions, and yet it is the primary repair pathway for induced DSBs in mammalian cells (Jeggo, 1990). Recent studies have begun to delineate the mechanism of a third pathway for DSB repair, called alternative end joining (alt-EJ) (Decottignies, 2013) (also called microhomology-mediated end joining (MMEJ) (Lee & Lee, 2007; Moore & Haber, 1996; Sharma et al., 2015) or theta-mediated end joining (TMEJ) (Beagan et al., 2017; Chan, Yu, & McVey, 2010; Kent, Chandramouly, McDevitt, Ozdemir, & Pomerantz, 2015; Mateos-Gomez et al., 2015; Wyatt et al., 2016; Yousefzadeh et al., 2014)). This pathway is believed to be a backup pathway for breaks that are unrepairable by HR or C-NHEJ (Nussenzweig & Nussenzweig, 2007; Sfeir & Symington, 2015). As this pathway appears to be used following the initiation of end resection (Seol, Shim, & Lee, 2017), alt-EJ is even more mutagenic than C-NHEJ, often causing hypermutation in flanking DNA (Sinha et al., 2017) and large deletions mediated by microhomology (Soni, Siemann, Pantelias, & Iliakis, 2015; van Schendel, van Heteren, Welten, & Tijsterman, 2016).

The choice between which repair mechanism to use, as well as the degree of mutagenicity, appears to be affected by a number of factors, including the presence of microhomologies in the DNA sequence. To further understand what factors affect the pathway used for DSB repair, and the types of mutations they cause, model cell systems have been developed that enable delineation of repair events at specific nuclease target sites (Chu, Wu, Xu, & Her, 2013; Gunn, Bennardo, Cheng, & Stark, 2011; Holmes & Haber, 1999; Liang, Han, Romanienko, & Jasin, 1998; Lin, Lukacsovich, & Waldman, 1999; Moynahan & Jasin, 1997; Nick McElhinny et al., 2005; Roth & Wilson, 1986; Rouet, Smih, & Jasin, 1994; Rudin & Haber, 1988; Xu, Wu, & Her, 2015), such as those generated by CRISPR-Cas9 technology (Vriend, Jasin, & Krawczyk, 2014; Vriend et al., 2016; Wyatt et al., 2016), or at ends of transfected linear DNA (Waters et al., 2014). Traditionally, these break sites have been induced, given time for repair, and then up to 100 individual isolates are chosen and sequenced. With the advent of next-generation sequencing (NGS) technology (Goodwin, McPherson, & McCombie, 2016), it is possible to sequence tens of thousands of events, from hundreds of samples, each with high fidelity and low cost. This shift in technology has led to a standard of high volume sequencing that accurately reveals the frequency of a broad spectrum of repair events including any specific mutations that occur (Huefner, Mizuno, Weil, Korf, & Britt, 2011; IJspeert et al., 2016; Liang, Sunder, Nallasivam, & Wilson, 2016; Mateos-Gomez et al., 2015; Soong et al., 2015; Waters et al., 2014; Wyatt et al., 2016), and even enabled single nucleotide resolution assessment of randomly occurring translocations (Chiarle et al., 2011). Because of the large number of events that are accessed, these methods are capable of detecting subtle differences in DSB repair junctions among model systems and how they are influenced by sequences around the break sites (Khodaverdian et al., 2017).

However, these large data sets demand robust methods for alignment of the sequences and analysis of the break junctions. The sequencing and alignment methods used to produce and analyze these data sets must be highly accurate to avoid misinterpretation of some repair events. For instance, sequencing errors near a break junction can cause misalignment of a repair product. Likewise, in downstream analysis sequencing error might be incorrectly called a base substitution, insertion, or deletion event.

In this chapter, we discuss methods for robustly sequencing and analyzing DSB repair junctions. We describe how to construct high-throughput sequencing libraries of DNA amplicons containing break-induced recombination events, how to align NGS sequence reads to a reference sequence, and ultimately how to analyze the alignments for base substitutions, insertions, and deletions at the repair junction (Fig. 1). The downstream analyses are largely performed by our custom script (written in Python3; https://github.com/Alexander-Brown13/Hi-FiBR), called Hi-FiBR (high-throughput fidelity of break repair) analysis. We describe how this tool works and compare alignment and sequencing strategies to determine how to maximize the accuracy and utility of the Hi-FiBR analysis.

2. CREATING COMPLEX LIBRARIES OF DSB REPAIR JUNCTIONS

High-throughput evaluation of the sequence of DSB repair junctions requires either the high efficiency induction of a site-specific DSB within chromosomal DNA, or the transfection of linearized episomal DNA in a large population of cells whose repair machinery can then rejoin the broken DNA ends (Soong et al., 2015; Waters et al., 2014; Wyatt et al., 2016). Nearly synchronous and uniform creation and repair of the break in the cell population is critical to obtain samples with a high degree of complexity in the repair products. Ultimately, this complexity is essential to effectively and robustly determine the roles of specific repair pathways or proteins in rejoining a DSB.

The first step in generating complex libraries of repair junction amplicons is to either isolate genomic DNA from a population of cells having undergone repair of a site-specific break or to harvest bulk circularized episomal DNA (i.e., linear DNA that has been joined into a circle) from

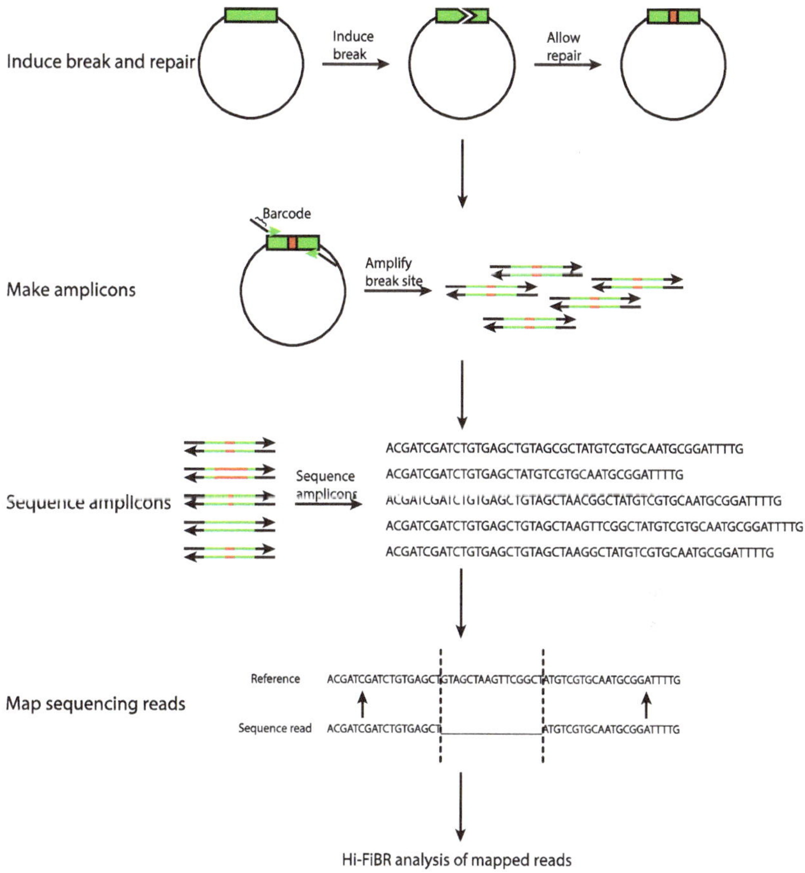

Fig. 1 Steps in building and sequencing an amplicon library of DSB repair events. First, linear or circular DNA with an inducible break site must undergo a break and be given time for repair. The sequence surrounding the break site (shown in *green*) may experience an error such as an insertion, deletion, or base substitution (shown in *red*). To capture the frequency and variety of these errors the sequence surrounding the break site is PCR amplified, optionally with primers that contain barcodes. The amplicon products are then sequenced, and the sequenced reads are mapped to the reference sequence. Finally, the mapped reads are analyzed by the Hi-FiBR analysis.

transfected or injected cells. For mammalian cells containing site-specific DSB repair events, a typical genomic DNA isolation kit, such as a Qiagen DNeasy Blood and Tissue kit, is sufficient.

Once isolated, the repair junctions must be PCR amplified using unique sequence surrounding the break site. The position of the primers plays a crucial role in dictating the type of repair events that can be observed. More distantly spaced primers allow larger deletion events to be recovered while also providing additional sequences that may be used in microhomology-mediated events. The maximum length of the amplicon, however, is determined by the type of high-throughput sequencing technology that is to be employed. For example, paired-end sequencing of amplicons on an Illumina HiSeq platform would limit analysis to amplicons in the 250–300 bp product size (150 bp of sequence is acquired from each end), while MiSeq sequencing would maximally allow analysis of 500 bp fragments. Other sequencing platforms like Ion Proton or PacBio can accommodate even larger amplicons (up to several kilobases for PacBio); however, systematic sequencing error common to these platforms is problematic for analysis (see Section 5.2).

To produce amplicons we recommend following a protocol using a high-fidelity DNA polymerase, as standard DNA *Taq* polymerase may introduce errors. The following PCR protocol uses NEB's high-fidelity DNA polymerase. The reagents and their corresponding final concentrations are as follows: 1 × Phusion HF Buffer, 200 μ*M* dNTPs, 0.5 μ*M* forward primer, 0.5 μ*M* reverse primer, 1.0 U/50 μL PCR Phusion DNA polymerase, and ~250 ng template genomic DNA. For repair junctions induced by Cas9 cleavage of human genomic DNA, 250 ng of input DNA would correspond to ~25,000–50,000 independent repair events that could be assessed. The PCR mix should then be cycled in a thermocycler using the following steps: initial denaturation at 98°C for 30 s, 22–25 cycles of amplification which are 98°C for 5–10 s, 45–72°C for 10–30 s, and 72°C for 15–30 s per kb, the final extension at 72°C for 5–10 min, and an optional hold step at 4–10°C.

Following the PCR amplification, the amplicon must be purified to add adapters for NGS. The following Qiagen PCR-cleanup kit protocol works well for this purpose:

(1) Mix PCR with 5 reaction volumes of Buffer PB,
(2) Place spin column in collection tube, apply sample, centrifuge for 1 min, discard flow-through, and place column back into collection tube,
(3) Add 750 μL Buffer PE, let stand for 2–5 min, centrifuge for 1 min, discard flow-through, place column back into collection tube, and centrifuge for 1 min to remove residual wash buffer,
(4) Place column into clean 1.5-mL microcentrifuge tube,
(5) Add 30 μL Buffer EB or water to center of column membrane, let stand for 1–4 min, and centrifuge for 1 min.

Once a pure amplicon is obtained, one can proceed with a NGS library preparation protocol appropriate for the type of sequencing that will be conducted. These preparations involve addition of adapter sequences used as the template sequence to prime sequencing. Adapter sequences can either be appended to PCR primers, or ligated to the amplicon ends. If the adapters are to be added by ligation, amplicon products first need to be 5′ phosphorylated by a 15-min incubation with 10 U T4 Polynucleotide kinase (New England Biosciences) and 10 m*M* ATP at 37°C. Then a 3′ terminal A nucleotide is added by addition of 15 U Klenow exo$^-$ polymerase (New England Biosciences), 0.2 m*M* dATP, and an additional incubation at 30 min. For Illumina libraries, adapters (for sequences, see https://support.illumina.com/downloads/illumina-customer-sequence-letter.html) are annealed and ligated to amplicon DNA with a 10:1 ratio of linker to amplicon, 10 m*M* ATP, 2000 U T4 DNA ligase (New England Biosciences), and 30% PEG. About 1–2 μg of DNA is typically used for a library preparation, but this equates to far more molecules of PCR product than the original ~20,000 molecules of repair events placed in the original PCR. Consequently, sequencing of a library from a single sample will likely result in a massive oversampling of the repair events in the sample. For instance, an Illumina HiSeq 2500 is capable of generating ~150 million reads per lane; thus, sequencing of a single sample/library per lane would be expected to sample the same repair event 10,000 times. To maximize cost effectiveness, amplicons from different treatment conditions can be pooled together prior to the library preparation. To differentiate between amplicons from different treatments, barcoded primers need to be used during the PCR step. Barcodes can be designed in multiple ways and can be assigned to both the forward and reverse primer, which increases the permutations available for barcoding each amplicon. After sequencing, the barcodes are used to parse the sequence reads into separate samples, and then the reads are examined for specific repair events.

To remove possible PCR products of incorrect size and unligated adapters from the NGS library preparation, it is recommended to do a gel extraction or PEG purification. The following abridged gel extraction protocol uses Qiagen's QIAquick gel extraction kit:

(1) Excise the DNA fragment,
(2) Weigh the gel and add 3 volumes Buffer QG to 1 volume gel,
(3) Incubate at 50°C for 10 min,
(4) Add 1 gel volume isopropanol to sample and mix,

(5) Place spin column in collection tube, apply sample, centrifuge for 1 min, discard flow-through, and place column back into collection tube,
(6) Add 500 μL Buffer QG to column, centrifuge for 1 min, discard flow-through, and place column back into collection tube,
(7) Add 750 μL Buffer PE, let stand for 2–5 min, centrifuge for 1 min, discard flow-through, place column back into collection tube, and centrifuge for 1 min to remove residual wash buffer,
(8) Place column into clean 1.5-mL microcentrifuge tube,
(9) Add 30 μL Buffer EB or water to center of column membrane, let stand for 1–4 min, and centrifuge for 1 min.

Alternatively, small molecular weight contaminants can be separated efficiently from an otherwise clean amplicon by PEG purification. An abridged PEG purification protocol is as follows:

(1) Add an equal volume PEG-8000/2.5 *M* NaCl to the PCR product and mix by brief gentle vortex (the final percentage of PEG is based on the PCR fragment size: 10% for >100 bp, 8.3% for >200 bp, 6.7% for >650 bp, and 5.6% for >800 bp),
(2) Incubate PCR + PEG mix at room temperature for 5 min,
(3) Centrifuge mix at ~15,000 × *g* for 20 min (at room temperature),
(4) Remove and discard the supernatant,
(5) Add the same volume of cold 80% ethanol (as PEG solution),
(6) Centrifuge 5 min at max RCF,
(7) Remove and discard supernatant,
(8) Centrifuge tubes for 10 s at max RCF,
(9) Remove and discard residual ethanol,
(10) Place tubes open in a 37°C incubator for 10 min to remove ethanol,
(11) Add an appropriate volume of molecular grade water, heat for 5 min at 37°C, and then vortex briefly to suspend DNA.

We recommend following this protocol with an additional QIAquick PCR-cleanup column purification (outlined above) to remove contaminating enzymes and trace levels of adapter that may still be present in the sample.

3. ALIGNING SEQUENCE READS

3.1 Merging Paired Reads

If a sequencing technology is used that generates pair-end reads (i.e., HiSeq or MiSeq), the reads will need to be merged. This can be done using a variety of tools, including CLC Genomics Workbench or BBMerge. BBMerge is an open access tool, written in Java, and can be installed on any operating system that runs Java, but is also a plug-in for Geneious. Here, we describe its usage in Geneious 10.2.2, but if used as a stand-alone tool, refer to the BBMap guide on GitHub (https://github.com/BioInfoTools/BBMap) for installation and execution.

To run BBMerge in Geneious, first select the unmerged, paired reads file. Then select "Sequence" in the top bar, and choose "Merge Paired Reads." A prompt will appear asking for Merge Rate and Maximum memory to use. Normal merge rate and 500 MB of memory is sufficient. Select "OK." The tool will produce two output files with the same name as the input file labeled either "Merged" or "Unmerged" as a prefix.

3.2 Trimming and Padding fastq Files

Once the paired-end sequences have been merged, the resulting fastq reads need to be cropped so that they all start and end at the same 5′ and 3′ sequences. This provides common reference index points that are critical for downstream automated determination of the repair junction structure. We have designed a "Trim_and_Pad" script that searches for user-defined sequences at the 5′ and 3′ ends of the reads and trims the fastq reads prior to the 5′ matching sequence and after the 3′ matching sequence. The 5′ sequence is looked for starting at the 5′ end of the read and progresses to the 3′ end, while the 3′ sequence starts at the 3′ end and passes to the 5′ end. This progression ensures that short sequences are not found prematurely in the rest of the sequence. Also, if the 3′ and 5′ sequences are both found, but in the incorrect relative orientation (i.e., the 3′ sequence is 5′ of the 5′ sequence), then it is not considered a match and the read is ignored. Reads containing both the 5′ and 3′ sequences in correct orientation are trimmed and printed to an output file, "file_Match," in fastq format, where "file" is the original file name. In addition to the sequence of the read being trimmed, the corresponding quality scores will be trimmed as well; most aligners will not accept fastq files if the read sequence length is different than the quality score length. Alternatively, reads lacking either the 5′ or the 3′ sequence are printed to a different file, "file_NoMatch." Nonmatching reads are primarily due to PCR contaminants from off-target amplification and low level sequencing error that modifies the corresponding sequence in a specific read. This step of read processing typically removes ~3% of the total reads in a sequenced library.

The user must decide what an appropriate length of sequence to search for is; if the sequence is too short, it will likely appear by random chance, but the longer the sequence is, the greater the chance of a sequencing error occurring in the search sequence, which will result in the removal of a larger number of reads. We suggest scaling the length of the searching sequence by the read length, using longer sequences for longer reads.

Some repair amplicon libraries may contain reads with large deletions, or mismatches near the ends of the sequence. Many aligners will have difficulty properly aligning the ends of these sequences which is required to generate the appropriate position indexes. This problem can be rectified using the padding functionality, which adds user-defined sequence to the ends of the read that will help anchor these sequences to the appropriate index. When the script prompts for padding sequence, the user can define nothing and the script will continue without adding sequence, but if padding is included it will add it to the corresponding end of the sequence and add the appropriate number of quality scores "G" (a perfect score). Once again, the user should make sure the reference sequence has the same sequence added to it. If the padding is being used to replace what is already present in the reference, the aligner should work correctly. If it is instead being used to act as an anchor and includes a highly repetitive sequence such as "AAAAAAAAAA," then that sequence must also be added to the reference.

The Trim_and_Pad script runs on all the files in a directory, and will output two files for each file in the directory with the same name as the file being processed, but with the extensions "Match.fastq" and "NoMatch.fastq," as described earlier. When the script finishes running, compare the sizes of the two files to each other. If the size of the Match file is significantly larger, then the sequencing data were probably very clean and the file is ready for aligning to the reference. If the sizes are close or the NoMatch is significantly larger, the sequencing data may have high frequency of sequencing error or the library may be significantly contaminated with off-target PCR product.

3.3 Choosing an Aligner

After processing the raw fastq reads, they are now ready to be aligned to a reference sequence to determine the sequence structure of each repair junction. Because the Hi-FiBR analysis operates on SAM files (Li et al., 2009), which is a standard high-throughput sequencing read alignment format, nearly any commercial or open source alignment software can be used to align your sequencing reads to a reference sequence. Using an aligner with appropriate parameters, however, can have a dramatic impact on the accuracy of the data analysis. In the following sections we provide tutorials for a few of the more widely used aligning softwares: commercial products Geneious 10.2.2 (BioMatters) and CLC Genomics Workbench 7.5 (Qiagen) and open source aligners Bowtie2 (Langmead & Salzberg, 2012) and Burrows–Wheeler Aligner (BWA) (Li & Durbin, 2009).

3.3.1 Mapping With Geneious 10

The trimmed fastq files can be batch imported into Geneious by selecting them all and dragging them into a new, empty folder within the Geneious database. The program will try to recognize the sequencing that was done based on the quality scores from the file, and the user must confirm which they are. The reference sequence should also be imported as a fasta file (reference file). If all the fastq files can use the same reference sequence for mapping, they can be batch mapped by selecting all of them, then choosing "Align/Assemble," followed by "Map to Reference."

A window will open allowing several parameters to be changed. Start by unchecking all the boxes. Check "Do not trim" (under Trim Before Mapping) and "Save contigs" (under Results). Under Data (at the top) choose the appropriate reference sequence. Under Method, Geneious should be the default Mapper, but to obtain the most accurate alignments, the Sensitivity should be changed to "Highest Sensitivity/Slow" and Fine Tuning should be set to "None (fast/read mapping)." Assembly name can be chosen by the user, but a convenient format is to use the exact phrase "{Reads Name} assembled to {Reference Name}" which will autopopulate the {Name} and produce a file clearly stating which reads were assembled to which reference file. Select "OK" when all the parameters have been set to launch the mapping. The high sensitivity setting may trigger a warning such as "Sensitivity is too high," which can be ignored.

Once all the files have been mapped, they can be individually or batch exported as SAM files. They must be exported with the following settings for the Hi-FiBR breakpoint analysis to work properly. Choose the files and select export, then make sure to change the file format to SAM, not just the extension. When you choose export, there may be a warning "Potential Data Loss," which can be ignored by choosing "Proceed." A window will appear called "SAM sequences/alignments Export." Select "More Options." Then uncheck all the boxes except "Replace IUPAC ambiguities with N." Be sure not to export "padded" CIGAR strings as this option will include information comparing each read to each other (as opposed to just a pairwise comparison of each read to the reference) into the format of the CIGAR string (Compact Idiosyncratic Gapped Alignment

Report: an alphanumeric description of the alignment of each read to the reference). The CIGAR string is the primary field used in Hi-FiBR analysis to determine the sequence structure of repair junctions. Therefore, failure to properly format it will ultimately cause the breakpoint analysis to fail. Select OK and the files will begin exporting. Once the exporting is finished, the breakpoint analysis can be done.

3.3.2 Mapping With CLC Genomics Workbench 7.5

The fastq files containing merged and trimmed sequencing reads of DSB repair junctions can be batch imported into CLC by selecting "Import," followed by the appropriate sequencing platform (in this example, Illumina), then finding the location of the files and selecting all of them. Before choosing "Next," make sure there are no options selected under "General options" or "Illumina options"; however, Remove failed reads may optionally be chosen. Make sure the "Quality scores" are correct, which should be 1.8 and later. After choosing "Next" another screen will prompt the user for "Result handling," and Save should be chosen. The next window will show folders in CLC that are mapped, and one must be chosen for the data to be saved to, then select "Finish."

Alternatively, the directory containing the data can be mapped to CLC by choosing "Add File Location" near the top of the Navigation Area, and finding the directory and adding it. The reference file can be similarly imported, instead as a fasta file, and preferably in a separate folder for batching purposes.

Once the data have been imported, choose "Toolbox," "NGS Core Tools," and "Map Reads to Reference." Choose the Batch option near the bottom of the new window and select the folder containing the fastq files by moving them into "Selected elements" with the blue arrow. The next window allows you to choose or exclude particular files in the folder to analyze; if all files in the folder are to be mapped, simply proceed to the next screen. In the next screen, choose the reference file and "No masking." Following is a screen with alignment parameters which are important for determining how well the aligner opens gaps. This can be left with the default settings which use linear gap penalties (2, 3, 3, 0.5, 0.8), no global alignment, and "Ignore" nonspecific match handling. Newer versions of CLC Genomics Workbench include an option to utilize affine gap penalties designed to favor generation of long, continuous gaps. Affine gap penalties tend to produce more accurate alignments as long as the insertion and deletion extension costs are set to 1. In the final screen choose "Create stand-alone read mappings," and Save; you may also choose to "Create report" for additional details. After choosing Finish the program will run and map the reads to the reference one at a time.

As with mapping in Geneious, read alignments generated in CLC need to be exported as SAM files. First select Export near the top of the screen, then choose SAM and Open. As before, choose the Batch option and move the folder into Selected elements. In the next window, do not use a compression, and we recommend making a Custom file name with the exact phrase "{1}.{2}" which will produce a SAM file with the original file name and the SAM extension. The final window requests the location for where the files will be exported. Once the exporting is finished, the breakpoint analysis can be done.

3.3.3 Mapping With Bowtie2

Bowtie2 (Langmead & Salzberg, 2012) runs in Linux or Mac OS from the Command Line, and thus requires no data importing. However, Bowtie2 does require an index for the reference genome to run properly. This is easily done through the command line by entering the directory in which the reference file is located and running "bowtie2-build reference.fa reference" where "reference.fa" is the name of the file and "reference" is the output prefix used for the index files. Once finished, Bowtie2 will have generated a number of index files with the given prefix and a suffix ending with "bt2" (bowtie2). When launching the mapping portion of the program, it is easiest to do in the directory containing these indices, so keep them together and in a directory in which you want to run the program.

Once the index files are finished, the trimmed fastq files can be mapped. Enter the directory where the bt2 index files are, and run the command "bowtie2 -x reference -U read.fastq -S output.sam" where "reference" is the prefix for the bt2 index files, "read.fastq" is the fastq file to be mapped, and "output.sam" is the name of the SAM file where the results of the analysis will be written. It is highly recommended to give the output SAM the same name as the input file. Also be sure to include the extensions in the input/output names, but not the reference prefix. The program runs fast and automatically creates SAM files, so there is no need to export them. However, it is difficult to batch run on multiple fastq files; multiple files can be listed, but they export to the same output SAM file. It may be simpler to use other software when a large number of fastq files are being analyzed. Regardless, once the program is finished, it will produce percentages of reads that aligned 0, 1, and >1 times. The vast majority of reads should map exactly once.

3.3.4 Mapping With BWA

BWA (Li & Durbin, 2009) is run almost exactly the same as Bowtie2. Like Bowtie2, BWA runs in Linux or Mac OS from the Command Line. BWA also requires an index of the reference genome, and it is generated by entering the directory with the reference file, through the command line, and running "bwa index reference.fa" where "reference.fa" is the name of the reference file. The output index files will share the same prefix as the reference file name and should be kept in the same directory to conveniently run the mapping function.

Once the index files are complete enter the directory that contains the BWA index files and run the command "bwa mem reference.fa read.fastq > output.sam" where "reference.fa" is the reference file, "read.fastq" is the fastq file to be mapped, and "output.sam" is the output SAM file. Include the appropriate extensions in the file names and, preferably, give the SAM file the same name as the input files. The program runs quickly, but it suffers from the same difficulty as Bowtie2 with an inability to batch run multiple fastq files. Consequently, it may be simpler to use other software when many fastq files are being analyzed. After the mapping is finished, the breakpoint analysis can be done on the SAM files.

4. BREAKPOINT ANALYSIS

4.1 Running Hi-FiBR

Hi-FiBR requires six inputs from the user: (1) the directory of SAM files, (2) a reference sequence file in fasta format, (3) the number of nucleotides from the start of the reference sequence to the position in the reference sequence where the DSB was generated (left breakpoint), (4) the number of nucleotides from the DSB site in the reference to the end of the reference sequence (right breakpoint), and (5) and (6) the number of nucleotides left and right of the break site in which base substitutions will be looked for. For convenience, once the script starts running it will ask for these inputs. The file and directory inputs need to be in the same directory that the script is in. Once all the inputs are provided, the script will run from start to finish on all files in the directory. This allows a large number of samples that utilize the same reference sequence to be analyzed all at once.

All the output files are generated in the same directory as the input directory in order to keep the files grouped. Please note that if the analysis needs to be rerun on the same directory of files, the output files from the initial analysis will also be considered as new inputs, so be sure to separate these files from the old inputs prior to reinitiating a second analysis on the same directory.

4.2 How Hi-FiBR Works

Hi-FiBR utilizes the CIGAR string for each read within the SAM files alignments to determine the nature of the repair that has occurred. A CIGAR string is, as its name (Compact Idiosyncratic Gapped Alignment Report) suggests, a report for the gapped alignment performed on the read to the reference sequence. In the string, "M" represents matching nucleotides, "D" indicates deleted nucleotides, and "I" designates nucleotides present in the read, but not the reference, i.e., insertions. The numbers indicate how many nucleotides are attributed to M, I, or D. For example, in our data set of repair events from a 197-bp amplicon, a "perfect" repair event would be indicated with a CIGAR string of 197M, as the reference is 197 nucleotides long and no processing of the break occurred.

In the Hi-FiBR analysis, first, the CIGAR string is examined for the character "S," which is used in programs such as CLC to denote soft clips, "*" which is used in some programs like Bowtie2 to signify reads that did not map properly, and "H" which is used in BWA to denote hard clips. As reads containing any of these characters are unlikely to result from appropriate amplification of repair junctions, they are removed from further analysis. Next, the CIGAR is broken down into its respective components (e.g., 20M5D25M → 20M, 5D, 25M) and the matching lengths are used to determine the region of each read where processing of the DSB occurred (called the processing window). This window can result from relatively simple processing of the break, such as loss of nucleotides surrounding the breakpoint (a deletion), or insertion of nucleotides at the break site. However, more complex events also occur, involving both resection and DNA synthesis from the break site. Errors during the resynthesis of these ends can result in small insertions, deletions, or base substitutions in regions flanking break sites. Mathematically, the processing window of these complex events is determined by calculating the distance between the breakpoint and the distal most error (i.e., nonreference base) on both sides of the repair junction. Immediately after determining the processing window, the class of repair event is determined as "exact," "insertion," "deletion," or "complex."

If the amplification, sequencing, and alignment algorithms were all perfectly accurate, then the number of times each specific event (or read sequence) is observed in the sequencing library could be counted at this time. However, each of these steps produces error that can cause misclassification of a repair event. For example, PCR amplification and high-throughput sequencing technologies produce errors in the reported sequence of individual reads that are intrinsic to the preparation and sequencing of the DNA and are not related to the actual repair event.

Likewise, small insertions during the repair event (both templated and non-templated) are frequently misaligned by most alignment algorithms. Such reads are frequently miscategorized. To overcome these issues, Hi-FiBR utilizes a local, realignment of the sequence near the DSB repair junction to ensure accurate calling of the repair processing window. This process specifically checks that the ends of inserted sequences are not from the reference sequence and looks for base substitutions within a user-defined distance on each side of the repair junction. If any of these events occur, the size and location of the processing window is adjusted (as shown in Fig. 2).

Once adjusted, all remaining base substitutions are likely due to sequencing error. This is apparent as sequencing error should be evenly distributed across the amplicon, while sequence changes stemming from biological processing of the repair junction should be enriched near the breakpoint of the reference. Indeed, in MiSeq-sequenced amplicons, small insertions, small deletions, and base substitutions all cluster near repair breakpoints (Fig. 3).

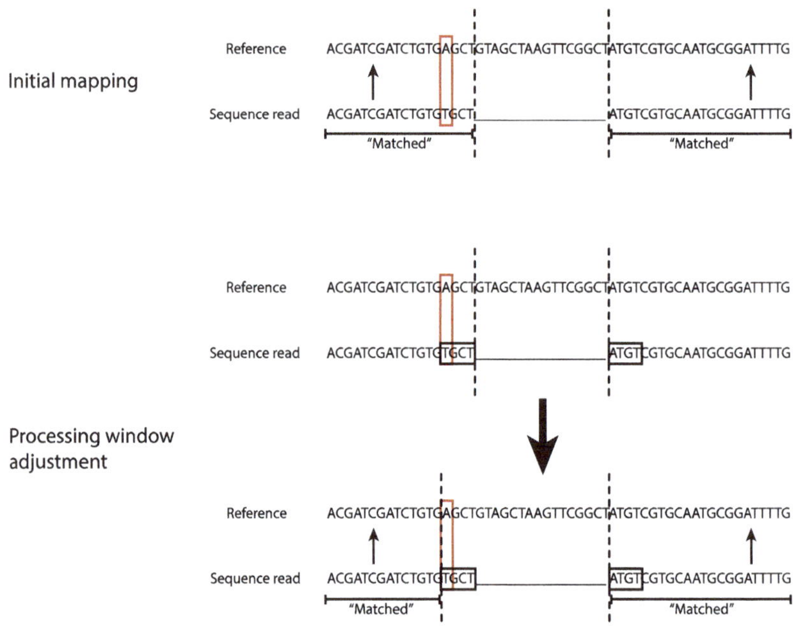

Fig. 2 Determining the processing window of a DSB repair event. Standard alignment tools will often treat base substitutions as sequencing error, which causes them to be called matches to the reference sequence. The Hi-FiBR analysis can recognize mismatches in a user-defined window around the break site (here shown as 4 bp on either side) and adjust the matching sequences. This also changes what the program considers as the processing window for repair event classification.

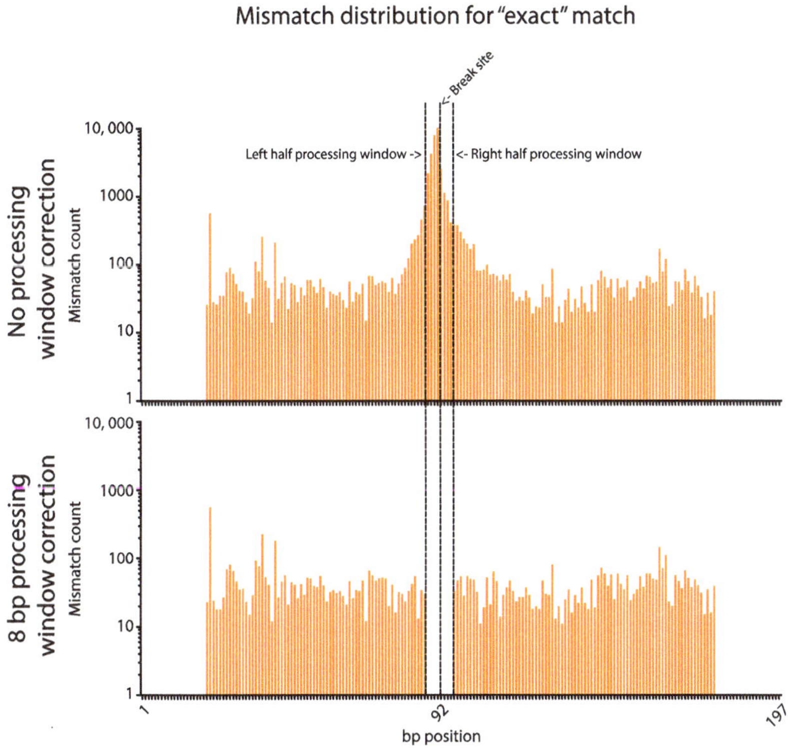

Fig. 3 Distribution of base substitutions in Illumina sequencing DSB repair amplicons. The distribution of mismatched bp is given for reads that were considered perfect repair events by Geneious, before and after the 8 bp search for base substitutions correction. The majority of mismatches occur homogenously across the read, as expected of sequencing error, with counts of ~10–100. However, the mismatches peak around the break site with a maximum of 11,982 immediately at the break site (~120–1200-fold higher than the surrounding sequence). The 8 bp processing window correction recategorizes these counts and leaves the rest that are likely sequencing error. No mismatches occur at the ends of the reads due to the "Trim_and_Pad" script used to filter reads.

However, after excluding base pair substitutions within eight nucleotides surrounding a repair junction, all remaining base substitutions are evenly distributed across the amplicon. This is consistent with the major source of Illumina sequencing error being random base substitutions. Hi-FiBR subsequently removes any base pair substitutions outside of the defined processing window by constructing a hybrid sequence for each read (called a "reconstructed read"), consisting of actual read sequence in the processing window surrounded by the corresponding reference sequence for the

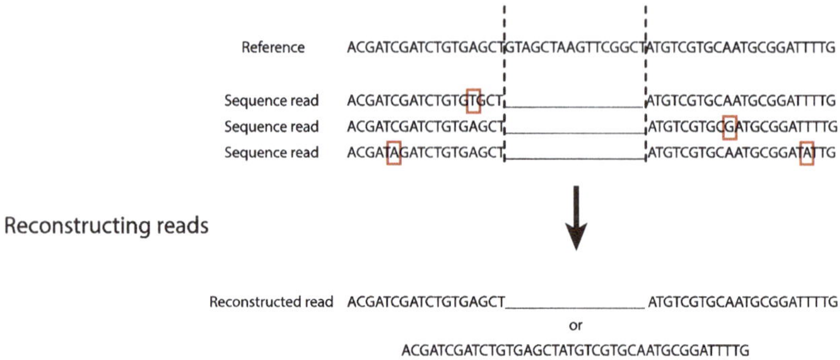

Fig. 4 Removing sequencing error by generating reconstructed reads. Sequencing error in the flanks of reads around the break site causes the same type of repaired read to appear as a different repair event. To correct for this discrepancy the reads are reconstructed in the regions of matching sequence to become the same as the reference sequence, thus creating a common, reconstructed read.

flanking regions (as shown in Fig. 4). The only variation in this reconstructed read should now be due to processing of the DSB during the repair event.

The number of nucleotides processed on each side of the breakpoint, deletion lengths, insertion lengths, insertion sequences, and reconstructed read sequences is added as additional columns for each read in the original SAM file and written as a new SAM with the same name plus the extension "_Extended.sam." Furthermore, a new data file containing the list of repair events observed is generated with the same name as the original SAM file plus the extension "_Final.sam." The "Final" output file contains a summary of all the analyses Hi-FiBR conducts. It is a tab-delimited table containing 22 columns, where each column contains unique information describing the repair event (see Table 1 for a detailed description of each column's contents). This file is generated by listing each reconstructed read observed in the data set once. The script then proceeds to count how many times each reconstructed read occurs in the full data set. It is also important to know the percent of mismatch between the reconstructed read and the original reads, as it may help identify systemic mapping errors or biological events that are being masked by the reconstruction. A small, nonzero value is always expected as sequencing error will occur at a low frequency, and thus, some portion of reads will always contain a mismatch. As the value increases, this suggests that there is some other issue, such as PCR cross-contamination, or real biological events being called sequencing error by the mapping tool. To help determine the cause, the script compares the original reads and the reconstructed read and outputs a read error distribution file which contains

Table 1 Descriptions Are Given for the Contents of Each Column in the "Final" Output File

Column	Description
1	Name of reference sequence to which read was mapped
2	Original CIGAR string
3	Length of read
4	Original CIGAR string; later may contain a reconstructed CIGAR string
5	Corrected length of sequence matching the left flank of the reference
6	Corrected length of sequence matching the right flank of the reference
7	Distance between the break and the end of the left matching sequence
8	Distance between the break and the end of the right matching sequence
9	Number of nucleotides deleted on the left side of the break
10	Number of nucleotides deleted on the right side of the break
11	Total number of deleted nucleotides
12	Start of inserted sequence
13	End of inserted sequence
14	Length of inserted sequence
15	Sequence of inserted nucleotides
16	Repair event class
17	Reconstructed read sequence
18	Number of times reconstructed read occurred in read population
19	Potential microhomology that mediated deletion
20	Length of microhomology
21	Number of times reconstructed read is observed in this file (in a row)
22	Percentage of reads that contain mismatches to reconstructed read

The output does not follow a conventional format, and so the descriptions should be carefully examined.

counts of the mismatches at each individual bp position among raw reads categorized as the same repair event (i.e., they produce the same reconstructed read). If the errors occur in a homogeneous distribution, it is likely just sequencing error. If there is a peak near the break site, then there is a biological event being ignored. In the event that the same reconstructed

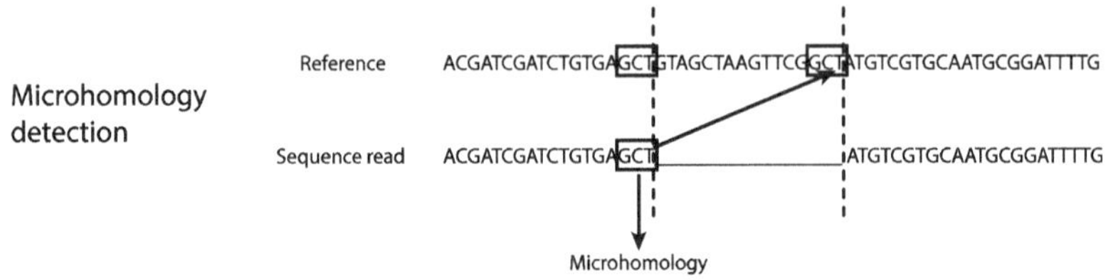

Fig. 5 Determination of microhomology usage. For reads classified as deletions, sequence microhomologies may mediate the repair event. These microhomologies are detected by examining the sequence to the left of the deletion in the read, and the sequence inside the deletion on the right side, as shown here. The same process is repeated for the sequence to the right of the deletion in the read (not shown here). The microhomologies detected are then combined into a single sequence.

read sequence can be generated by multiple alignments of the read sequence, all possible alignments are included in _Final.sam file; however, they will contain the same count for the number of times that event was observed due to them having the same reconstructed read sequence. Hi-FiBR marks these events in column 21 of the _Final.sam file (which indicates the number of times a sequence occurs in the _Final.sam file) so that the user can manually determine which alignment is most likely.

For repair events that are simple deletions, Hi-FiBR also determines whether microhomology mediated the event. This is done by comparing the matching indices to the position of the breakpoint and then moving along the working sequence and reference sequence in single base iterations to search for homology (see Fig. 5). Specifically, the sequence on the left side of the repair junction is compared to the rightmost portion of the deleted sequence. Similarly, the sequence to the right of repair junction is compared to the leftmost portion of the deleted sequence. If microhomologies are found on each half, they are added together to make the complete sequence and the homology length is calculated.

Finally, each line in the newly written file is reread by the script and the insert sequences are exported into fasta files based on their class as "complex" or "insertion," and all the working (reconstructed) sequences are exported into a single fasta file.

5. RESULTS OF ANALYSIS AND DISCUSSION

5.1 Effects of Using Different Aligners for Hi-FiBR Analysis

Many alignment tools besides those described here are used by the scientific community to analyze sequencing data. However, scoring penalties associated with features such as gap creation, extension, and mismatching nucleotides are not universal, neither are the methods for alignment (local vs global, gapped vs ungapped, etc.), and, ultimately, lead to differences in mapping. Here, we compared four of the most commonly used aligners to see if differences in their algorithms made a significant impact on the number and diversity of events they called, as denoted by the CIGAR strings. Additionally, to test the versatility of each aligner, we used a data set that had a highly varied array of sequences (suggesting many different repair event types) (Fig. 6). This helps show how individual alignment tools handle events with a high degree of complexity, and how efficaciously Hi-FiBR corrects errors that the aligners make when aligning complex reads.

The data set we used was first trimmed, then mapped using each alignment tool, individually, and the SAM files from each mapper were analyzed using the Hi-FiBR analysis script with a 0- and an 8-bp window around the break site, creating eight permutations. The count and percentage of reads remaining after each step are listed in Table 2, and show that most of the reads are kept throughout the entire analysis. Importantly, the change in window size around the break site does not remove reads, but reclassifies them. It is worth noting, though, that the final outputs for Geneious and BWA retain the highest number of reads at ~97% and ~96.7%, respectively, and CLC and Bowtie2 retain the least at ~93% and ~94%, respectively. For CLC, this is because ~4% of the reads are mapped using methods such as soft clipping that the script does not process and removes from the analysis. For Bowtie2, this is because ~3% of the reads do not map, and never make it to the analysis (even if the reads are left in the SAM file, the Hi-FiBR analysis will skip them as they are denoted with a "*" symbol). Based on the percent of reads retained, all aligners appear to be highly similar; however, Geneious would maintain the highest number of reads to use in conjunction with the Hi-FiBR analysis.

Pearson correlation coefficients were calculated between the final reconstructed reads for each of the alignment tools to determine how similarly they processed the data. This showed high correlation between all mappers (see Fig. 7). Interestingly, when ignoring base substitutions near the break site, a small percentage of reads show vastly different counts among the mappers, with Geneious tending to produce higher counts per read than the other alignment tools. Many of these reads can be traced back to reads that were lost during the mapping and processing of CLC and Bowtie2 (as noted by the 0 counts with these aligners). Most of the remaining

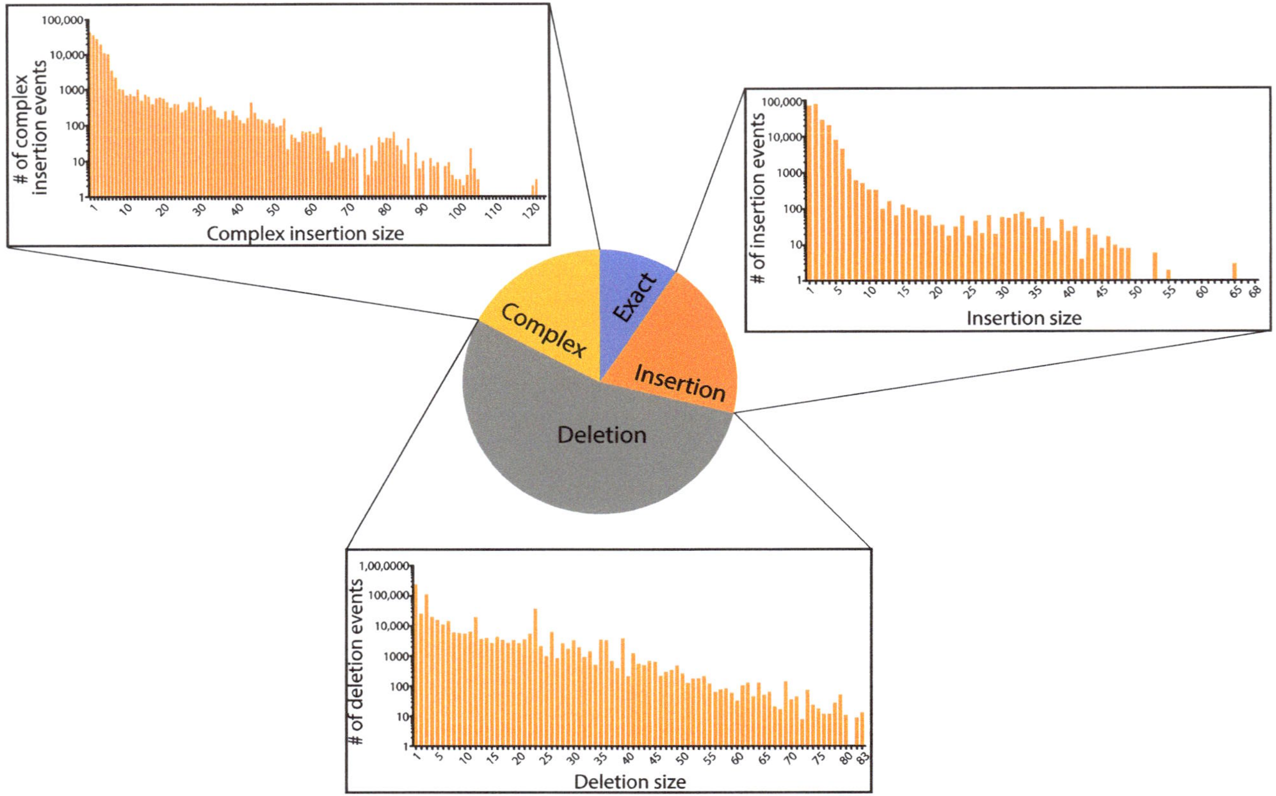

Fig. 6 Overview of the data set's complexity. The data set used here mostly consisted of deletion events, followed by insertion events, complex events, and finally exact matches (i.e., error-free repair). A cursory examination shows that in each event there is a diverse number of repair events (based on size of insertion or deletion) and the sizes tend to follow a normal distribution from smallest size to largest (although the scale here is logarithmic and, consequently, is not immediately obvious as normal).

Table 2 The Number of Sequencing Reads Retained Following Each Processing Step in the Analysis

Processing Step	Counts	Percent Retained
fastq	1134957	100
Trimmed	1101132	97.01971088
Geneious	1100535	96.96710977
Hi-FiBR Geneious (No Processing Window Correction)	1100535	96.96710977
Hi-FiBR Geneious (8 bp Processing Window Correction)	1100535	96.96710977
CLC	1099030	96.83450562
Hi-FiBR CLC (No Processing Window Correction)	1054561	92.91638362
Hi-FiBR CLC (8 bp Processing Window Correction)	1054561	92.91638362
Bowtie2	1070588	94.3285076
Hi-FiBR Bowtie2 (No Processing Window Correction)	1070588	94.3285076
Hi-FiBR Bowtie2 (8 bp Processing Window Correction)	1070588	94.3285076
BWA	1100458	96.96032537
Hi-FiBR BWA (No Processing Window Correction)	1097364	96.68771592
Hi-FiBR BWA (8 bp Processing Window Correction)	1097364	96.68771592

About 3% of reads are lost to the trimming step, and the remaining lost are due to the alignment tools. The order of most reads retained from highest to lowest is Geneious (~97%), BWA (~96.7%), Bowtie2 (~94%), and CLC (~93%). No difference occurs from the size of the processing window adjustment.

differentially counted reads appear to be caused by mismatches near the break site that influence how each mapper aligns the read. Once the processing windows of each repair event were enlarged to include base substitutions occurring within 8 bp of the break site, nearly all reads are counted equally, regardless of the aligner used, and the Pearson correlation coefficients increased (see Table 3).

To determine if specific aligners, particularly CLC and Bowtie2 which mapped the lowest number of reads, had difficulty mapping specific types of repair events, we compared (using a nonparametric one-way ANOVA; i.e., Kruskal–Wallis test) the mappers ability to count repair events parsed by class

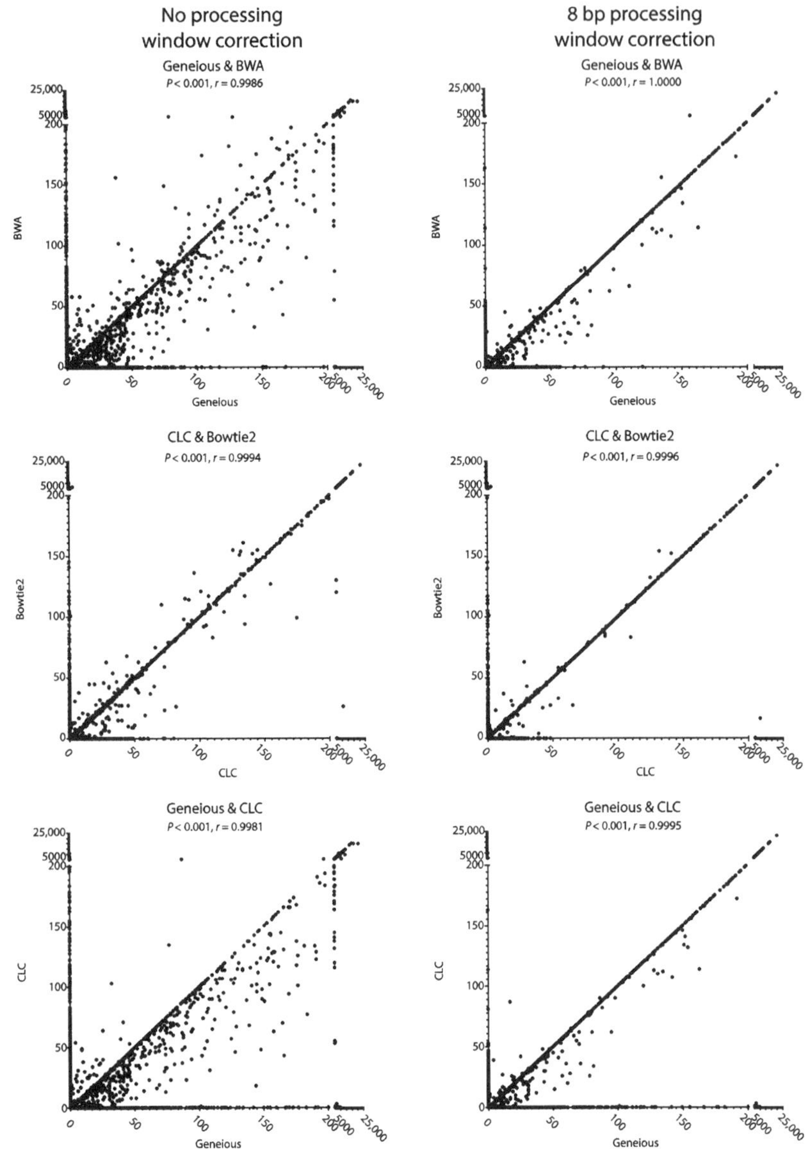

Fig. 7 Comparison of NGS alignment softwares for analyzing DSB repair junctions. Example Pearson correlations are given for Geneious, CLC, Bowtie2, and BWA. After processing window correction the aligning tools are more similar. However, there are still many reads that Geneious calls that tools, like CLC, do not catch, as shown by the many 0 reads. This is unsurprising, as CLC retained fewer reads than Geneious.

($P=0.9778$, Fig. 8), the size of read deletions ($P<0.0001$, Fig. 9), the size of read insertions ($P=0.0098$, Fig. 10), the size of read complex insertions ($P=0.4905$, Fig. 11), the size of microhomologies used in repair ($P=0.9207$, Fig. 12), and the types of microhomologies used in repair ($P=0.0108$, Fig. 13).

This analysis indicated that the aligners differed primarily based upon the size of deletion and insertion events as well as the specific microhomology sequences utilized during repair. Post-ANOVA Dunn's multiple comparison tests showed that Geneious and BWA called events with larger deletion sizes more efficiently than CLC and Bowtie2 ($P<0.05$, see Table 4). The

Table 3 Pearson Correlation Coefficients for Pairwise Comparison of NGS Alignment Softwares

Aligner Pairs	No Processing Window Correction	8 bp Processing Window Correction
Geneious & CLC	0.9981	0.9995
Geneious & Bowtie2	0.9985	0.9999
Geneious & BWA	0.9986	1
CLC & Bowtie2	0.9994	0.9996
CLC & BWA	0.9994	0.9995
Bowtie2 & BWA	0.9994	0.9999

After processing window adjustment, all aligners become more similar to each other. However, Geneious and BWA remain the most similar.

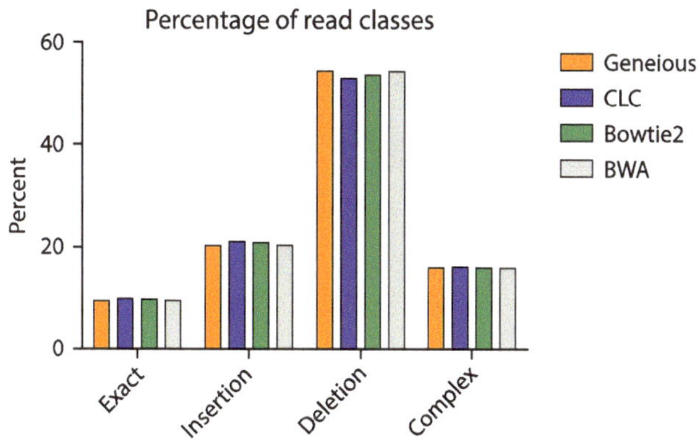

Fig. 8 Distribution of repair classes identified by Hi-FiBR analysis of an example data set. The different aligners have the same distribution of repair class types ($P=0.9778$). For our data set, deletion events occurred the most frequently, followed by insertions, complex events, and exact matches (i.e., error-free repair).

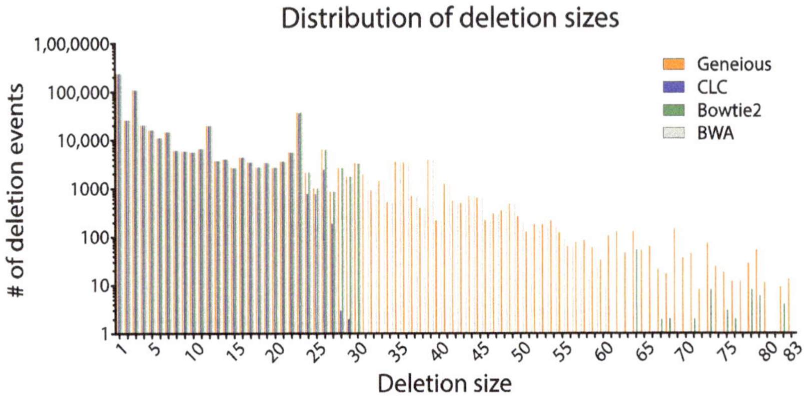

Fig. 9 Distribution of observed deletion events based on deletion size. The different aligners produced different distributions of deletion sizes ($P < 0.0001$). Geneious and BWA called larger deletion sizes. However, they are a small percentage of the total deletions.

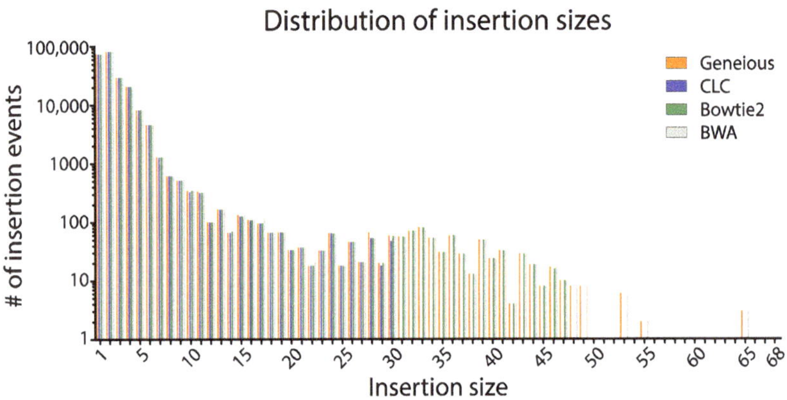

Fig. 10 Distribution of observed insertion events based on insertion size. The different aligners produce statistically different distributions of insertion sizes ($P = 0.0098$). Manual observation shows that Geneious, BWA, and Bowtie2 called larger insertion sizes, but only a small percent of the reads are larger.

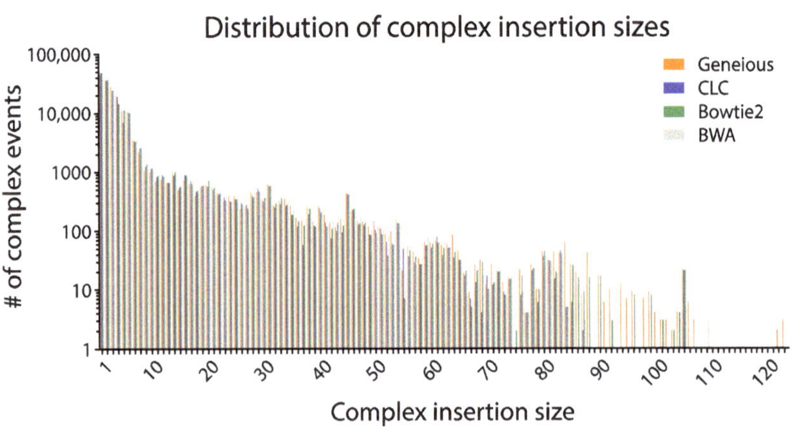

Fig. 11 Distribution of observed complex repair events based on insertion size. The complex insertion sizes are the same for all alignment tools ($P = 0.4905$), unlike the deletion and insertion sizes.

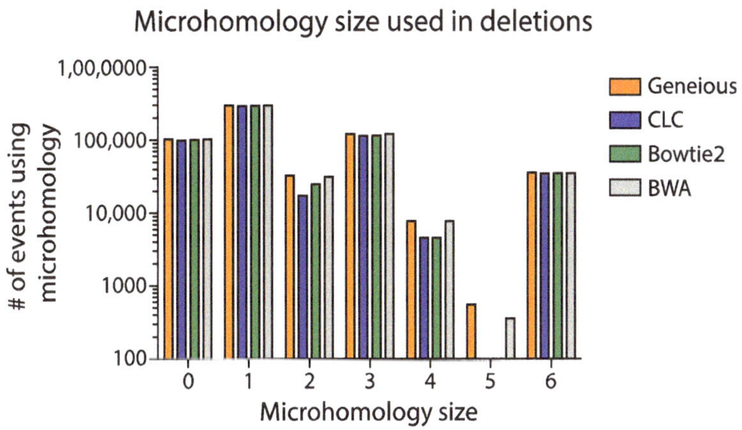

Fig. 12 Distribution of microhomology size observed in deletion events. The size of the microhomologies that mediate deletion events was the same for all aligners ($P = 0.9207$). The sizes tended to be small (0–3 bp), but this would likely shift depending on the sequence context around the break site.

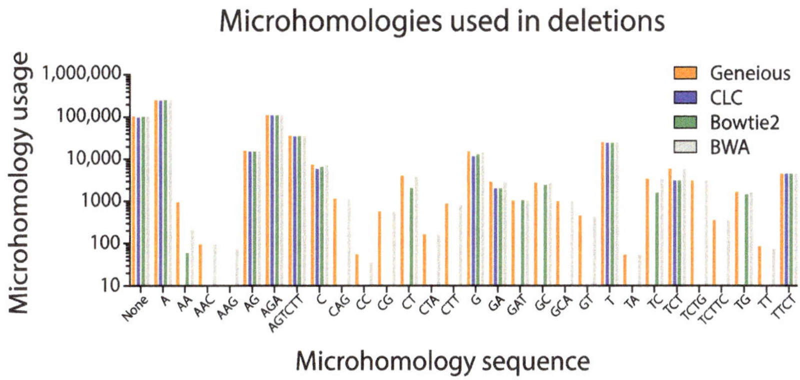

Fig. 13 Microhomology sequence usage. The sequence of the microhomologies used differs between all alignment tools ($P = 0.0108$). Since Geneious and BWA detect events with larger deletion sizes, the difference in microhomologies are likely linked to the larger deletions.

Table 4 The Deletion Size Dunn's Multiple Comparison *P* values Are Listed Here Between the Different Aligners

Aligner Pairs	$P < 0.05$
Geneious & CLC	Yes
Geneious & Bowtie2	Yes
Geneious & BWA	No
CLC & Bowtie2	No
CLC & BWA	Yes
Bowtie2 & BWA	No

Table 5 The Insertion Size Dunn's Multiple Comparison *P* values Are Listed Here Between the Different Aligners

Aligner Pairs	$P < 0.05$
Geneious & CLC	Yes
Geneious & Bowtie2	No
Geneious & BWA	No
CLC & Bowtie2	No
CLC & BWA	Yes
Bowtie2 & BWA	No

ability of Geneious and BWA to accurately map these larger deletion events also accounts for the difference in the specific microhomology sequences observed by the different mapping methods, as these additional reads increased the number of sequences that can potentially be used to find microhomology. The distribution of insertion sizes between mappers was also statistically different (see Table 5). Manual observation of the insertion graph indicates Geneious, Bowtie2, and BWA call events with larger insertion sizes than CLC; however, these events generally constitute a small fraction of the total number of repair events that occur. Cumulatively, this may suggest that Geneious and BWA are better at recognizing biological events in the flanks of DNA sequences around the break site.

5.2 Using Different Sequencing Platforms With Hi-FiBR

The data set used here was generated on an Illumina sequencing platform; however, because Hi-FiBR operates on a standard SAM file format, other sequencing methods could theoretically be used for sequencing repair junction amplicons. PacBio (Eid et al., 2009) and Ion Proton sequencing both have potential benefits that may make them particularly appealing for such analyses. They both produce longer reads, with PacBio easily able to accommodate multiple kb amplicons, which would allow even longer deletion events to be assessed. Moreover, these platforms cost less per SMRT cell or semiconductor, operate on scales that produce less reads per run (50,000 and 80,000,000 reads, respectively), and run faster than Illumina platforms (providing results sooner). Consequently, these sequencing methods would allow smaller scale experiments to be easily and cheaply conducted. This is opposed to experiments utilizing Illumina sequencing, where many libraries need to be accumulated and sequenced on a single lane to gain the greatest cost effectiveness.

As mentioned previously however, both Ion Proton and PacBio can experience systematic sequencing error that significantly inhibits sequence analysis of repair junctions. In particular, both platforms can produce frequent deletions or insertions in homopolymer runs of greater than three nucleotides in length. All NGS alignment tools identify these sequencing errors within the CIGAR string of the SAM file. This will ultimately cause Hi-FiBR to propagate the error and inaccurately report the sequence between the error and the break site as part of the repair processing window. If the sequencing reads can be corrected to account for these errors, namely, in the CIGAR string, Hi-FiBR will function normally. In the future, it may be desirable to design additional functions that examine a population of sequencing reads for highly recurrent errors, and that cleanse these sequences prior to mapping and sequence analysis.

6. SUMMARY AND CONCLUSION

The large data sets generated by NGS-based analysis of DSB repair events require robust alignment and downstream analyses. We report here an automated method for quantifying the relative frequency of repair events from amplicons sequenced with NGS technologies. The analysis is accurate and highly efficient. The 615 MB data set analyzed in this chapter was processed by the Hi-FiBR analysis on average in 8 min 13 s (for five runs), on a 3.50-GHz processor with 32 GB RAM available (although the RAM usage only peaked at ~0.7 GB, not including other background processes). Thus, it should easily run on most computers with Python3 installed.

Our study suggests that aligners map reads differently in regard to mismatches near breaks sites, but the Hi-FiBR analysis effectively corrects for this alignment error. These differences are normally small and the analysis for the majority of events strongly correlates between mappers even without Hi-FiBR's alignment correction. However, in data sets where base substitution near break sites is high, Hi-FiBR can identify events that would otherwise be improperly classified.

Despite the fact that Hi-FiBR can produce universally robust results regardless of the alignment software used to map the NGS reads to the reference sequence, Geneious's and BWA's alignment algorithms inherently allow them to call larger deletion/insertion events than CLC and Bowtie2. Most of these events are lost from the analysis due to failed or improper mapping of the specific reads, which may result from penalties associated with gap creation and extension. We therefore recommend using Geneious or BWA in association with Hi-FiBR analysis to preserve as many repair events as possible. However, it is likely that use of a version of CLC Genomics Workbench that allows affine gap penalties or adjusting the alignment parameters in CLC or Bowtie2 would increase the mappability of reads with larger deletion and insertions.

We believe that expanded utilization of NGS sequencing of DSB repair events will greatly enhance our understanding of the mechanisms employed in cells to repair these extremely deleterious lesions. Moreover, as inappropriate repair of DSBs commonly underlies potentially carcinogenic genome rearrangements (Arlt, Casper, & Glover, 2003; Byrne et al., 2014; Huebner & Croce, 2001), developing highly detailed signatures of error-prone DSB repair processes will greatly facilitate our ability to determine what enzymatic processes contribute to cancer genetic heterogeneity. NGS sequencing of DSB repair junctions provides an easy means to produce these signatures. Already, it is clear that the translocation junctions in many different cancer types bear resemblance to sequences used by alt-EJ (Chiarle et al., 2011; Hromas et al., 2016; Nussenzweig & Nussenzweig, 2007; Soni et al., 2015). In particular, BRCA1- and BRCA2-deficient breast cancers maintain a mutation signature similar to the spectra of sequences produced by pol theta-mediated repair of DSBs (Wyatt et al., 2016). It will be important to further outline what initiates the usage of these pathways and the genomic environments in which it is most mutagenic. This will be accomplished through the high data production of NGS technologies and the accurate analyses of these data, as shown here by the Hi-FiBR analysis.

ACKNOWLEDGMENTS

We would like to thank Drs. Varandt Khodaverian, Mitch McVey, and David Weinstock, as well as Jacob Layer for their input on this analysis pipeline as it was being developed. This research was supported by grants from the National Institute of Environmental Health Sciences (NIEHS) (R00ES022633 to S.A.R.), the National Cancer Institute (NCI) (R01ES002614 to S.A.R.), and the Department of Defense Congressionally Directed Medical Research Programs (BC141727 to S.A.R.).

REFERENCES

Aparicio, T., Baer, R., & Gautier, J. (2014). DNA double-strand break repair pathway choice and cancer. *DNA Repair (Amst)*, *19*, 169–175. https://doi.org/10.1016/j.dnarep.2014.03.014.

Arlt, M. F., Casper, A. M., & Glover, T. W. (2003). Common fragile sites. *Cytogenetic and Genome Research*, *100*(1–4), 92–100. https://doi.org/10.1159/000072843.

Beagan, K., Armstrong, R. L., Witsell, A., Roy, U., Renedo, N., Baker, A. E., et al. (2017). Drosophila DNA polymerase theta utilizes both helicase-like and polymerase domains during microhomology-mediated end joining and interstrand crosslink repair. *PLoS Genetics*, *13*(5), e1006813. https://doi.org/10.1371/journal.pgen.1006813.

Bennett, C. B., Lewis, A. L., Baldwin, K. K., & Resnick, M. A. (1993). Lethality induced by a single site-specific double-strand break in a dispensable yeast plasmid. *Proceedings of the National Academy of Sciences of the United States of America, 90*(12), 5613–5617. Retrieved from http://www.ncbi.nlm.nih.gov/pubmed/8516308.

Byrne, M., Wray, J., Reinert, B., Wu, Y., Nickoloff, J., Lee, S. H., et al. (2014). Mechanisms of oncogenic chromosomal translocations. *Annals of the New York Academy of Sciences, 1310*, 89–97. https://doi.org/10.1111/nyas.12370.

Chan, S. H., Yu, A. M., & McVey, M. (2010). Dual roles for DNA polymerase theta in alternative end-joining repair of double-strand breaks in Drosophila. *PLoS Genetics, 6*(7), e1001005. https://doi.org/10.1371/journal.pgen.1001005.

Chiarle, R., Zhang, Y., Frock, R. L., Lewis, S. M., Molinie, B., Ho, Y. J., et al. (2011). Genome-wide translocation sequencing reveals mechanisms of chromosome breaks and rearrangements in B cells. *Cell, 147*(1), 107–119. https://doi.org/10.1016/j.cell.2011.07.049.

Chu, Y. L., Wu, X., Xu, Y., & Her, C. (2013). MutS homologue hMSH4: Interaction with eIF3f and a role in NHEJ-mediated DSB repair. *Molecular Cancer, 12*(51). https://doi.org/10.1186/1476-4598-12-51.

Ciccia, A., & Elledge, S. J. (2010). The DNA damage response: Making it safe to play with knives. *Molecular Cell, 40*(2), 179–204. https://doi.org/10.1016/j.molcel.2010.09.019.

Decottignies, A. (2013). Alternative end-joining mechanisms: A historical perspective. *Frontiers in Genetics, 4*(48). https://doi.org/10.3389/fgene.2013.00048.

Eid, J., Fehr, A., Gray, J., Luong, K., Lyle, J., Otto, G., et al. (2009). Real-time DNA sequencing from single polymerase molecules. *Science, 323*(5910), 133–138. https://doi.org/10.1126/science.1162986.

Goodwin, S., McPherson, J. D., & McCombie, W. R. (2016). Coming of age: Ten years of next-generation sequencing technologies. *Nature Reviews. Genetics, 17*(6), 333–351. https://doi.org/10.1038/nrg.2016.49.

Gunn, A., Bennardo, N., Cheng, A., & Stark, J. M. (2011). Correct end use during end joining of multiple chromosomal double strand breaks is influenced by repair protein RAD50, DNA-dependent protein kinase DNA-PKcs, and transcription context. *The Journal of Biological Chemistry, 286*(49), 42470–42482. https://doi.org/10.1074/jbc.M111.309252.

Holmes, A., & Haber, J. E. (1999). Physical monitoring of HO-induced homologous recombination. *Methods in Molecular Biology, 113*, 403–415. https://doi.org/10.1385/1-59259-675-4:403.

Hromas, R., Williamson, E., Lee, S. H., & Nickoloff, J. (2016). Preventing the chromosomal translocations that cause cancer. *Transactions of the American Clinical and Climatological Association, 127*, 176–195. Retrieved from http://www.ncbi.nlm.nih.gov/pubmed/28066052.

Huebner, K., & Croce, C. M. (2001). FRA3B and other common fragile sites: The weakest links. *Nature Reviews. Cancer, 1*(3), 214–221. https://doi.org/10.1038/35106058.

Huefner, N. D., Mizuno, Y., Weil, C. F., Korf, I., & Britt, A. B. (2011). Breadth by depth: Expanding our understanding of the repair of transposon-induced DNA double strand breaks via deep-sequencing. *DNA Repair (Amst), 10*(10), 1023–1033. https://doi.org/10.1016/j.dnarep.2011.07.011.

IJspeert, H., Rozmus, J., Schwarz, K., Warren, R. L., van Zessen, D., Holt, R. A., et al. (2016). XLF deficiency results in reduced N-nucleotide addition during V(D)J recombination. *Blood, 128*(5), 650–659. https://doi.org/10.1182/blood-2016-02-701029.

Jeggo, P. A. (1990). Studies on mammalian mutants defective in rejoining double-strand breaks in DNA. *Mutation Research, 239*(1), 1–16. Retrieved from http://www.ncbi.nlm.nih.gov/pubmed/2195330.

Kent, T., Chandramouly, G., McDevitt, S. M., Ozdemir, A. Y., & Pomerantz, R. T. (2015). Mechanism of microhomology-mediated end-joining promoted by human DNA polymerase theta. *Nature Structural & Molecular Biology*, *22*(3), 230–237. https://doi.org/10.1038/nsmb.2961.

Khodaverdian, V. Y., Hanscom, T., Yu, A. M., Yu, T. L., Mak, V., Brown, A. J., et al. (2017). Secondary structure forming sequences drive SD-MMEJ repair of DNA double-strand breaks. *Nucleic Acids Research*, *45*(22), 12848–12861. https://doi.org/10.1093/nar/gkx1056.

Langmead, B., & Salzberg, S. L. (2012). Fast gapped-read alignment with Bowtie 2. *Nature Methods*, *9*(4), 357–359. https://doi.org/10.1038/nmeth.1923.

Lee, K., & Lee, S. E. (2007). Saccharomyces cerevisiae Sae2- and Tel1-dependent single-strand DNA formation at DNA break promotes microhomology-mediated end joining. *Genetics*, *176*(4), 2003–2014. https://doi.org/10.1534/genetics.107.076539.

Li, H., & Durbin, R. (2009). Fast and accurate short read alignment with Burrows-Wheeler transform. *Bioinformatics*, *25*(14), 1754–1760. https://doi.org/10.1093/bioinformatics/btp324.

Li, H., Handsaker, B., Wysoker, A., Fennell, T., Ruan, J., Homer, N., et al. (2009). The sequence alignment/map format and SAMtools. *Bioinformatics*, *25*(16), 2078–2079. https://doi.org/10.1093/bioinformatics/btp352.

Liang, F., Han, M., Romanienko, P. J., & Jasin, M. (1998). Homology-directed repair is a major double-strand break repair pathway in mammalian cells. *Proceedings of the National Academy of Sciences of the United States of America*, *95*(9), 5172–5177. Retrieved from http://www.ncbi.nlm.nih.gov/pubmed/9560248.

Liang, Z., Sunder, S., Nallasivam, S., & Wilson, T. E. (2016). Overhang polarity of chromosomal double-strand breaks impacts kinetics and fidelity of yeast non-homologous end joining. *Nucleic Acids Research*, *44*(6), 2769–2781. https://doi.org/10.1093/nar/gkw013.

Lieber, M. R. (2010). The mechanism of double-strand DNA break repair by the non-homologous DNA end-joining pathway. *Annual Review of Biochemistry*, *79*, 181–211. https://doi.org/10.1146/annurev.biochem.052308.093131.

Lin, Y., Lukacsovich, T., & Waldman, A. S. (1999). Multiple pathways for repair of DNA double-strand breaks in mammalian chromosomes. *Molecular and Cellular Biology*, *19*(12), 8353–8360. Retrieved from http://www.ncbi.nlm.nih.gov/pubmed/10567560.

Mateos-Gomez, P. A., Gong, F., Nair, N., Miller, K. M., Lazzerini-Denchi, E., & Sfeir, A. (2015). Mammalian polymerase theta promotes alternative NHEJ and suppresses recombination. *Nature*, *518*(7538), 254–257. https://doi.org/10.1038/nature14157.

Moore, J. K., & Haber, J. E. (1996). Cell cycle and genetic requirements of two pathways of nonhomologous end-joining repair of double-strand breaks in Saccharomyces cerevisiae. *Molecular and Cellular Biology*, *16*(5), 2164–2173. Retrieved from http://www.ncbi.nlm.nih.gov/pubmed/8628283.

Moynahan, M. E., & Jasin, M. (1997). Loss of heterozygosity induced by a chromosomal double-strand break. *Proceedings of the National Academy of Sciences of the United States of America*, *94*(17), 8988–8993. Retrieved from http://www.ncbi.nlm.nih.gov/pubmed/9256422.

Nick McElhinny, S. A., Havener, J. M., Garcia-Diaz, M., Juarez, R., Bebenek, K., Kee, B. L., et al. (2005). A gradient of template dependence defines distinct biological roles for family X polymerases in nonhomologous end joining. *Molecular Cell*, *19*(3), 357–366. https://doi.org/10.1016/j.molcel.2005.06.012.

Nickoloff, J. A. (2017). Paths from DNA damage and signaling to genome rearrangements via homologous recombination. *Mutation Research*, *806*, 64–74. https://doi.org/10.1016/j.mrfmmm.2017.07.008.

Nussenzweig, A., & Nussenzweig, M. C. (2007). A backup DNA repair pathway moves to the forefront. *Cell*, *131*(2), 223–225. https://doi.org/10.1016/j.cell.2007.10.005.

Resnick, M. A., & Martin, P. (1976). The repair of double-strand breaks in the nuclear DNA of Saccharomyces cerevisiae and its genetic control. *Molecular & General Genetics*, *143*(2), 119–129. Retrieved from http://www.ncbi.nlm.nih.gov/pubmed/765749.

Roth, D. B., & Wilson, J. H. (1986). Nonhomologous recombination in mammalian cells: Role for short sequence homologies in the joining reaction. *6*, 4295–4304. http://www.ncbi.nlm.nih.gov/pubmed/3025650.

Rouet, P., Smih, F., & Jasin, M. (1994). Introduction of double-strand breaks into the genome of mouse cells by expression of a rare-cutting endonuclease. *Molecular and Cellular Biology*, *14*(12), 8096–8106. Retrieved from http://www.ncbi.nlm.nih.gov/pubmed/7969147.

Rudin, N., & Haber, J. E. (1988). Efficient repair of HO-induced chromosomal breaks in Saccharomyces cerevisiae by recombination between flanking homologous sequences. *Molecular and Cellular Biology*, *8*(9), 3918–3928. Retrieved from http://www.ncbi.nlm.nih.gov/pubmed/3065627.

Seol, J. H., Shim, E. Y., & Lee, S. E. (2017). Microhomology-mediated end joining: Good, bad and ugly. In *Mutation Research*, pii: S0027-5107(17)30041-6. https://doi.org/10.1016/j.mrfmmm.2017.07.002.

Sfeir, A., & Symington, L. S. (2015). Microhomology-mediated end joining: A back-up survival mechanism or dedicated pathway? *Trends in Biochemical Sciences*, *40*(11), 701–714. https://doi.org/10.1016/j.tibs.2015.08.006.

Sharma, S., Javadekar, S. M., Pandey, M., Srivastava, M., Kumari, R., & Raghavan, S. C. (2015). Homology and enzymatic requirements of microhomology-dependent alternative end joining. *Cell Death & Disease*, *6*, e1697. https://doi.org/10.1038/cddis.2015.58.

Sinha, S., Li, F., Villarreal, D., Shim, J. H., Yoon, S., Myung, K., et al. (2017). Microhomology-mediated end joining induces hypermutagenesis at breakpoint junctions. *PLoS Genetics*, *13*(4), e1006714. https://doi.org/10.1371/journal.pgen.1006714.

Soni, A., Siemann, M., Pantelias, G. E., & Iliakis, G. (2015). Marked contribution of alternative end-joining to chromosome-translocation-formation by stochastically induced DNA double-strand-breaks in G2-phase human cells. *Mutation Research, Genetic Toxicology and Environmental Mutagenesis*, *793*, 2–8. https://doi.org/10.1016/j.mrgentox.2015.07.002.

Soong, C. P., Breuer, G. A., Hannon, R. A., Kim, S. D., Salem, A. F., Wang, G., et al. (2015). Development of a novel method to create double-strand break repair fingerprints using next-generation sequencing. *DNA Repair (Amst)*, *26*, 44–53. https://doi.org/10.1016/j.dnarep.2014.12.002.

van Schendel, R., van Heteren, J., Welten, R., & Tijsterman, M. (2016). Genomic scars generated by polymerase theta reveal the versatile mechanism of alternative end-joining. *PLoS Genetics*, *12*(10), e1006368. https://doi.org/10.1371/journal.pgen.1006368.

Vriend, L. E., Jasin, M., & Krawczyk, P. M. (2014). Assaying break and nick-induced homologous recombination in mammalian cells using the DR-GFP reporter and Cas9 nucleases. *Methods in Enzymology*, *546*, 175–191. https://doi.org/10.1016/B978-0-12-801185-0.00009-X.

Vriend, L. E., Prakash, R., Chen, C. C., Vanoli, F., Cavallo, F., Zhang, Y., et al. (2016). Distinct genetic control of homologous recombination repair of Cas9-induced double-strand breaks, nicks and paired nicks. *Nucleic Acids Research*, *44*(11), 5204–5217. https://doi.org/10.1093/nar/gkw179.

Waters, C. A., Strande, N. T., Pryor, J. M., Strom, C. N., Mieczkowski, P., Burkhalter, M. D., et al. (2014). The fidelity of the ligation step determines how ends are resolved during nonhomologous end joining. *Nature Communications*, *5*, 4286. https://doi.org/10.1038/ncomms5286.

Wyatt, D. W., Feng, W., Conlin, M. P., Yousefzadeh, M. J., Roberts, S. A., Mieczkowski, P., et al. (2016). Essential roles for polymerase theta-mediated end joining in the repair of chromosome breaks. *Molecular Cell*, *63*(4), 662–673. https://doi.org/10.1016/j.molcel.2016.06.020.

Xu, Y., Wu, X., & Her, C. (2015). hMSH5 facilitates the repair of camptothecin-induced double-strand breaks through an interaction with FANCJ. *The Journal of Biological Chemistry*, *290*(30), 18545–18558. https://doi.org/10.1074/jbc.M115.642884.

Yousefzadeh, M. J., Wyatt, D. W., Takata, K., Mu, Y., Hensley, S. C., Tomida, J., et al. (2014). Mechanism of suppression of chromosomal instability by DNA polymerase POLQ. *PLoS Genetics*, *10*(10), e1004654. https://doi.org/10.1371/journal.pgen.1004654.

CHAPTER EIGHT

Assaying Mutations Associated With Gene Conversion Repair of a Double-Strand Break

Gajendrahar Dwivedi✠, James E. Haber[1]
Rosenstiel Basic Medical Sciences Research Center, Brandeis University, Waltham, MA, United States
[1]Corresponding author: e-mail address: haber@brandeis.edu

Contents

Abstract

DNA double-strand break (DSB) is a cytotoxic lesion and needs to be repaired immediately. There are several metabolic pathways evolved to repair a DSB. Gene conversion is one of the least error-prone pathway for repair of a DNA DSB. Despite this there is nearly 1000-fold increase in mutation rate associated with gene conversion. Not only higher mutation rate is associated with gene conversion but also there is a very distinct mutation profile compared to spontaneous mutation events. Gene conversion is characterized by the presence of very high frameshift mutation events and other complex mutations that are not present during regular DNA replication. Another DNA DSB repair pathway widely studied is "break-induced replication" (BIR). BIR has been shown to be highly mutagenic in nature. BIR may lead to chromosomal rearrangement and has potential to cause cluster mutations with serious disease implications. In this chapter,

✠ This paper celebrates the life of Gajendrahar Dwivedi, who passed away suddenly in December, 2017.

Methods in Enzymology, Volume 601
ISSN 0076-6879
https://doi.org/10.1016/bs.mie.2017.11.029

the design of assay systems to study various mutation types and experimental procedures to measure specific mutation frequency associated with gene conversion are discussed.

1. INTRODUCTION

Repair of DNA double-strand breaks (DSBs) is critical to preserve genome integrity. Two sets of distinct pathways have evolved to repair such breaks: nonhomologous end joining (NHEJ) or homologous recombination (HR). Whereas NHEJ—either by "classic" or microhomology-mediated end joining (McVey & Lee, 2008; Moore & Haber, 1996)—often lead to the creation of insertions or deletions (indels) that are intrinsically mutagenic, HR is generally thought to be relatively free from associated errors, as the break is repaired by copying an intact template such as a sister chromatid, a homologous chromosome, or a homologous sequence at an ectopic location (Krogh & Symington, 2004; Paques & Haber, 1999). However, DNA repair by the presumably most accurate of these processes—gene conversion without an accompanying crossover—proves to be 1000 times more likely to contain new mutations than if the same sequences were simply replicated (Hicks, Kim, & Haber, 2010). Here we review methods to study such recombination-associated mutations.

Gene conversion (GC) was classically defined as a nonreciprocal transfer of genetic information from an unbroken chromosome to repair a broken chromosome. Initially such studies examined meiotic recombination, where DSBs are frequent and GC could be seen as the inheritance of alleles in 3:1 pattern instead of the expected 2:2. Meiotic GC at "hotspots" can occur in several percent of cells, whereas in mitotic cells spontaneous events are quite rare (Lichten & Haber, 1989; Nicolas & Petes, 1994). However, with the advent of site-specific endonucleases, it has been possible to examine analogous repair events in mitotic cells. In the studies we will focus on here, a DNA sequence is cleaved by a site-specific endonuclease such as the HO endonuclease (HO), I-SceI, or RNA-directed CRISPR/Cas9 (Gallagher & Haber, 2017; Jasin & Haber, 2016). The broken DNA is repaired by a donor sequence that has a mutation that prevents it from being cleaved.

By a combination of genetic and molecular biological analysis, DSB repair by GC has been described in detail; readers are directed to a number of review articles (Haber, 2016; Symington, Rothstein, & Lisby, 2014). Here we briefly outline the key steps and note in what ways these steps may contribute to the formation of mutations during the repair process. In most of these studies, DSBs are formed by conditional expression of

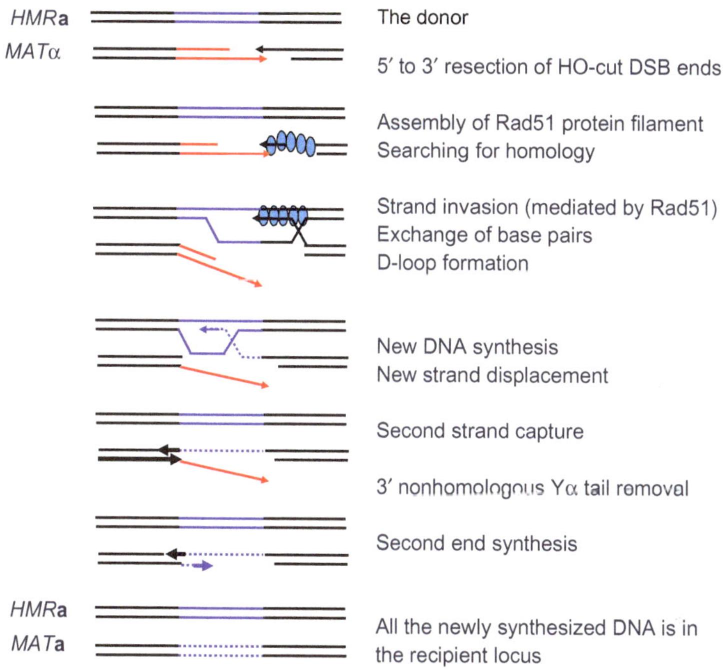

Fig. 1 Repair of a DSB by gene conversion in the context of *MAT* switching. An HO endonuclease-induced DSB at the *MAT*α locus is processed by 5′ to 3′ exonucleases, allowing assembly of the Rad51 protein as a filament on the 3′-ended ssDNA. Rad51 mediates a search for homology, resulting in the formation of base pairs between the invading strand and the complementary strand of a homologous donor sequence (*HMR***a**), resulting in a displacement loop (D-loop). New DNA synthesis is initiated from the invading 3′ end (*blue dashed line*) but is displaced from the donor, allowing this end to be captured by homologous, complementary ssDNA sequences from the second end of the DSB. Nonhomolgous Yα sequences (*red line*) are clipped off and new DNA synthesis fills in the gap, using the new 3′ end, completing the replacement of Yα with Y**a** and the switch from *MAT*α to *MAT***a**.

the endonuclease, for example, by placing the HO or I-SceI genes under the control of a galactose-inducible promoter. Cleavage is highly efficient and synchronous. All HR pathways rely on 5′ to 3′ end resection of the DSB to create long 3′-ended single-strand DNA (ssDNA) tails, which are bound first by the single-strand DNA-binding proteins, RPA, and then by the Rad51 recombinase to form a nucleoprotein filament that is the principal actor in DSB repair (Fig. 1).

Rad51 filament formation occurs with the aid of mediators including Rad52 and the Rad55/Rad57 paralogs.[a] The Rad51 filament engages in

[a] In metazoans, Rad52 plays a minor role and is superseded by BRCA2, along with the Rad51 paralogs Rad51B, Rad51C, Rad51D, XRCC2, and XRCC3.

a search for homologous sequences in a process that can embrace the entire genome but favors sites that have a high probability of contact, as measured by chromosome conformation capture (Lee et al., 2016; Wang, Lee, & Haber, 2017). This search initially may test as few as 6–8 consecutive bases for a suitable match but usually requires about 70 bp for efficient recombination to proceed (Anand, Beach, Li, & Haber, 2017; Danilowicz, Yang, Kelley, Prevost, & Prentiss, 2015; Greene, 2016; Lee et al., 2015; Ragunathan, Liu, & Ha, 2012). The ssDNA::Rad51 filament can synapse with a homologous double-strand DNA donor and promotes strand exchange, forming heteroduplex DNA in a displacement loop (D-loop). Correction of mismatches in this structure appears to depend principally on the 3′ to 5′ exonucleolytic "proofreading" activity of DNA polymerase δ, rather than on mismatch repair proteins (Anand et al., 2017). This strand invasion step requires the chromatin remodeler, Rad54, to form a stable-enough intermediate that can be used to continue the repair process (Hicks, Yamaguchi, & Haber, 2011). The 3′ end of the invading strand within this intermediate is then extended by DNA polymerase(s) to copy the template. This step is quite different from normal replication, in that the newly copied DNA is displaced from the donor rather than remaining in a semiconservatively replicated state (Ira, Satory, & Haber, 2006). It is this displacement step that apparently is at the root of the high rate of mutation seen in DSB repair. The copied and displaced new strand can then be annealed by the 3′-ended, resected end of the DSB (called second-end capture), requiring at least 100 nt of homology to be efficient (Mehta, Beach, & Haber, 2017). The 3′ end of the second end then apparently recruits DNA polymerase to convert the copied DNA into dsDNA and seal the break, so that all the newly copied DNA is found in the recipient and the donor remains unaltered.

The discovery that DSB repair in budding yeast was strikingly error-prone came from Jeffery Strathern's lab who measured the frequency of reversion of a homozygous point mutation in a *trp1* gene placed adjacent to a DSB that was repaired by gene conversion in a diploid, between homologous chromosomes. There was 300- to 1000-fold increase in Trp^+ revertants compared to replication (McGill, Holbeck, & Strathern, 1998; Rattray, McGill, Shafer, & Strathern, 2001; Strathern, Shafer, & McGill, 1995). The reversions occurred on the allele adjacent to the DSB rather than at the donor. For *trp1* base substitution alleles, reversion was dependent on the error-prone DNA polymerase ζ (Rev3) that is involved in translesion DNA synthesis after UV exposure; however, frameshift mutations showed high reversion but were largely independent of Rev3 (Holbeck & Strathern, 1997). Subsequent work in

Strathern's lab showed that break repair-induced mutations were predominantly base substitutions. However, there was significant increase in frameshift mutations during gene conversion as compared to regular replication, and all the frameshift mutations were observed in homonucleotide runs with a nucleotide being repeated three or more times (Rattray, Shafer, McGill, & Strathern, 2002). Rad55 and Rad57 proteins were observed to be important not only for recombinational repair of a DNA DSB but also for the fidelity of break repair DNA synthesis (Rattray et al., 2002).

Increased mutations associated with DSB formation and repair have also been seen in the transient long stretches of ssDNA near uncapped telomeres (Yang, Sterling, Storici, Resnick, & Gordenin, 2008). Mutations associated with DSB have been hypothesized to be responsible for localized hypermutability leading to clusters of multiple mutations or Kataegis (Burch et al., 2011; Chan & Gordenin, 2015; Sakofsky et al., 2014). Recently, Sinha et al. (Sinha et al., 2017) observed that a DSB flanked with microhomologies that can mediate deletion formation has very high mutagenic potential to induce changes in DNA sequences that are many kilobases further away from the sites that are rejoined. These regions may be transiently single-stranded as exonucleases continue to degrade DNA in a 5′ to 3′ direction until the deletion is successfully generated and these ssDNA regions filled in.

In the assays described below, we exploited the *MAT* switching system that is the normal target of HO in budding yeast (Lee & Haber, 2015). Here, HO can cut the *MAT*α locus, but not cleave its homologous donor, *HMR***a** because these sequences are heterochromatic, silenced by highly positioned nucleosomes in a Sir2-dependent process. The normal Y**a** sequences of *HMR* were modified to include the presence of a *Kluyveromyces lactis URA3* gene, as well as to ablate the HO cleavage site that would be cut if the locus were expressed (Fig. 2).

When HO was induced, repair was efficient, creating Ura$^+$ cells, since the *K. lactis URA3* gene was now copied into the euchromatic *MAT* locus. It was then possible to find 5-fluoroorotic acid (5-FOA)-resistant ura3 mutations that had arisen during DSB repair. Inhibiting Sir2 with nicotinamide further confirmed that the mutations were de novo, at *MAT* and not preexisting at the donor. Hicks et al. observed a nearly 1400-fold increase in the mutation rate associated with gene conversion as compared to the spontaneous mutation rate of *MAT::Kl-URA3* sequences (Hicks et al., 2010). Sequencing of the repair-associated mutations showed a spectrum quite different from spontaneous events. About 60% of the mutations were single base-pair changes, but the remaining 40% suggested that the repair DNA

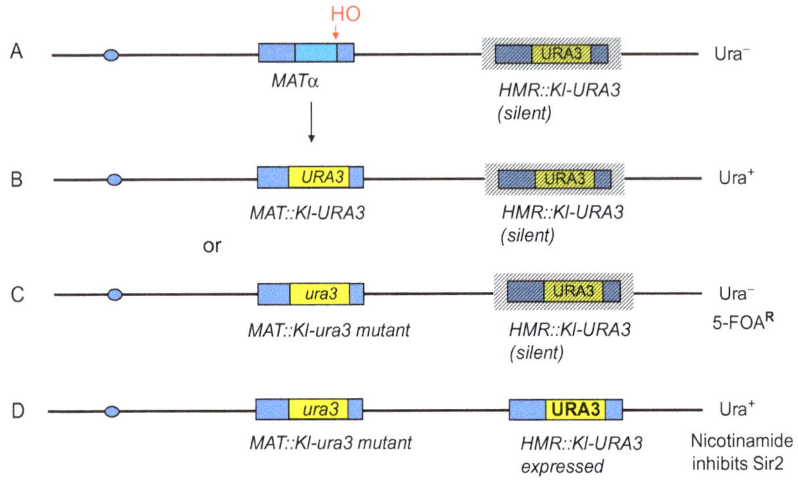

Fig. 2 Modified *MAT* switching system (*HMR::Kl-URA3*) to study errors made during gene conversion. (A) On chromosome 3, *HMLα* has been deleted (not shown). The Y**a** sequence in *HMR***a** has been partially replaced by the *Kluyveromyces lactis URA3* gene (*Kl-URA3*), but this gene is inactive because it sits within silenced heterochromatin embracing *HMR*. (B) This phenotypically Ura$^-$ strain becomes Ura$^+$ when the *Kl-URA3* sequences are copied into the *MAT* locus, replacing Yα after induction of the HO endonuclease. (C) Mutations arising during gene conversion can result in a 5-fluoorotic acid-resistant Ura$^-$ phenotype. (D) A demonstration that the mutations were not preexisiting at *HMR::Kl-URA3* was to grow cells on medium containing 5 m*M* nicotinamide, which inhibits the Sir2 histone deacetylase and unsilences the (still Ura$^+$) *HMR::Kl-URA3* locus.

polymerase was highly unstable. There were frequent frameshift mutations in homonucleotide runs, almost all of which were −1 (i.e., CCCC → CCC). There were also two classes of more complex mutations. First, many events involved both changes in sequence length and base pair substitution; these could be explained by quasipalindrome (QP)-associated template switching events in which it would be necessary for the partially copied new strand to dissociate from its template and pair intramolecularly, so that continued DNA synthesis would "perfect the palindrome" and created complex changes (Hicks et al., 2010). QP sequences-mediated mutagenesis have been observed in *Escherichia coli* to human genomes (Greenblatt, Grollman, & Harris, 1996; Lovett, 2004; Ripley, 1982). To be able to study QP mutations in detail we have devised a specific assay in which virtually all of the mutations arise by this process, as described below).

The second type of complex mutations that occurred at about 1% of all the 5-FOA-resistant events were template switches involving a highly divergent (72% identical) *Saccharomyces cerevisiae ura3-52* allele located on a different chromosome. This copy of *Sc-URA3* is intact but expression is prevented by the presence of a large transposon insertion near the 5′ end of the gene. For these interchromosomal template switching (ICTS) events to occur, the process of copying the initial template must stop, the partially copied strand must again dissociate from the template and now engage in copying the divergent sequences; however, for the events to be recovered in our assay, there had to be a second template switch back to the initial *HMR::Kl-URA3* template. ICTS events (Hicks et al., 2010) were infrequent when the donor was homeologous (only 72% identical), but—using the assay described later—such events can occur as often as 0.1% of all events if the distant donor is made 100% identical to the *Kl-URA3* sequences that are copied at (Tsaponina & Haber, 2014).

Similarly high rates of mutation are also found in another DSB repair process, break-induced replication (BIR), as described in chapter "Investigation of break-induced replication in yeast" by Elango et al. We note only that BIR also shows very high rates of template switching, as described by Anand et al. (2014) and Smith, Llorente, and Symington (2007).

In this chapter, we discuss construction of strains to study specific types of mutations arising during DSB repair by GC.

2. SCORING INTERCHROMOSOMAL TEMPLATE SWITCHING (ICTS) EVENTS

To study specifically ICTS during DNA DSB repair in *S. cerevisiae*, Tsaponina and Haber (2014) modified the strain containing *MAT*α and *HMR::Kl-URA3* described in Fig. 2, so that DSB repair events other than ICTS would be excluded. The *SIR3* gene was deleted to unsilence *HMR::Kl-URA3* so that the phenotype of the strain would be Ura$^+$. A 32-bp deletion was inserted into *Kl-URA3* by transformation, thereby creating a Ura$^-$ cell that could be selected by resistance to 5-fluoroorotic acid (Tsaponina & Haber, 2014). In this derivative, a simple GC copying the *Y***a***::Kl-ura3-Δ32* sequences from *HMR* to *MAT* yields only Ura$^-$ events. The strain used in these experiments also contained the *S. cerevisiae ura3-52* allele, which has a wild-type *URA3* open reading frame but whose expression is prevented by the insertion of a Ty1 retrotransposon near the 5′ end of the gene. The sequence of the *Sc-URA3* gene, on chromosome 5, is only 72% identical to that of the *Kl-URA3* sequence at *HMR*, on chromosome 3. During an ICTS event, the 32-bp deletion in *Kl-ura3-Δ32* is corrected by a jump of the DNA polymerase (perhaps aided by the recombination machinery) to the highly diverged *ura3-52* template, followed by a second jump back to the original *HMR* template, without which the repair event cannot be recovered. This results in a functional, chimeric *Kl-Sc-Kl-URA3* product (Fig. 3) (Tsaponina & Haber, 2014).

Since the entire DSB repair process take place during a single cell division, the frequencies are equivalent to rates of recombination. ICTS is a relatively rare event ($\sim 5 \times 10^{-7}$) when the jump is between 72% identical sequences, but it increases to more than 1×10^{-3} when the chromosome five donor sequence is changed from *S. cerevisiae ura3-52* to the *K. lactis URA3* sequence, lacking a promoter.

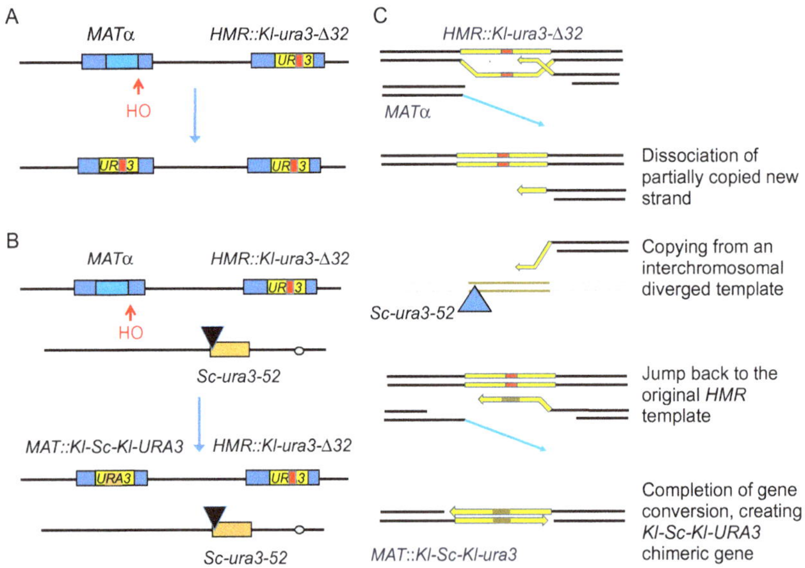

Fig. 3 Interchromosomal template switching (ICTS) between divergent templates. (A) HO endonuclease-mediated gene conversion repair using *HMR::Kl-ura3-Δ32* as the donor sequence usually results in copying of *Kl-ura3-Δ32* into *MAT*, yielding Ura^- cells. (B) A rare ICTS event incorporates part of 72% identical *Sc-ura3-52* sequence from chromosome 5, resulting in a functional, chimeric *Kl-Sc-Kl URA3* sequence at *MAT*, yielding Ura^+ cells. (C) For ICTS to occur, the right DSB end at *MAT* invades *HMR* homologous sequences and initiates copying of the *Kl-ura3-Δ32 donor*, carrying a 32-bp deletion within the open reading frame. The partially copied sequence must then dissociate from this template and jump to the *Sc-ura3-52* sequence on a different chromosome, copying enough sequence to include the region deleted in *Kl-ura3-Δ32*. Subsequently, DNA polymerase must again dissociate and return to the original *HMR* donor in order to complete the repair event, which is found as a chimeric, *Kl-Sc-Kl-URA3* sequence in the switched *MAT* locus.

2.1 Equipment

- Nanodrop or spectrophotometer
- Light microscope
- Hemocytometer or counting chamber

2.2 Solutions and Reagents

- YEP-dextrose (YEPD) solid media
  - Yeast extract—10 g
  - Peptone—20 g
  - Agar—25 g (do not add when preparing liquid media)
  - 900 mL ddH_2O
  - 8 mL 0.5% adenine in 0.05 N HCL
  - 5.5 mL 1 N HCL
  - After autoclaving and cooling, add 100 mL 20% dextrose
- YEP-raffinose (YEP-Raf)/YEP-galactose (YEP-Gal)
  - Preparation would be same as YEPD, except that after autoclaving add 100 mL of 20% filter sterilized raffinose or galactose, respectively
- YEP-lactate (YEP-Lac)
  - 10 g yeast extract
  - 20 g bacto peptone
  - 37 g lactic acid 35%
  - 30 mL 10 *M* NaOH until pH 5.5
  - 8 mL 0.5% adenine in 0.05 N HCl, if needed
  - 1000 mL total

2.3 Protocol for Assaying ICTS Mutation Frequency

1. Pick up a single yeast colony from a freshly grown YEPD plate and inoculate in liquid YEP-Raf media or YEP-Lac.
2. Grow until mid-log phase (5×10^6 to 2×10^7 cells/mL), as determined either by optical density or by counting cells, using a hemocytometer by light microscopy. The OD_{600} (spectrophotomer or nanodrop) reading should be 0.4–0.6.
3. Induce DSB by adding galactose to the final concentration of 2%. Since the repair process will eliminate the HO endonuclease cleavage site at *MAT*, each cell can experience only a single cut.
4. After 4 h of galactose induction wash cells with YEP-Raf/YEP-Lac three times to ensure complete removal of galactose.

5. Collect all the cells and plate appropriate dilutions on uracil dropout and YEPD.
6. Plate appropriate serial dilutions of cells on YEPD and YEP-Gal for survival analysis.
7. Incubate all the plates at 30°C and count colony number after 3 days.

3. SCORING QUASIPALINDROME (QP)-MEDIATED MUTATIONS DURING DSB REPAIR

In collaboration with Susan Lovett (Brandeis University) we modified the aforementioned *HMR::Kl-URA3* gene conversion strain to be able to characterize genetic pathways affecting QP-mediated mutations. QPs are palindromic sequences whose self-complementary hairpin of approximately 40–150 bases is interrupted by a short nonpalindromic element (Lisnic, Svetec, Stafa, & Zgaga, 2009; Pierce, Kong, & Masker, 1991). We introduced a 40-nucleotide QP sequence containing a 4-bp nonpalindromic element (NPE) (Fig. 4A) into the *HMR::Kl-URA3* sequence, by selecting for FOA-resistant, Ura^- transformants (the experimental procedure is given below).

The design of the sequence is such that a perfect (36-bp) hairpin sequence would result in a functional Ura3 protein, but the presence of the 4-bp NPE disrupts the *Kl-URA3* ORF by creating a frameshift. A QP-mediated mutation during the HO-induced copying of these sequences to *MAT* would result in the deletion of the NPE, thus perfecting the hairpin and resulting in a $Ura3^+$ gene convertant (Fig. 4). Subsequent DNA sequence analysis has shown that all of the corrected sequences are at *MAT*, with those at the donor being unchanged. Thus, every colony growing on solid media lacking uracil represents one mutation event, making it very easy to calculate the mutation rate.

Two different QP sequences (QP-L and QP-R) were designed that differ in the position of the NPE from each other. QP-L has its NPE in the left arm of hairpin sequence, whereas QP-R has the NPE in the right arm of the hairpin. Different locations of the NPE can indicate at what stage the NPE is removed. In case of QP-L, DNA, DNA polymerase would encounter NPE while copying the second half of a QP sequence, before which it will have copied sufficient sequence to create a stable hairpin if the newly copied DNA were to dissociate from the template. However, in case of QP-R, the NPE would already be copied before the polymerase generated a self-complementary segment. In this case, the correction of the QP may only happen during second-end synthesis and be a more complex

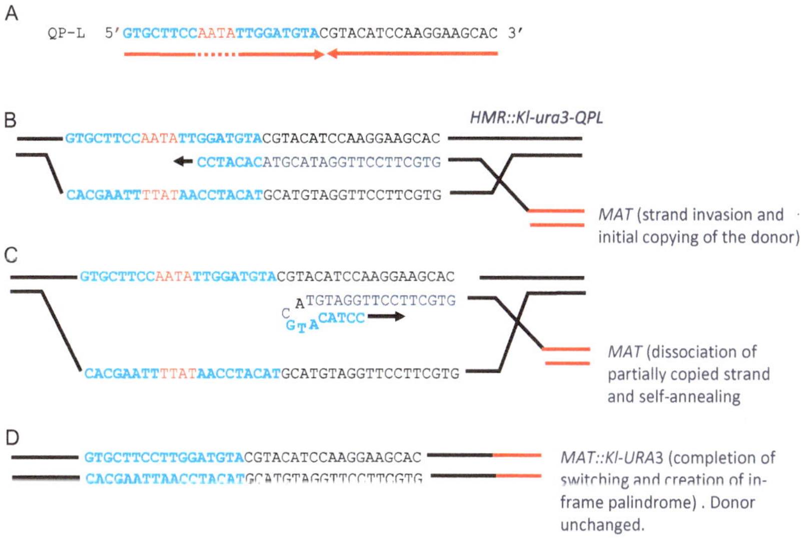

Fig. 4 Quasipalindrome (QP)-mediated mutagenesis. (A) A QP sequence. QP-L represents palindrome sequence with a 4-nt (*red*) nonpalindromic element (NPE) inserted in the *left side* (*blue*) of the otherwise 36-bp perfect palindrome. A similar construct can be made with the NPE at an equivalent location in the *right side* of the palindrome (QP-R). (B) Strand invasion of the DSB end from MAT initiates copying the *HMR::Kl-ura3-QP-L* donor. (C) Dissociation of at least the end of the newly copied sequence from the template allows self-annealing to initiate copying of the newly copied strand. This dissociation may lead to the self-annealing of the template strand as well, creating a cruciform structure (not shown). (D) Completion of the DSB repair event requires that copying resume in the normal direction through the rest of the Kl-ura3 template into the region homologous between *HMR* and *MAT*. This event places a now-perfect palindrome (and Ura3^{+}) *Kl-URA3* sequence at *MAT*.

event, as the NPE will still be present in the top strand and likely requires mismatch repair of heteroduplex DNA before the cells will be Ura^{+}.

G blocks (500 bp in length) were procured containing respective QP sequences with flanking homology to *Kl-URA3* ORF. A *HMR::Kl-URA3* strain containing the *sir3Δ* mutation (the same as ICTS background strain) was transformed with QP G-block sequences. Successful integration of a QP sequence was screened by selecting Ura^{-} colonies on 5-FOA-supplemented plates. A bottleneck in the screening procedure was a very high occurrence of false-positive colonies. To solve this problem G-block sequences were cotransformed with a centromeric plasmid carrying *LEU2* (pRS315). Transformants were first selected on plates lacking leucine and subsequently replica plated on 5-FOA plates. Colonies growing on 5-FOA plate were confirmed by sequencing. GC-associated QP-mediated mutation frequency was estimated by counting colonies on URA dropout (d/o) plate after HO endonuclease induction (see Section 5). The spontaneous QP mutation frequency was estimated by fluctuation test analysis.

4. SCORING FRAMESHIFT MUTATIONS DURING DSB REPAIR

We also modified the gene conversion strain *HMR::Kl-URA3* to construct a study frameshift mutations. Hicks et al. (2010) had observed that during DSB repair there was a very high occurrence of frameshift mutations in homonucleotide runs. Almost all were −1 frameshift mutations rather than +1. To focus on −1 frameshifts arising during repair, we modified a frequently mutated site in the *Kl-URA3* ORF, where CCCC was mutated to CCC during repair (Hicks et al., 2010) (Fig. 5).

We introduced one extra "C" at this site, making it a "5C" motif (CCCCC), resulting in a Ura$^-$ phenotype that can revert to Ura$^+$ by a −1 frameshift (Fig. 5). Although other compensatory −1 mutations can be imagined to occur, all of the Ura$^+$ events (~40) we have analyzed were indeed −1 deletions in the 5C sequence, at the *MAT* locus, with the *HMR* donor unchanged. Again every colony growing on uracil dropout solid media represents one mutation event and the frequency is equivalent to the rate.

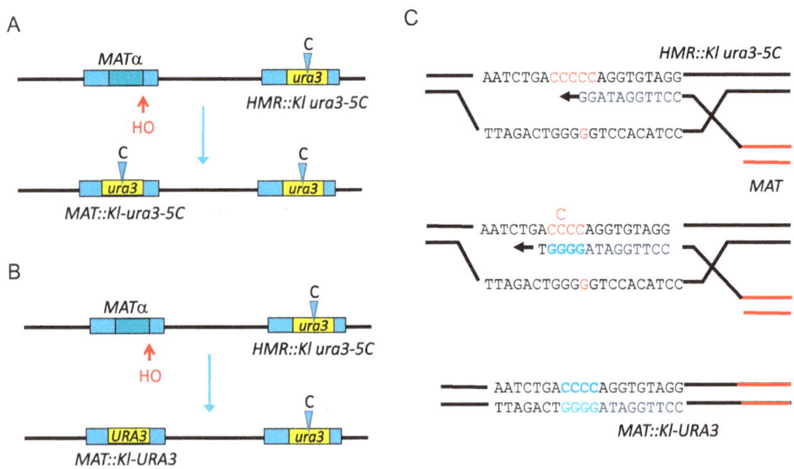

Fig. 5 An assay to detect −1 frameshifts during DSB repair. A run of 4Cs in Kl-URA3 was notably "hot" for −1 frameshifts in an assay for Ura$^-$ mutations arising during DSB repair (Hicks et al., 2010). This sequence was mutated to have 5Cs, creating a Ura$^-$ mutation that could revert to Ura$^+$ by loss of 1C. (A) Most DSB repair events result in Ura$^-$ outcomes. (B) −1 frameshift mutations create a Ura$^+$ event. (C) A possible mechanism of generating a −1 frameshift in a homonucleotide run.

Frameshift mutation strain was constructed in the same way as QP strain except that the G-block sequence was synthesized with "5C" motif instead of the QP sequence. DSB repair-associated frameshift mutation frequency was determined as discussed in Section 5. Spontaneous mutation frequency was determined by fluctuation test analysis (Hicks et al., 2010; Tsaponina & Haber, 2014).

4.1 Equipment

- Nanodrop or spectrophotometer
- Light microscope
- Hemocytometer or counting chamber

4.2 Solutions and Reagents

- YEPD solid media
- YEP-Raf/YEP-Lac
- YEP-Gal solid media

5. PROTOCOL FOR ASSAYING QP AND FRAMESHIFT MUTATION FREQUENCY

1. Pick up a single yeast colony from a freshly grown YEPD plate and inoculate in YEP-Raf or YEP-Lac media.
2. Grow until mid-log phase, as described above.
3. Collect cells and plate appropriate dilutions on YEP-Gal and YEPD media. Cells growing on YEP-Gal will suffer a DSB at the *MAT* locus.
4. Since these mutation events are rare, nearly 10^8–10^9 cells are plated to get statistically reliable numbers of mutants per YEP-galactose plate. Dilutions are used to count total cells on YEPD.
5. Incubate plates at 30°C for 48h.
6. YEP-Gal and YEPD plates with higher concentration of cells (10^8–10^9) show a lawn of growth after incubation. Replica-plate these plates on uracil dropout solid media to identify Ura^+ cells.
7. Incubate at 30°C for 24h
8. Colonies growing on uracil dropout plates after replica-plating from YEP-Gal represent the mutation frequency associated with DSB repair. Colonies growing on uracil d/o plates after replica-plating from YEPD gives a count of spontaneous mutation frequency. However, an accurate assessment of the spontaneous rate requires a fluctuation test based on a minimum of 10 independent colonies.

ACKNOWLEDGMENTS

Work on mutagenesis in DSB repair has been supported by NIH Grants R37 GM20056, R01 GM76020, and P01 GM105473.

REFERENCES

Anand, R., Beach, A., Li, K., & Haber, J. (2017). Rad51-mediated double-strand break repair and mismatch correction of divergent substrates. *Nature*, *544*, 377–380.

Anand, R. P., Tsaponina, O., Greenwell, P. W., Lee, C. S., Du, W., Petes, T. D., et al. (2014). Chromosome rearrangements via template switching between diverged repeated sequences. *Genes & Development*, *28*, 2394–2406.

Burch, L. H., Yang, Y., Sterling, J. F., Roberts, S. A., Chao, F. G., Xu, H., et al. (2011). Damage-induced localized hypermutability. *Cell Cycle*, *10*, 1073–1085.

Chan, K., & Gordenin, D. A. (2015). Clusters of multiple mutations: Incidence and molecular mechanisms. *Annual Review of Genetics*, *49*, 243–267.

Danilowicz, C., Yang, D., Kelley, C., Prevost, C., & Prentiss, M. (2015). The poor homology stringency in the heteroduplex allows strand exchange to incorporate desirable mismatches without sacrificing recognition in vivo. *Nucleic Acids Research*, *43*, 6473–6485.

Gallagher, D. N., & Haber, J. E. (2017). Repair of site-specific DNA cleavage: Old-school lessons for Cas9-mediated gene editing. *ACS Chemical Biology*. https://doi.org/10.1021/acschembio.7b00760. [Epub ahead of print].

Greenblatt, M. S., Grollman, A. P., & Harris, C. C. (1996). Deletions and insertions in the p53 tumor suppressor gene in human cancers: Confirmation of the DNA polymerase slippage/misalignment model. *Cancer Research*, *56*, 2130–2136.

Greene, E. C. (2016). DNA sequence alignment during homologous recombination. *The Journal of Biological Chemistry*, *291*, 11572–11580.

Haber, J. E. (2016). A life investigating pathways that repair broken chromosomes. *Annual Review of Genetics*, *50*, 1–28.

Hicks, W. M., Kim, M., & Haber, J. E. (2010). Increased mutagenesis and unique mutation signature associated with mitotic gene conversion. *Science*, *329*, 82–85.

Hicks, W. M., Yamaguchi, M., & Haber, J. E. (2011). Real-time analysis of double-strand DNA break repair by homologous recombination. *Proceedings of the National Academy of Sciences of the United States of America*, *108*, 3108–3115.

Holbeck, S. L., & Strathern, J. N. (1997). A role for REV3 in mutagenesis during double-strand break repair in *Saccharomyces cerevisiae*. *Genetics*, *147*, 1017–1024.

Ira, G., Satory, D., & Haber, J. E. (2006). Conservative inheritance of newly synthesized DNA in double-strand break-induced gene conversion. *Molecular and Cellular Biology*, *26*, 9424–9429.

Jasin, M., & Haber, J. E. (2016). The democratization of gene editing: Insights from site-specific cleavage and double-strand break repair. *DNA Repair*, *44*, 6–16.

Krogh, B. O., & Symington, L. S. (2004). Recombination proteins in yeast. *Annual Review of Genetics*, *38*, 233–271.

Lee, C. S., & Haber, J. E. (2015). Mating-type gene switching in *Saccharomyces cerevisiae*. *Microbiology Spectrum*, *3*(2). MDNA3-0013-2014. https://doi.org/10.1128/microbiolspec.

Lee, J. Y., Terakawa, T., Qi, Z., Steinfeld, J. B., Redding, S., Kwon, Y., et al. (2015). DNA RECOMBINATION. Base triplet stepping by the Rad51/RecA family of recombinases. *Science*, *349*, 977–981.

Lee, C. S., Wang, R. W., Chang, H. H., Capurso, D., Segal, M. R., & Haber, J. E. (2016). Chromosome position determines the success of double-strand break repair. *Proceedings of the National Academy of Sciences of the United States of America, 113*, E146–54.

Lichten, M., & Haber, J. E. (1989). Position effects in ectopic and allelic mitotic recombination in *Saccharomyces cerevisiae*. *Genetics, 123*, 261–268.

Lisnic, B., Svetec, I. K., Stafa, A., & Zgaga, Z. (2009). Size-dependent palindrome-induced intrachromosomal recombination in yeast. *DNA Repair, 8*, 383–389.

Lovett, S. T. (2004). Encoded errors: Mutations and rearrangements mediated by misalignment at repetitive DNA sequences. *Molecular Microbiology, 52*, 1243–1253.

McGill, C. B., Holbeck, S. L., & Strathern, J. N. (1998). The chromosome bias of misincorporations during double-strand break repair is not altered in mismatch repair-defective strains of *Saccharomyces cerevisiae*. *Genetics, 148*, 1525–1533.

McVey, M., & Lee, S. E. (2008). MMEJ repair of double-strand breaks (director's cut): Deleted sequences and alternative endings. *Trends in Genetics, 24*, 529–538.

Mehta, A., Beach, A., & Haber, J. E. (2017). Homology requirements and competition between gene conversion and break-induced replication during double-strand break repair. *Molecular Cell, 65*, 515–526, e3.

Moore, J. K., & Haber, J. E. (1996). Cell cycle and genetic requirements of two pathways of nonhomologous end-joining repair of double-strand breaks in *Saccharomyces cerevisiae*. *Molecular and Cellular Biology, 16*, 2164–2173.

Nicolas, A., & Petes, T. D. (1994). Polarity of meiotic gene conversion in fungi: Contrasting views. *Experientia, 50*, 242–252.

Paques, F., & Haber, J. E. (1999). Multiple pathways of recombination induced by double-strand breaks in *Saccharomyces cerevisiae*. *Microbiology and Molecular Biology Reviews, 63*, 349–404.

Pierce, J. C., Kong, D., & Masker, W. (1991). The effect of the length of direct repeats and the presence of palindromes on deletion between directly repeated DNA sequences in bacteriophage T7. *Nucleic Acids Research, 19*, 3901–3905.

Ragunathan, K., Liu, C., & Ha, T. (2012). RecA filament sliding on DNA facilitates homology search. *eLife, 1*, e00067.

Rattray, A. J., McGill, C. B., Shafer, B. K., & Strathern, J. N. (2001). Fidelity of mitotic double-strand-break repair in *Saccharomyces cerevisiae*: A role for *SAE2/COM1*. *Genetics, 158*, 109–122.

Rattray, A. J., Shafer, B. K., McGill, C. B., & Strathern, J. N. (2002). The roles of *REV3* and RAD57 in double-strand-break-repair-induced mutagenesis of Saccharomyces cerevisiae. *Genetics, 162*, 1063–1077.

Ripley, L. S. (1982). Model for the participation of quasi-palindromic DNA sequences in frameshift mutation. *Proceedings of the National Academy of Sciences of the United States of America, 79*, 4128–4132.

Sakofsky, C. J., Roberts, S. A., Malc, E., Mieczkowski, P. A., Resnick, M. A., Gordenin, D. A., et al. (2014). Break-induced replication is a source of mutation clusters underlying kataegis. *Cell Reports*, 7, 1640–1648.

Sinha, S., Li, F., Villarreal, D., Shim, J. H., Yoon, S., Myung, K., et al. (2017). Microhomology-mediated end joining induces hypermutagenesis at breakpoint junctions. *PLoS Genetics, 13*, e1006714.

Smith, C. E., Llorente, B., & Symington, L. S. (2007). Template switching during break-induced replication. *Nature, 447*, 102–105.

Strathern, J. N., Shafer, B. K., & McGill, C. B. (1995). DNA synthesis errors associated with double-strand-break repair. *Genetics, 140*, 965–972.

Symington, L. S., Rothstein, R., & Lisby, M. (2014). Mechanisms and regulation of mitotic recombination in Saccharomyces cerevisiae. *Genetics*, *198*, 795–835.

Tsaponina, O., & Haber, J. E. (2014). Frequent interchromosomal template switches during gene conversion in S. cerevisiae. *Molecular Cell*, *55*, 615–625.

Wang, R. W., Lee, C. S., & Haber, J. E. (2017). Position effects influencing intrachromosomal repair of a double-strand break in budding yeast. *PLoS One*, *12*, e0180994.

Yang, Y., Sterling, J., Storici, F., Resnick, M. A., & Gordenin, D. A. (2008). Hypermutability of damaged single-strand DNA formed at double-strand breaks and uncapped telomeres in yeast Saccharomyces cerevisiae. *PLoS Genetics*, *4*, e1000264.

CHAPTER NINE

Investigation of Break-Induced Replication in Yeast

Rajula Elango, Zachary Kockler, Liping Liu, Anna Malkova[1]
University of Iowa, Iowa City, IA, United States
[1]Corresponding author: e-mail address: anna-malkova@uiowa.edu

Contents

Methods in Enzymology, Volume 601
ISSN 0076-6879
https://doi.org/10.1016/bs.mie.2017.12.010

Abstract

Break-induced replication (BIR) is an important mechanism aimed to repair one-ended double-strand DNA breaks. BIR is initiated by invasion of a broken DNA end into a homologous template followed by DNA synthesis that can proceed for hundreds of kilobases to the end of the chromosome. Unlike S-phase replication, BIR is carried out by a migrating DNA bubble and is associated with conservative inheritance of newly synthesized DNA. The unusual mode of DNA synthesis during BIR leads to an increased level of genetic instabilities including increased mutagenesis and chromosomal rearrangements. Here, we describe our experimental system in yeast *Saccharomyces cerevisiae* where BIR is initiated by a site-specific DNA break and where the repair involves two copies of chromosome III. This system allows investigation of BIR using genetic and molecular biology approaches, and can be used for characterization of the BIR mechanism, roles of individual proteins in BIR, and for the analysis of genetic instabilities associated with BIR.

ABBREVIATIONS

BIR break-induced replication
CHEF contour-clamped homogenous electric field
DSB double-strand break
GC gene conversion
GCRs gross chromosomal rearrangements
HR homologous recombination
NHEJ nonhomologous end joining
ssDNA single-strand DNA

1. INTRODUCTION

Double-strand breaks (DSBs) are a dangerous type of DNA damage that can result from various sources including exposure of DNA to radiation or chemicals, from the collapse of DNA replication, or from problems occurring during chromosome segregation (reviewed in Mehta & Haber, 2014). The repair of DSBs is critical for cell survival and for maintaining genetic integrity. Two major pathways of DSB repair include non-homologous end joining (NHEJ) and homologous recombination (HR). During NHEJ, the broken DNA ends are ligated back together and this often leads to the loss or alteration of genetic information (reviewed in

Rodgers & McVey, 2016). In contrast, HR utilizes a homologous DNA template for repair of DSBs. HR repair can be further divided into several pathways including: gene conversion (GC), which can occur via gap repair or by synthesis-dependent strand annealing, single-strand annealing, and break-induced replication (BIR) (reviewed in Mehta & Haber, 2014).

In this review, we focus on BIR pathway, which is employed only when one end of the DNA break can find homology in the genome for invasion and repair (Fig. 1A) (reviewed in Sakofsky & Malkova, 2017). This situation typically arises at eroded telomeres or at collapsed replication forks. BIR is initiated by a 5′–3′ resection of the broken DNA end (Fig. 1B) followed by invasion of 3′ single-strand end into a homologous DNA template (Fig. 1C) and copying of DNA template that can proceed for more than 100 kb through the telomere (Fig. 1D). Unlike S-phase DNA replication, which is carried out by a replication fork, BIR proceeds via a migrating DNA

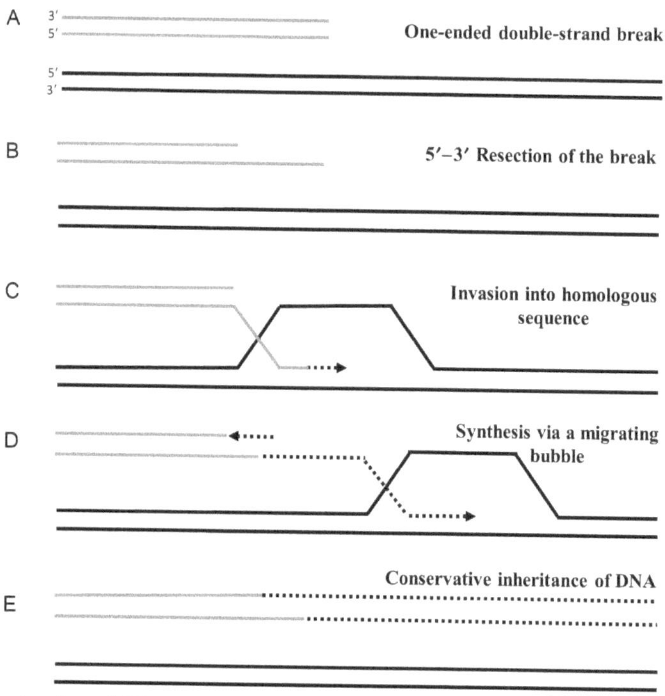

Fig. 1 Mechanism of BIR initiation and progression. (A) One-ended DSB. (B) 5′–3′ resection of the break results in exposed 3′ ssDNA. (C) The ssDNA end invades into the homologous template initiating DNA synthesis. (D) Leading strand synthesis progresses by a migrating bubble. ssDNA accumulates behind the migrating bubble due to uncoupled leading and lagging strand synthesis. (E) Conservative inheritance of DNA.

bubble with asynchronous synthesis of the leading and lagging strands, and leads to conservative inheritance of newly synthesized DNA (Fig. 1E). This unusual type of DNA synthesis leads to accumulation of large amounts of single-strand DNA (ssDNA), which promotes mutagenesis associated with BIR (Saini et al., 2013; Sakofsky et al., 2014). In addition, BIR leads to frequent genomic rearrangements similar to those associated with cancer in humans (Sakofsky et al., 2015; Smith, Lam, & Symington, 2009; Vasan, Deem, Ramakrishnan, Argueso, & Malkova, 2014).

For a long time budding yeast remained the only eukaryotic organism where BIR could be studied, and several different approaches have been employed to investigate BIR in the budding yeast *Saccharomyces cerevisiae*. First, a site-specific HO endonuclease was used to initiate one DSB in chromosome III in a way that only one broken end could invade the homologous chromosome and initiate BIR that can proceed for approximately 100 kb through the telomere (Malkova, Ivanov, & Haber, 1996; Malkova, Naylor, Yamaguchi, Ira, & Haber, 2005). Second, a linearized fragment was transformed into yeast, where it invaded the homologous template chromosome and initiated BIR (Davis & Symington, 2004; Morrow, Connelly, & Hieter, 1997). Third, BIR was initiated in telomerase-deficient yeast strains to maintain their telomeres by alternative lengthening of telomeres (ALT) pathway (Lydeard, Jain, Yamaguchi, & Haber, 2007; Teng, Chang, McCowan, & Zakian, 2000; Teng & Zakian, 1999).

Recently, development of new experimental approaches has allowed detection of BIR in higher eukaryotes including human cells (Costantino et al., 2014; Hashimoto & Costanzo, 2011; Minocherhomji et al., 2015). It has been demonstrated that at mammalian fragile sites, BIR-like repair synthesis is initiated following replication collapse (Bhowmick, Minocherhomji, & Hickson, 2016; Minocherhomji et al., 2015). Also, it was shown that overexpression of oncogenes in human cells leads to the initiation of BIR promoting chromosomal rearrangements (Costantino et al., 2014; Sotiriou et al., 2016). Finally, several important observations have recently been made investigating ALT (that is carried out by BIR-like pathway) studied in human telomerase-defective cells (Cho, Dilley, Lampson, & Greenberg, 2014; Dilley et al., 2016).

In this review we focus on one experimental system in yeast where BIR is initiated by a site-specific HO endonuclease and involves two copies of chromosome III (Deem et al., 2008). This system allows for a detailed investigation of BIR by using genetic and molecular biology techniques. Using this system, we were able to describe the mechanism of BIR, including its

kinetics, molecular intermediates, and also to identify proteins that carry out and regulate BIR (Deem et al., 2008; Saini et al., 2013; Vasan et al., 2014). In addition, this experimental system allowed us to study mutagenesis and chromosomal rearrangements associated with BIR (Saini et al., 2013; Sakofsky et al., 2015, 2014). In this review we describe this BIR system as well as various genetic and molecular biology techniques that we use to study BIR in yeast.

2. CALCULATING BIR EFFICIENCY

The efficiency of BIR is determined using our yeast strain disomic for chromosome III AM1003 and its derivatives (Fig. 2) with the genotype *hmlΔ::ADE1/hmlΔ::ADE3 MATa-LEU2-tel/MATα-inc hmrΔ::HPH FS2Δ::NAT/FS2 leu2/leu2-3112 thr4 ura3-52 ade3::GAL::HO ade1 met13* (Deem et al., 2008). In this strain, HO endonuclease is under the regulation of a galactose-inducible promoter and the HO recognition site is at the *MAT***a** locus of the truncated copy of chromosome III acting as the recipient chromosome. The full-length chromosome III, acting as the donor copy, contains an uncleavable *MATα-inc* allele and serves as the template for DSB repair. The HO endonuclease gene is inserted at *ADE3* gene located at chromosome VII and is placed under the control of *GAL1/GAL10* (galactose-inducible) promoter. Almost all sequence homology (except for 46 bp) between the recipient and donor on centromere-distal side of the break was eliminated via replacement with *LEU2* and telomere sequences, allowing for efficient DSB repair through BIR. The outcomes resulting from DSB repair by BIR and alternative repair pathways can be followed based on maintenance of the markers located at all arms of the two recombining chromosome IIIs (Deem et al., 2008; Vasan et al., 2014).

The efficiency of DSB repair in AM1003 can be assayed by plating cells on a YEP-Gal (galactose-containing) medium to induce expression of HO endonuclease leading to the DSB formation. After plating on YEP-Gal followed by replica plating of the colonies on a series of dropout media, approximately 80% of the colonies show the expected BIR phenotype of Ade^+ Leu^- (Deem et al., 2008; Fig. 3). A small fraction of colonies (approximately 10%) show the Ade^+ Leu^+ phenotype indicative of GC. Approximately 2%–5% of the colonies are Ade^- Leu^- indicative of failed repair resulting in the loss of the broken (recipient) chromosome. These chromosome loss (CL) events are easily distinguished by accumulation of red pigment on adenine dropout media due to the loss of a functional *ADE1*

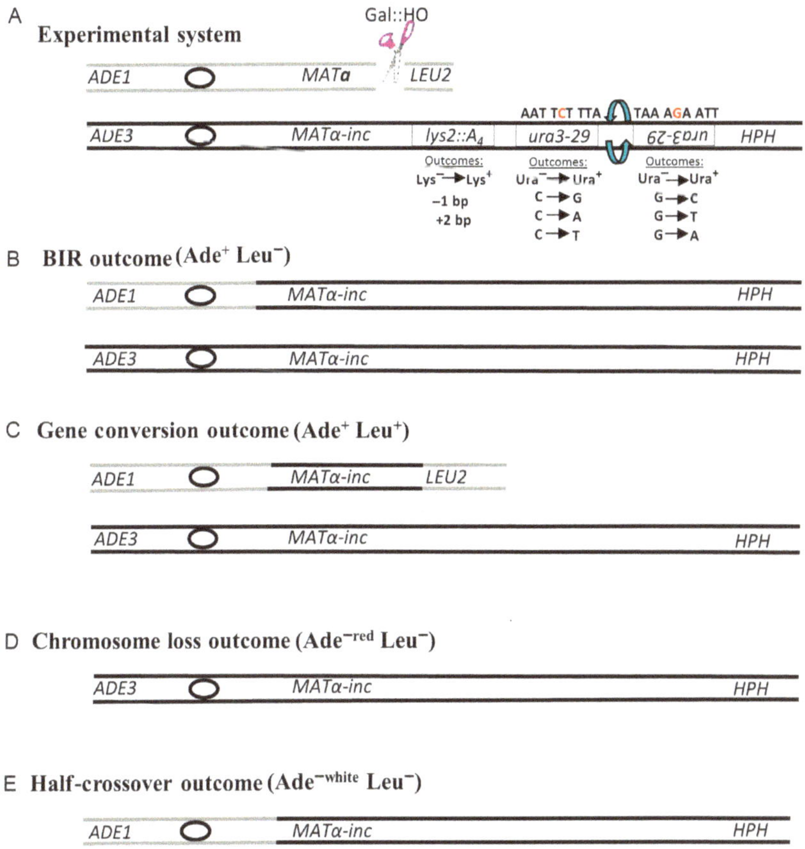

Fig. 2 Schematic of the AM1003 disome and DSB repair outcomes. (A) Depiction of the AM1003 experimental disome system that contains two copies of chromosome III. One copy is truncated and contains a *GAL::HO* site at the *MAT***a** locus. The other, full length, chromosome III (*MATα-inc*) serves as a donor template for repair. *Rectangles* denote insertions of mutation reporters, including the frameshift (*lys2::A$_4$*) or base-substitution (*ura3-29*) reporters with *red letters* corresponding to mutant bases in the reporter. (B) BIR repair outcome (Ade$^+$ Leu$^-$). (C) Gene conversion repair outcome (Ade$^+$ Leu$^+$). (D) Chromosome loss outcome (Ade^{-red} Leu$^-$). (E) Half-crossover outcome (Ade^{-white} Leu$^-$).

gene (this phenotype is indicated as Ade^{-red} Leu$^-$) (Fig. 3; Deem et al., 2008). Another rare type of aberrant repair, called half-crossovers, constitutes 2%–5% of all events and leads to the Ade^{-white} Leu$^-$ phenotype where the colonies fail to grow on adenine dropout media but appear white due to the absence of any pigment accumulation resulting from the loss of *ADE3* gene. These colonies also fail to grow on histidine dropout media because

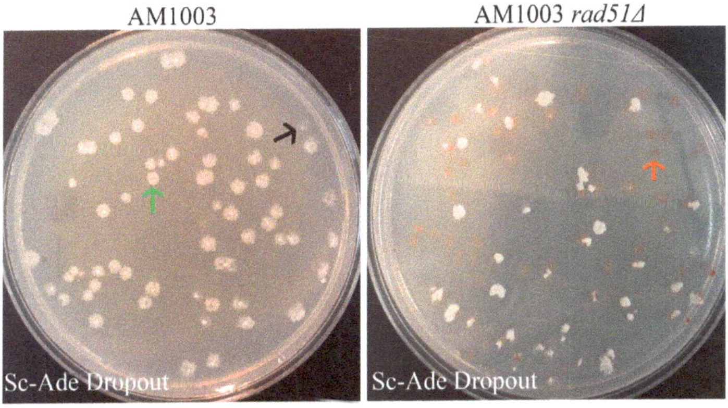

Fig. 3 BIR repair plating assay. The cells were pregrown in YEP-lactate liquid media followed by plating on YEP-galactose plates. The colonies were replica-plated on the Sc-Ade dropout plates. The *left panel* shows the replica of AM1003 colonies on Sc-Ade media. The *right panel* is the replica of a *rad51Δ* AM1003 derivative. *Arrows* point to different repair outcomes, *green arrow* indicates a BIR outcome, *black arrow* indicates a half-crossover, and the *red arrow* indicates a chromosome loss outcome.

the protein encoded by *ADE3* plays also an important role in the biosynthesis of histidine. Half-crossovers are formed due to interruption of BIR leading strand synthesis leading to resolution of BIR intermediates producing a fusion of one arm of the donor chromosome and one arm of the recipient chromosomes with all other arms lost (Deem et al., 2008).

Because the repair by BIR occurs in G2/M cells (Malkova et al., 2005), two copies of each chromosome are present. Therefore, each colony grown on YEP-Gal plate is considered to be a result of two separate repair events that occurred in two broken chromosomes. Therefore, while calculating the total number of BIR events, the number of full Ade^{+} Leu^{-} colonies is multiplied by two, while for the colonies containing one Ade^{+} Leu^{-} sector, the number of BIR events is counted as one. For example, for the colony where one half is Ade^{+} Leu^{-}, while another half is Ade^{-red} Leu^{-}, we conclude that it represents two separate events: the repair of one broken chromatid by BIR, while another sister chromatid failed to repair and was lost (see Deem et al., 2008 for details). Also, in some DNA repair mutants, we observed the formation of sectored colonies that were smaller than a half of the colony, which resulted from the retention of an unrepaired chromatid for several cell divisions (Malkova et al., 1996; Vasan et al., 2014). Most of the cells in such colonies eventually stabilize their broken chromatids by BIR, GC, or half-crossover, while in others the repair fails, leading to a CL.

The formation of colonies with small and often multiple repair sectors has been observed in a number of different mutants, including *rad9Δ*, *rad24Δ*, and *rad51Δ* (Fig. 3; Malkova et al., 1996; Vasan et al., 2014).

2.1 Equipment and Materials

- Hemocytometer (Hausser scientific, #3100)
- 1.5-mL microcentrifuge tubes
- Tabletop vortex
- Incubator (30°C)
- 30°C incubator with shaking table
- Olympus Microscope (with 40 × ocular) (or a similar microscope)
- Petri dishes (100 × 15 mm)

2.2 Solutions and Reagents

- Sc-Leu dropout liquid: 2% D-glucose (RPI, #50-99-7), 0.67% yeast nitrogen base without amino acids (US Biological, #Y2025), 0.087% Leu dropout amino acid mix, pH 5.5
- YEPD plates: 2% D-glucose, 2.5% agar (Difco, #214520), 1% yeast extract (Difco, #212750), 2% peptone (Difco, #211677), pH 5.5
- YEP-lactate liquid: 3.15% lactic acid (Fisher, #A162-500), 1% yeast extract, 2% peptone pH 5.5
- YEP-galactose plates: 2% galactose (Sigma-Aldrich, #59-23-4), 2.5% agar, 1% yeast extract, 2% peptone pH 5.5
- MilliQ water
- Sc-Ade dropout plates: 2% glucose, 2.5% agar, 0.67% yeast nitrogen base without amino acids, 0.087% Ade dropout amino acid mix pH 5.5
- Sc-His dropout plates: 2% glucose, 2.5% agar, 0.67% yeast nitrogen base without amino acids, 0.087% His dropout amino acid mix pH 5.5
- Sc-Leu dropout plates: 2% glucose, 2.5% agar, 0.67% yeast nitrogen base without amino acids, 0.087% Leu dropout amino acid mix pH 5.5
- Yeast strain AM1003 (Deem et al., 2008) or its derivative (available upon request from the corresponding author)

2.3 Procedure

1. Inoculate a single colony of AM1003 in liquid Sc-Leu dropout media and grow for 24 h at 30°C to saturation (approximately to 1×10^8 cells/mL).
2. Transfer 5 mL of this culture (inoculum) to 45 mL of YEP-lactate liquid media in 250 mL flask and grow for 16 h at 30°C with constant aeration.

3. Count cells using a hemocytometer to calculate the cell density. The expected cell density is approximately 2×10^7 cells/mL.
4. Serially dilute the cell culture to a density that allows plating of approximately 60 cells/plate by initially adding 6×10^5 cells to 1 mL of sterile water. Make three serial 1:10 dilutions in sterile water to achieve a final concentration of 6×10^2 cells/mL.
5. Plate 0.1 mL of cell suspension (containing approximately 60 cells) onto 1 plate of YEPD and 1 plate of YEP-galactose media. Plate at least 3 YEPD plates and 5–10 YEP-galactose plates.
6. Incubate YEPD and YEP-galactose plates at 30°C for 3–5 days. Typically, the colonies will form in 3 days on YEPD and in 5 days on YEP-galactose.
7. Replica plate the colonies onto Sc-Ade and Sc-Leu, dropout media plates.
8. Score repair outcomes based on auxotrophic marker phenotype (Figs. 2 and 3).
   a. BIR: Ade^+ Leu^-
   b. GC: Ade^+ Leu^+
   c. Half-crossover: Ade^{-white} Leu^-
   d. CL: Ade^{-red} Leu^-
9. Calculate the frequency of BIR outcomes using the following formula:

$$\text{BIR frequency} = \frac{2 \times (\text{\# of full BIR colonies}) + \text{\# of colonies with a BIR sector}}{2 \times (\text{total colonies})}$$

2.4 Notes

1. Initial inoculum can be started from a patched colony, or a single colony.
2. Sc-Leu dropout media is used for the inoculum to eliminate cells that underwent spontaneous BIR and therefore became Leu^- prior to the beginning of experiment.
3. The lack of glucose in the media enables the efficient galactose uptake ensuring the maximum HO induction and therefore DSB initiation. It is expected that more than 90% of cells will undergo HO-induced DSB in less than 20 min following galactose addition.
4. The comparison of the number of colonies grown on YEP-galactose vs YEPD allows to determine cell viability following DSB induction. The cell viability can be calculated using the formula:

$$\text{Cell viability (\%)} = \frac{\text{c.f.u. per mL on YEP-Gal}}{\text{c.f.u. per mL on YEPD}} \times 100$$

where c.f.u. = colony-forming units.

Inability to form colonies, or loss of viability after the DSB, is rare due to the design of the disomic system that contains two chromosomes III (Fig. 2A). These cells remain viable even when repair of the one (broken) chromosome is unsuccessful and it is lost. However, the loss of viability can occur due to more complex events that alter both recombining chromosomes.

5. It is important to note that not all Ade^{+} Leu^{-} events have been proven to be BIR. Some Ade^{+} Leu^{-} outcomes contain gross chromosomal rearrangements (GCRs). BIR can be distinguished from GCR outcomes using contour-clamped homogenous electric field (CHEF) gel electrophoresis (see Section 3). The frequency of GCRs among Ade^{+} Leu^{-} in each strain background should be taken into account when calculating BIR efficiency (see Vasan et al., 2014 for details).

3. ANALYSIS OF BIR KINETICS

BIR kinetics is analyzed by separating BIR products from other chromosomes using CHEF gel electrophoresis. Using this technique, the BIR product (~300 kb) can be distinguished from the original truncated chromosome III that is approximately 200-kb-long before HO-induced DSB (Fig. 4). BIR repair time-course is performed by collecting 50 mL samples for CHEF before and after DSB induction. BIR typically takes 4 h for initiation and 5–10 h for completion. Therefore, time points are collected every 2 h following BIR induction for the total of 8–10 h. Cells collected at each time point are spun down and embedded in 1% low melting agarose plugs after treatment with zymolyase. The resulting DNA plugs are then used for CHEF gel electrophoresis using the Bio-Rad-CHEF-DRII or other CHEF machines.

3.1 Determining BIR Kinetics in a Time-Course Experiment

3.1.1 Equipment

- 50-mL Conical tubes
- Tabletop centrifuge for spinning 50-mL conical tubes (Beckman Coulter Allegra 25R or similar model)
- 30°C Incubator with shaking table
- −80°C freezer

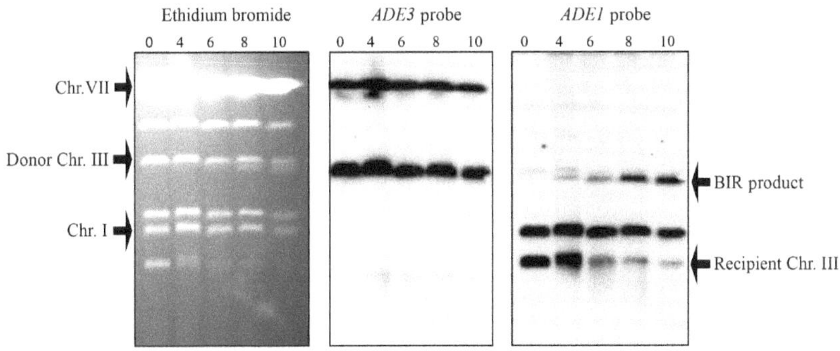

Fig. 4 BIR kinetics assay using CHEF gel electrophoresis. The aliquots were taken from the liquid cultures undergoing BIR repair before (0 h) and following the addition of galactose to the media every 2 h. The genomic DNA was separated by CHEF. The *left panel* is an ethidium bromide-stained gel of the separated chromosomes. The *middle panel* represents the results of Southern blot analysis using *ADE3*-specific probe that highlights the donor chromosome III. The *right panel* represents the results of Southern blot analysis using *ADE1*-specific probe highlighting the recipient chromosome III. *Chr.*, chromosome.

3.1.2 Solutions and Reagents

- YEP-lactate liquid
- Sc-Leu dropout liquid
- 20% Galactose: 20 g galactose dissolved in 100 mL of MilliQ water and filter sterilized
- 10% Sodium azide: 10 g sodium azide dissolved in 100 mL MilliQ water (Sigma-Aldrich, #26628-22-8)
- 50 m*M* EDTA: 14.6 g dissolved in 1 L MilliQ water pH 7.5 (Fisher, BP120-1)

3.1.3 Procedure

1. Inoculate AM1003 in 50 mL of Sc-Leu dropout media and incubate at 30°C with shaking for 24 h. Culture should be of a final density of ~10^8 cells/mL.
2. Add ~5 mL of inoculum to 500 mL of YEP-lactate to obtain a cell density of 1×10^6 cells/mL. The culture was incubated at 30°C with shaking for approximately 16 h, to reach a cell density of 2×10^7 cells/mL.
3. Collect 50 mL of culture for a 0-h sample (sample taken before galactose addition) in a 50-mL conical tube, then add 0.5 mL of 10% sodium azide to this aliquot. Mix thoroughly.
4. Pellet the sample using a tabletop centrifuge by spinning at 3000 rpm for 5 min at 4°C.

5. Pour off supernatant into hazardous waste container and resuspend in 50 mL of 50 m*M* EDTA pH 7.5 solution (chilled).
6. Pellet the sample using a tabletop centrifuge by spinning at 3000 rpm for 5 min. Discard supernatant.
7. Store pellet in −80°C freezer until needed for plug preparation.
8. Induce the DSB by adding 50 mL of 20% galactose to the YEP-lactate culture.
9. Every hour (or every 2 h) after the galactose addition, collect 50 mL of culture and repeat steps 3 through 7.

3.1.4 Notes

- DSB formation should occur within the first hour following addition of galactose.
- 4 h after the break, repair products of BIR usually appear.

3.2 Preparing Agarose Plugs With Embedded DNA for CHEF

3.2.1 Equipment

- 40°C water bath
- 37°C water bath
- 50°C water bath
- 30°C incubator with shaking table
- 4°C refrigerator
- Plug molds (Bio-Rad, #1703713)
- 50-mL conical tubes
- 1.5-mL microcentrifuge tubes
- Serological pipettes (5 and 10 mL)
- Centrifuge for 50-mL conical tubes (Beckman Coulter Allegra 25R centrifuge or a similar brand)
- Scoopula

3.2.2 Solutions and Reagents

- SCE solution: 1 *M* sorbitol (RPI, #10043-35-3), 0.1 *M* sodium citrate dihydrate (RPI, #6132-04-3), 50 m*M* EDTA pH 7.5
- 50 m*M* EDTA pH 7.5 solution in water
- LMP agarose (Fisher, 9012-36-6) 1% LMP agarose in a mix of 0.75 mL water and 0.25 mL 0.5 *M* EDTA pH 7.5
- Zymolyase buffer: 0.005 g 20T zymolyase (MP#320921), 1 mL SCE solution, 50 μL 2-mercaptoethanol (AMRESCO, #60-24-2)

- Incubation buffer 1: 50 m*M* EDTA pH 7.5, 0.01 *M* Tris–HCl (RPI, #77-86-1), pH 7.5
- RNase A 10 mg/mL: (Thermo Fisher Scientific #EN0531)
- Incubation buffer 2: 0.01 *M* Tris–HCl pH 7.5, 0.25 *M* EDTA pH 7.5, 1% Sarkosyl (Fisher, #137-16-6)
- Wash buffer: 10 m*M* Tris pH 7.5, 50 m*M* EDTA pH 7.5
- Storage buffer: 40 m*M* EDTA pH 7.5, 55% glycerol (RPI, #56-81-5)
- Proteinase K: (RPI, #39450-01-6)
- MilliQ water

3.2.3 Procedure

1. Thaw pellets from the time-course assay on ice.
2. During thawing, LMP agarose is prepared by adding 0.01 g of LMP agarose to 0.75 mL of MilliQ water then adding 0.25 mL 0.5 *M* EDTA pH 7.5. Heat LMP agarose on heating block to 100°C to dissolve the LMP agarose. Zymolyase buffer is prepared by adding 50 μL of 2-mercaptoethanol and 0.005 g 20T zymolyase to 1 mL of SCE where it then stored on ice.
3. Place LMP agarose in 40°C water bath to equilibrate temperature to 40°C.
4. Loosen the pellet using tabletop vortex and equilibrate in a 40°C water bath.
5. Mix 0.415 mL of agarose with 0.085 mL zymolyase buffer quickly and then add 0.25 mL cells to this mixture.
6. Quickly transfer the cell agarose mixture into plug molds by pipetting.
7. Chill plug molds containing cell agarose mixture in refrigerator at 4°C for 5 min.
8. Prepare incubation buffer 1 mix by adding 750 μL mercaptoethanol and 10 μL 10 mg/mL RNase into 10 mL of incubation buffer 1. Dispense 2.5 mL of incubation buffer 1 mixture into 50-mL conical tubes.
9. Use a toothpick to remove solidified plugs from the plug molds and into conical tubes with incubation buffer 1 mix from the step 8, and incubate in 37°C water bath for 1 h.
10. Decant incubation buffer from conical tubes, while using scoopula to prevent losing the plugs.
11. Rinse plugs in MilliQ water and decant water while using scoopula to prevent losing the plugs.
12. Prepare incubation buffer 2 mix by adding 0.005 g proteinase K into 10 mL of incubation buffer 2.

13. Add 2.5 mL of incubation buffer 2 mixture from step 12 into each conical tube with plugs and incubate at 50°C for 16 h.
14. Rinse plugs in MilliQ water and decant water while using scoopula to prevent losing the plugs.
15. Rinse plugs in 2.5 mL of wash buffer.
16. Plugs can be stored in wash buffer for 2 days at 4°C and used for CHEF gel electrophoresis. Long-term storage requires decanting wash buffer, using scoopula to prevent losing the plugs, and adding 50 mL of storage buffer and keeping at −20°C for up to 6 months.

3.2.4 Notes

1. Colonies for analysis can be patched beforehand to allow for further studies to be performed.
2. LMP agarose mix should not be hot at the time of mixing with zymolyase to prevent degradation of zymolyase.
3. When transferring cell mixture, with LMP agarose and zymolyase, into plug molds avoid air bubbles in the mold so as to have a solid plug.
4. A pressurized stream of air can also be used to remove plugs from the plug mold as an alternative to a toothpick.
5. For long-term storage of the plugs, the volume of storage solution should be at least half of the storage tube to prevent freezing of the plugs as well as to protect against breaking of the plugs.

3.3 Analysis of BIR Products Using CHEF Gel Electrophoresis

3.3.1 Equipment

- CHEF-DRII (Bio-Rad) apparatus or similar brands.
- CHEF gel-casting tray.

3.3.2 Solutions and Reagents

- 5× TBE: 64 g Tris base (RPI, #77-86-1), 31 g boric acid (RPI, #10043-35-3), 3.35 g EDTA disodium salt dihydrate (Fisher, BP120-1) for 1 L of water
- 1% Pulse-field certified agarose gel (Bio-Rad, #L004315B): 2 g pulse-field certified agarose in 200 mL 0.5× TBE
- Wash buffer: 10 m*M* Tris pH 7.5, 50 m*M* EDTA pH 7.5

3.3.3 Procedure

1. Make 3 L of 0.5× TBE by adding 300 mL of 5× TBE and 2.7 L of MilliQ water to a 4-L beaker.

2. Assemble gel-casting tray (a part of CHEF Bio-Rad equipment) with black base plate on the bottom and dams tightened on each side.
3. Using a scoopula remove each plug and rinse in wash buffer. Then place the plug against the long wall of the gel-casting tray.
4. Remove 200 mL of the previously made 0.5× TBE and make a 1% pulse-field agarose gel, and microwave until it dissolves. Let the 1% pulse-field agarose gel cool to about 60°C then pour into gel-casting tray.
5. Transfer the remaining 0.5× TBE buffer into CHEF machine to circulate and precool buffer to 14°C.
6. Once the gel solidifies, disassemble the gel-casting tray while leaving the gel adhered to the black plate.
7. Place the gel and the black base plate into frame in the middle of CHEF apparatus.
8. Close the lid of the CHEF machine.
9. On the control module enter the parameters of the initial and final switch times, the duration, and the voltages. For the CHEF-DRII apparatus to separate the donor chromosome III (356 kb) and the repaired recipient chromosome III (346 kb) of AM1003, the parameters are: initial switch time of 10 s, final switch time of 35 s, running time of 40 h, and the voltage is 6 V/cm.

3.3.4 Notes

1. When pouring the gel into the casting tray, be sure that the holes in each corner of the black plate are filled without air bubbles. Failure for the holes to be filled could lead to the gel breaking away and floating in the buffer.
2. When removing the gel from the casting tray, remove as much excess gel from the edges and bottom of the black plate so as to not clog the CHEF apparatus.
3. For different CHEF machines the optimal running parameters to separate the chromosomes of interest vary.

3.4 Capillary Transfer and Hybridization

3.4.1 Equipment

- Glass baking dish 19 × 28 cm (or similar dish)
- UV Cross-linker wavelength 254 nm (Stratagene Stratalinker UV 2400)
- UV Gel imager (Gel DOC Bio-Rad or similar brand)
- Glass plate 20.5 × 32 cm

- Whatman paper 45 × 45 cm
- Scott C-fold paper towels (Fisher, #06-666-32B)
- Amersham Hybond Membrane-N+ (GE Healthcare, #RPN303B)
- 1.5-mL microcentrifuge tubes
- 1.5-mL microcentrifuge tube locker (VWR, #14229-941)
- 37°C dry bath
- 100°C dry bath
- Hybridization tubes
- Hybridization Oven (BigShot 3 or similar brand)
- Geiger counter (Ludlum 5000)
- Typhoon Phosphorimager FLA7000 (or similar brand)

3.4.2 Solutions and Reagents

- Ethidium bromide (Sigma-Aldrich, #1239-45-8)
- 20× SSC pH 7: 175.3 g NaCl, 88.2 g sodium citrate, 900 mL MilliQ water
- Alkaline solution: 20 g NaOH (RPI, #1310-73-2), 87.7 g NaCl (RPI, #7647-14-5) in 1 L of MilliQ water
- Neutralization solution: 87.7 g NaCl (RPI, #7647-14-5), 121.4 g Tris base (RPI, #77-86-1) in 1 L of MilliQ water, pH 7
- 1× TBE: dilute 5× TBE by adding 200 mL 5× TBE in 1 L of MilliQ water
- 1% agarose gel: 1 g of agarose (Sigma-Aldrich, #A0169) in 100 mL of 1× TBE
- QIAGEN gel extraction kit: (QIAGEN, #28704)
- P^{32}-dCTP (Perkin Elmer, #BLU513H250)
- RmT Random primer kit (Agilent, #300392)
- Hybridization buffer: 50 mL of 0.5 *M* sodium phosphate buffer pH 7.2 (Fisher, #S374-1), 25 mL of 20% SDS (Fisher, #BP166-5), 200 μL of 0.5 *M* EDTA pH 8. Make up to 100 mL using MilliQ water
- Low stringency wash buffer: 2× SSC, 0.1% SDS
- MilliQ water

3.4.3 Procedure

1. Add 300 mL MilliQ water into glass baking dish.
2. Remove the CHEF gel from the CHEF apparatus and place into the glass baking dish that contains 0.5× TBE.
3. Add 9 μL of ethidium bromide solution into the glass dish containing the CHEF gel and incubate for 30 min with constant gentle shaking.
4. Carefully remove gel from black base plate.

5. Using UV gel imager, take a picture of the CHEF gel with separated chromosomes.
6. Remove buffer with ethidium bromide and place gel into UV cross-linker and set machine to 600 μJ total exposure at 254 nm wavelength light.
7. After cross-linking, submerge the gel in alkaline solution and incubate for 30 min with constant gentle shaking.
8. Decant alkaline solution.
9. Submerge gel in neutralization solution and incubate for 30 min with constant gentle shaking.
10. Transfer gel via capillary transfer methods described in (Sambrook & Russell, 2006) using 10 × SSC and let the transfer occur for 20 h.
11. Cross-link membrane (DNA side up) in UV cross-linker at 254 nm wavelength light using the autocrosslink function (1200 μJ).
12. *ADE1*-specific DNA probe targets not only the recipient copy of chromosome III (before and after the DSB, and also following DSB repair) but also chromosome I containing a native *ADE1* gene. This probe is prepared by PCR amplification of the *ADE1* fragment from the yeast genome (using genomic DNA prepared from AM1003) and using the following PCR primers:
    FP: 5′-GGTTTGAAACAACCTCAAGGACTT-3′
    RP: 5′-AAGTCCTTGAGGTTGTTTCAAACC-3′.
13. Run the PCR probe fragment on a 1% agarose 1 × TBE gel and excise the fragment.
14. Using the QIAGEN Gel extraction kit extract the probe PCR probe fragment.
15. Resuspend and dissolve the contents of one tube of the RmT Random primer kit in 37 μL of MilliQ water, and transfer to a 1.5-mL microcentrifuge tube.
16. Add 5 μL (~70 ng) of the gel-extracted PCR probe fragment to the reaction mixture in step 18.
17. Incubate RmT Random primer kit and probe fragment mixture at 100°C for 5 min. Be sure to add a microcentrifuge tube locker to keep the lid closed.
18. After incubation period, place reaction mixture on ice for 1 min and briefly spin down (to collect all the liquid on the bottom of the tube).
19. Add 5 μL of P^{32}-dCTP to the cooled reaction mixture, and add 3 μL of magenta polymerase from the RmT Random primer kit, and incubate at 37°C dry bath for 10 min.

20. Add 2 μL of stop mix from the RmT Random primer kit to stop the reaction.
21. Add 200 μL of MilliQ water to the reaction mixture and denature at 100°C dry bath for 10 min and snap cool on ice for 1 min.
22. Prehybridize the cross-linked membrane from step 14 in 25 mL of hybridization buffer at 65°C for at least 10 min.
23. Hybridize the membrane at 65°C using 25 mL of hybridization buffer for 24 h (Church & Gilbert, 1984).
24. Wash the membrane using 65°C low stringency wash buffer in an open tray. Gently rock the wash buffer and membrane for 10 min. Discard waste under the guidelines of your radiation safety officer.
25. Repeat wash two more times or until background is washed away, as being checked using a Geiger counter.
26. Using 20 × 25 cm GE Phosphor screen (or similar screen), expose membrane for 1 day and image on a Typhoon Phosphorimager FLA7000 (or similar brand). See Fig. 4 for the example of Southern blot hybridization images.

3.4.4 Notes

1. Follow precautions while using P^{32}. The laboratory staff using P^{32} has to undergo radiation safety training prior to usage of these materials.

4. ANALYSIS OF MUTAGENESIS ASSOCIATED WITH BIR

DNA synthesis during BIR is associated with a very high level of mutagenesis which could be 100- to 1000-fold higher as compared to S-phase DNA replication (Deem et al., 2011). Both frameshifts and base substitutions are drastically increased during BIR (Deem et al., 2011; Saini et al., 2013; Sakofsky et al., 2014). Mutagenesis associated with BIR can be assayed using our reporter systems. The level of frameshift mutagenesis during BIR can be assessed using the *lys2::A*$_4$ reporter (Fig. 2; Deem et al., 2011). This reporter can be inserted at several positions on the donor chromosome, on the track of BIR to determine the level of mutagenesis during its entire course of synthesis (Fig. 2). The level of base-substitution mutations formed during BIR could be determined by using the *ura3-29* reporter (Saini et al., 2013) inserted at various positions along the track of BIR (Fig. 2).

4.1 Analysis of Frameshift Mutagenesis

The reporter *lys2::A*$_4$ is an allele of the *LYS2* gene with an insert of 61 bp resulting in a +1-bp frameshift in the *LYS2* reading frame, leading to a Lys$^-$ phenotype (Tran, Keen, Kricker, Resnick, & Gordenin, 1997). A Lys$^+$ phenotype can be restored by a frameshift mutation that can occur in an approximately 71-bp region of the allele (that includes the inserted sequence) and restores the reading frame. Several isogenic strains were created by inserting the *lys2::A*$_4$ reporter into three positions on the donor (*MATα-inc*-containing) copy of chromosome III along the track of BIR synthesis. In our previous work (Deem et al., 2011) three insertion positions of *lys::A*$_4$ reporter were employed: at *MATα-inc* ("*MAT*" or "0 kb" position), 16 kb centromere-distal from *MATα-inc*, in the region between *RSC6* and *THR4* ("16 kb" position), and 36 kb centromere-distal to *MATα-inc*, in the region between *SED4* and *ATG15* ("36 kb" position). In all these strains, *LYS2* copy was fully deleted from its native location in chromosome II to prevent recombination with the reporter at chromosome III.

4.1.1 Equipment

- 30°C incubator
- 30°C incubator with shaker table (New Brunswick Scientific I2500 series or similar model)
- Hemocytometer (Hausser scientific, #3100)
- 50-mL conical tubes
- Tabletop centrifuge for 50-mL conical tubes (Beckman Coulter Allegra 25R centrifuge or similar model)

4.1.2 Solutions and Reagents

- Sc-Leu dropout liquid: 2% D-glucose, 0.67% yeast nitrogen base without amino acids, 0.087% Sc-Leu dropout amino acid mix
- YEP-lactate liquid media: 3.15% lactic acid, 1% yeast extract, 2% peptone, pH 5.5
- 20% Galactose: 20 g galactose dissolved in 100 mL MilliQ water and filter sterilized
- Sc-Lys dropout plates: 2% D-glucose, 2.5% agar, 0.67% yeast nitrogen base without amino acids, 0.087% Sc-Lys dropout amino acid mix pH 5.5

- YEPD plates: 2% D-glucose, 2.5% agar, 1% yeast extract, 2% peptone pH 5.5
- MilliQ water

4.1.3 Procedure

1. Inoculate a single colony of a strain (isogenic to AM1003) containing *lys2::A₄* reporter into 50 mL of Sc-Leu dropout media and incubate with constant shaking at 30°C 24 h.
2. Transfer ~10 mL of inoculum into 100 mL YEP-lactate media (in a 500-mL flask), with a starting cell density of $\sim 1 \times 10^6$/mL. Incubate at 30°C for ~16–20 h with constant shaking, to a final concentration of $\sim 1\text{–}2 \times 10^7$/mL.
3. Collect 50 mL of culture from step 2 into a conical tube. Take this as "0 h" time point. This sample is used to assess frameshift mutagenesis associated with S-phase replication (as a control).
4. Serially dilute these samples by transferring 0.1 mL of culture into 0.9 mL water. Typically, four 10-fold dilutions are made to obtain ~100 cells/100 μL of water.
5. Plate 0.1 mL per YEPD plate and use three plates in total (approximately 100 cells per plate). This will allow to measure the concentration of cells in the culture.
6. To estimate the number of mutations accumulated in the culture at 0 h (before induction of DSB by HO) collect the "0 h" culture and centrifuge at 5000 rpm for 2 min at 25°C. Decant the supernatant and keep the pellet. Resuspend the pellet with water to a final volume of 0.5 mL and spread equal amounts onto Sc-Lys dropout plates (0.1 mL per plate). Be cautious to not plate $>1 \times 10^8$ cells per plate to avoid formation of a lawn due to residual growth.
7. Add 5.5 mL 20% galactose (filter sterilized) to the remaining 50 mL of the culture from step 2, to obtain a final concentration of 2% galactose. Addition of galactose initiates formation of DSB. Incubate with constant aeration at 3000 rpm at 30°C for 7 h.
8. After 7 h of incubation, collect 5 mL of culture from step 7 and centrifuge at 5000 rpm for 2 min at 25°C. Decant the supernatant and resuspend the pellet with water to a final volume of 0.5 mL and spread them evenly onto Sc-Lys dropout plates (0.1 mL per plate). Be cautious to not plate $>1 \times 10^8$ cells per plate to avoid formation of a lawn.
9. Incubate all plates at 30°C for 5 days and then count the colonies on each plate.

10. Calculate mutation rate during BIR using the following equation (the example for the volume and dilutions mentioned here):

Mutation Rate associated with BIR = Mutation frequency (7 h)–Mutation frequency (0 h).

$$\text{Mutation frequency (0 h)} = \frac{\text{Total number of colonies on all Sc} - \text{Lys plates (0 h)}}{\text{Average number of colonies on YEPD plates (0 h)} \times 5 \times 10^{6}}$$

$$\text{Mutation frequency (7 h)} = \frac{\text{Total number of colonies on all Sc} - \text{Lys plates (7 h)}}{\text{Average number of colonies on YEPD plates (7 h)} \times 5 \times 10^{5}}$$

4.1.4 Notes

1. The best way to determine mutation rate during S-phase is by using a no-DSB control strain. A no-DSB strain has a genotype similar to AM1003 except it contains *MAT**α**-inc-LEU2-tel*, and in these strains DSBs cannot be induced by HO endonuclease.
2. Due to the accumulation of mutations over several generations, the rate of spontaneous mutagenesis (the rate of mutations during S-phase replication) is calculated using Drake equation: $\mu = 0.4343\ f/\log(N\mu)$, where μ is the rate of spontaneous mutagenesis, *f* is mutation frequency, and *N* is the number of cells in yeast culture (see Deem et al., 2011 for details). However, during BIR, the cells are arrested at G2/M by DSB induction until they are repaired, and then they recover from the arrest. This means that there are no cell divisions between 0 and 7 h time points following galactose addition. This is why the mutation rate during BIR can be calculated using a modified Drake equation. (Mutation Rate during BIR = Mutation frequency (7 h)–Mutation frequency (0 h)).
3. Typically for wild-type strains, 5 mL samples are collected at 7 h. However, for certain mutants, more cells need to be collected due to a lower mutation rate.

4.2 Analysis of Base-Substitution Mutagenesis Associated With BIR

The *ura3-29* reporter (used to assay the frequency of base substitutions associated with BIR) is an allele of the *URA3* gene containing a mutation at the position 257 resulting in an amino acid change from phenylalanine to serine rendering the strain Ura^{-} (Shcherbakova & Pavlov, 1996). This reporter can be inserted in two orientations—Ori1 (AGA in the mutant position in the leading strand of BIR) and Ori2 (TCT in the leading strand). CG → AT,

CG → TA, and CG → GC base-substitution events in the mutant position leads to restoration of a Ura^+ phenotype. In two strains isogenic to AM1003, AM2161, and AM2820, the *ura3-29* reporter is inserted in Ori1 and Ori2 orientations, respectively, at the position located 16 kb centromere-distal from *MATα-inc* in the region between *RSC6* and *THR4* (16 kb) (Saini et al., 2013). In these strains, *URA3* was fully deleted from its native location in chromosome V.

4.2.1 Equipment

- 30°C incubator
- 30°C incubator with shaker table (New Brunswick Scientific I2500 series or similar model)
- Hemocytometer (Hausser scientific, #3100)
- 50-mL conical tubes
- Tabletop centrifuge for 50-mL conical tubes (Beckman Coulter Allegra 25R centrifuge or similar model)

4.2.2 Solutions and Reagents

- Sc-Leu dropout liquid: 2% D-glucose, 0.67% yeast nitrogen base without amino acids, 0.087% Sc-Leu dropout amino acid mix
- YEP-lactate liquid: 3.15% lactic acid, 1% yeast extract, 2% peptone, pH 5.5.
- 20% Galactose: 20 g galactose dissolved in 100 mL of water and filter sterilized
- Sc-Ura dropout plates: 2% D-glucose, 2.5% agar, 0.67% yeast nitrogen base without amino acids, 0.087% Sc-Ura dropout amino acid mix
- YEPD plate media: 2% D-glucose, 2.5% agar, 1% yeast extract, 2% peptone, pH 5.5
- MilliQ water

4.2.3 Procedure

1. Inoculate a single colony of a strain containing *ura3-29* reporter into 10 mL of Sc-Leu liquid dropout media and incubate with constant shaking at 30°C for 24 h.
2. Transfer ~10 mL inoculum into 100 mL YEP-lactate media, with a starting cell density at $\sim 1 \times 10^6$/mL. Incubate at 30°C for ~16–20 h with constant shaking, to a final concentration of $\sim 1\text{–}2 \times 10^7$/mL.

3. Collect 50 mL of culture from step 2 into a conical tube. Take this as "0 h" time point. This sample is used to assess base-substitution mutagenesis associated with S-phase replication (used as a control).
4. Serially dilute these samples by transferring 0.1 mL of culture into 0.9 mL water. Typically, four 10-fold dilutions are made to obtain ~100 cells/100 μL of water.
5. Use 0.1 mL per plate and use 3 YEPD plates in total (approximately 100 cells per plate). This will allow to measure the concentration of cells in the culture.
6. To estimate the number of mutations accumulated in the culture at 0 h (before induction of DSB by HO) collect the "0 h" culture and centrifuge at 5000 rpm for 2 min at 25°C. Decant the supernatant and keep the pellet. Resuspend the pellet with water to a final volume of 0.5 mL and spread equal amounts onto Sc-Ura dropout plates (0.1 mL per plate). Be cautious to not plate $>1 \times 10^8$ cells per plate.
7. Add 5.5 mL of 20% galactose (filter sterilized) to the remaining 50 mL of the culture from step 2, to obtain a final concentration of 2% galactose. Incubate with constant aeration at 3000 rpm at 30°C for 7 h.
8. After 7 h, collect 5 mL of culture from step 7 and centrifuge at 5000 rpm for 2 min at 25°C. Decant the supernatant and resuspend the pellet in water to a final volume of 0.5 mL and spread them evenly onto Sc-Ura dropout plates (0.1 mL per plate).
9. Incubate all plates at 30°C for 5 days and then count the colonies on each plate.
10. Calculate mutation rate during BIR using the following equation.

Mutation Rate associated with BIR = Mutation frequency (7 h)–Mutation frequency (0 h).

See Section 4.1.3 and the notes in Section 4.1.4 for the details of calculating mutation rates. (Here, use the number of colonies grown on Sc-Ura dropout plates instead of Sc-Lys dropout plates that are used in the Section 4.1.3.)

4.2.4 Notes

1. Typically for wild-type strains, 5 mL samples are collected at 7 h. However, for certain mutants, more cells are collected due to a lower mutation rate.

4.3 Analysis of Lys^+ and Ura^+ Mutations by Sanger Sequencing

4.3.1 Equipment

- Thermal cycler (Bio-Rad T100 thermal cycler) or a similar model.

4.3.2 Solutions and Reagents

- PCR reaction kit containing dNTPs (SibEnzymes, #N026), PCR buffer and Taq Polymerase (SibEnzymes, #E331).

4.3.3 Procedure

1. Pick Lys^+ or Ura^+ colonies from Sc-Lys dropout and Sc-Ura dropout plate, respectively, and streak for singles to isolate single colonies.
2. Take the singles from last step and patch them on YEPD.
3. Replica plate these patches on Sc-Leu and Sc-Ade to confirm that the colonies have completed BIR repair. (They should be Ade^+ Leu^-).
4. To amplify the mutant Ura^+ or Lys^+ alleles, use colony PCR.
5. A 50-μL PCR reaction is set up for each isolate and a small amount of cells are scooped using a sterile toothpick and resuspended in the PCR reaction mixture.

   The primers used to amplify the mutant fragment from Lys^+ BIR isolates are:
   FP: 5′-GTTCGTACCCCTCTCGAGAATA-3′ and
   RP: 5′-ATTTGAGGCAAATTTTTCGTTCCAA-3′
   The primers used to amplify the mutant fragment from Ura^+ BIR isolates are:
   FP: 5′-GTGTGCTTCATTGGATGTTCGTAC-3′ and
   RP: 5′-AAAAGGCCTCTAGGTTCCTTTGTT3′
   The PCR reaction condition used for colony PCR is:

a. Initial denaturation	95°C–2 min
b. Denaturation	95°C–30 s
c. Annealing	55°C–1 min
d. Extension	72°C–1 min
e. Repeat steps **b, c, d**	30–40 times
f. Final extension	72°C–10 min

6. Once the PCR fragments are obtained, they are sent to the DNA sequencing facility along with the following sequencing primers.

   Sequencing primer for Lys^+ BIR isolates: 5′-GTTCGTACCCCT CTCGAGAATA-3′

   Sequencing primer for Ura^+ BIR isolates: 5′-GTTCGTACCCCT CTCGAGAATA-3′.

4.3.4 *Notes*

1. Be careful not to scoop out too many cells for colony PCR. The cellular debris interferes with PCR amplification.

4.4 Sequencing of *URA3* and *LYS2* Sequences Located on Either Donor or Recipient Chromosomes

Following BIR, the repair outcomes contain two copies of the reporter gene (*URA3* or *LYS2*). When mutation occurs during BIR, only one copy of the reporter contains mutation, while another one does not. Here, we describe the method allowing to determine which copy of chromosome III (the donor or the recipient (the BIR product)) contains mutation.

4.4.1 *Equipment*

- 68°C water bath
- 40°C water bath
- CHEF-DRII apparatus (Bio-Rad)
- Razor blades

4.4.2 *Solutions and Reagents*

- Beta-Agarase enzyme (Lonza, #BMA58005)
- 50× Lonza Beta-Agarase buffer (Lonza, #BMA58005)
- Ethidium bromide (Sigma, #1239-45-8)
- 5× TBE: 64 g Tris base (RPI, #77-86-1), 31 g boric acid (RPI, #10043-35-3), 3.35 g EDTA disodium salt dihydrate (Fisher, BP120-1). Start from the new lane here and insert the new bullet. Wash buffer: 10 m*M* Tris pH 7.5, 50 m*M* EDTA pH 7.5
- Low melting agarose (Fisher, #9012-36-6 (BP165-25))
- PCR reaction kit containing dNTPs (SibEnzymes, #N026), PCR buffer, and Taq Polymerase (SibEnzymes, #E331)

4.4.3 *Procedure*

1. Take agarose plugs containing DNA isolated from cells undergoing BIR repair and wash with wash buffer similar to what is described in Section 3.2.3.
2. Make 3 L of 0.5× TBE by adding 300 mL of 5× TBE and 2.7 L of MilliQ water to a 4-L beaker.
3. Equilibrate plugs in 0.5× TBE for 1 h.
4. Run CHEF (refer to Section 3.3 for details) using 1% low melting agarose gel and 0.5× TBE.

5. CHEF conditions: 6 V/cm, 41 h, initial switch time—29.7, final switch time—30.9, 14°C.
6. After 41 h, stain the gel with 3 μL ethidium bromide per 100 mL of 0.5× TBE buffer for about an hour.
7. Using new razor blade excise donor and recipient chromosome bands from the gel and incubate it 1× Lonza beta-agarase buffer for 1 h.
8. Remove buffer and add fresh 2× beta-agarase buffer for 1 h.
9. Decant the 2× beta-agarase buffer after the 1 h incubation.
10. Melt the gel at 68°C for 10 min.
11. Equilibrate the melted gel at 40°C for 5 min.
12. Add 1 μL of beta-agarase per 100 μL of gel and incubate at 40°C for 2 h.
13. The gel can be stored at 4°C until further use.
14. To PCR a desired fragment from either the donor chromosome III or recipient chromosome III, add 2 μL of DNA to 50 μL of PCR reaction mixture.

5. ANALYSIS OF BIR STRUCTURAL INTERMEDIATES BY TWO-DIMENSIONAL GEL ELECTROPHORESIS

While previous methods described in this review allowed for the detection of BIR products, 2D electrophoresis allows for the analysis of the DNA intermediates as BIR progresses. This method involves two separate electrophoresis steps described later. The first-dimension (1D) gel electrophoresis separates samples on the basis of molecular weight, whereas the second dimension separates DNA on the basis of both size and shape. Here, we describe a modified procedure for 2D gel electrophoresis (Brewer & Fangman, 1987) to separate the DNA intermediates formed during BIR.

5.1 Preparation of Yeast Samples

5.1.1 Equipment

- Shaking incubator 30°C
- 50-mL conical tubes
- 4 L Erlenmeyer flask
- Hemocytometer (Hausser scientific, #3100)
- 500 mL centrifuge bottle
- Ultracentrifuge (Beckman Coulter J2-21 centrifuge or a similar one)

5.1.2 Solutions and Reagents

- Sc-Leu liquid dropout media: 2% D-glucose (RPI, #50-99-7), 0.67% yeast nitrogen base without amino acids (US Biological, #Y2025), 0.087% Sc-Leu dropout amino acid mix, pH 5.5.
- YEP-raffinose: 3% raffinose (MP, #4010-022), 1% yeast extract (BD, #212750), 2% peptone (BD, #211677) pH 5.5.
- 50 m*M* EDTA (Fisher Bioreagents, #BP120-1).
- 20% Galactose: 20 g of galactose (Sigma-Aldrich, #G0750, 59-23-4) dissolved in 100 mL MilliQ water and filter sterilized.
- 0.2 *M* EDTA pH 7.5: 400 mL of 0.5 *M* EDTA pH 7.5 in 600 mL of MilliQ water.
- MilliQ water chilled on ice (4°C).
- 10% Sodium azide (Sigma-Aldrich, #26628-22-8).
- Liquid nitrogen.
- Nocodazole 10 mg/mL (US Biological, #N3000).

5.1.3 Procedure

1. Inoculate 50 mL Sc-Leu dropout media and incubate at 30°C for 24 h.
2. Transfer 20 mL of the inoculum into 850 mL of YEP-raffinose to a final concentration of $\sim 1 \times 10^6$ cells/mL.
3. Incubate YEP-raffinose for 16 h at 30°C with constant shaking.
4. After 16 h calculate cell density. Cell density should reach 1×10^7 cells/mL.
5. Add 2.5 mL of 10% sodium azide to a 500-mL autoclaved centrifuge bottle containing 81 mL of 0.2 *M* EDTA pH 7.5 frozen in a slant.
6. Collect 0 h 2D sample (before addition of galactose) by collecting 250 mL of log-phase culture (a total of $\sim 10^9$ cells), into the centrifuge bottle containing sodium azide and EDTA slants.
7. Shake bottles (like a martini) until the EDTA dislodges from the side of the bottle.
8. Spin bottles at 6000 rpm for 5 min at 4°C and discard supernatant.
9. Resuspend the pellet in 10 mL of cold sterile H_2O and transfer into prechilled 50 mL conical tube.
10. Spin down at 6000 rpm for 5 min.
11. Decant supernatant and freeze pellet in liquid nitrogen for approximately 2 min and store in −80C.
12. Take 50 mL sample for a 0-h time point in 50 mL conical tube (for CHEF analysis similar to Section 3.3).

a. Spin down and discard supernatant.
b. Rinse with 50 mL 50 m*M* EDTA pH 7.5 containing 0.5 mL 10% sodium azide.
c. Spin down at 3000 rpm at 4°C. Decant supernatant.
d. Store in −80°C until use.

13. Add 60 mL of 20% galactose to the remaining culture from step 6 (to a working concentration of 2%).
14. Three hours after addition of galactose, add 1.2 mL nocodazole to the final concentration of 0.015 mg/mL.
15. At 7 h after addition of galactose, collect 250 mL of culture following steps 5–11 to collect the 2D samples. These samples can be used for further 2D analysis.

5.1.4 Notes

1. In parallel with the samples for 2D, samples are always collected for CHEF to detect the presence of final BIR product (see Section 3.3).
2. Nocodazole prevents cell progression into the cell cycle after BIR repair.
3. The samples are always placed on ice (until mentioned otherwise).

5.2 DNA Cross-linking

5.2.1 Equipment

- DNA cross-linker wavelength 365 nm (Stratagene Stratalinker UV 2400 or similar model)
- Tabletop centrifuge for spinning 50-mL conical tubes (Beckman Coulter Allegra 25R or similar model)
- Petri dishes (100 × 15 mm)
- 10-mL serological pipettes
- 50-mL conical tubes
- 15-mL conical tubes

5.2.2 Solutions and Reagents

- Cross-link wash buffer: 50 m*M* Tris and 50 m*M* EDTA in ddH_2O
- Psoralen solution: 4.8 mg of psoralen (Sigma-Aldrich, #T6137-500MG) in 9 mL of 200 proof ethanol (Deacon, #64-17-5)
- Cross-link buffer: 2.8 mL of psoralen solution in 11.2 mL of cross-link wash buffer

5.2.3 Procedure

1. Thaw the frozen pellet on ice for ~1 h (keep cold throughout the protocol).

2. While thawing the samples make cross-link wash buffer:
   Dissolve 4.8 mg of psoralen in 9 mL of 200 proof ethanol by constant vortexing. Psoralen may take up to 1 h to dissolve.
3. Make fresh cross-link buffer by adding 2.8 mL of psoralen solution to 11.2 mL of cross-link wash buffer and resuspend the pellet by vortexing with this buffer.
4. Prewarm cross-linking bulbs (monochromatic white light at wavelength 365 nm) for 2 min.
5. Spread a thin layer of the suspended pellet in cross-link buffer onto six Petri dishes.
6. Place plates in the UV cross-linker at wavelength 365 nm and cross-link for 4 min. After the 4 min swirl the cell suspension on the plate to expose more cells. Repeat the cross-linking step twice more at 3 min cycles (total of 10 min).
7. Wash Petri dishes with chilled cross-linking wash buffer and add to conical tube and keep on ice.
8. Repeat wash two times till the Petri dish is clear of cells.
9. Spin down 6000 rpm for 4 min and discard supernatant.

5.2.4 Notes

1. Psoralen solution (ethanol and psoralen) should be wrapped in a paper towel/aluminum foil due to psoralen's light sensitivity and vortexed for approximately 1 h to allow the psoralen to dissolve into solution.
2. Note, normally cross-link buffer will form a white precipitate instead of being transparent.
3. Washing of Petri dishes done with serological pipette and automatic pipetter to collect as many cells as possible.

5.3 Extraction of DNA

5.3.1 Equipment

- Tabletop vortex
- Phase-contrast microscope (Leica DMRBE trinocular research microscope or similar model)
- 37°C water bath
- 0.5 mm diameter acid-washed glass beads (BioSpec products, #11079105)
- 5.75 in. disposable Pasteur pipettes (Fisher, #13-678-20A)
- Round bottom 50 mL centrifuge tube (Thermo scientific, #3119-0050)
- Opti-seal tubes (Beckman Coulter, #361627)

- 4°C ultracentrifuge with 50 mL round bottom tube rotor tabletop (Beckman model J2-21 centrifuge or similar model)

5.3.2 Solutions and Reagents

- NIB solution: 17% glycerol (RPI, #56-81-5), 50 m*M* MOPS (RPI, #1132-61-2), 150 m*M* Potassium acetate (RPI, #127-08-2), 2 m*M* Magnesium chloride (Fisher Scientific, #7791-18-6), 500 μ*M* Spermidine (Sigma-Aldrich, #S2626-5G), 150 μ*M* Spermine (Sigma-Aldrich, #S3256-1G), filter sterilize
- TEN solution pH 8.0 : 50 m*M* Tris base, 50 m*M* EDTA, 100 m*M* NaCl, filter sterilize
- 10% Sarkosyl (Fisher, #137-16-6)
- Proteinase K (RPI, #39450-01-6)
- Cesium chloride (Fisher, #7647-17-8)
- Hoechst 33258 dye 5 mg/mL (Life technologies, #23491-44-3)

5.3.3 Procedure

1. Resuspend pellet from Section 5.2.3 (step 9) with 4 mL NIB solution (keep on ice).
2. Add 0.5 mm acid-washed glass beads (~1/3 the volume of the pellet).
3. Vortex for 1 min and then place on ice for 1 min.
4. Repeat for 20 cycles then using phase-contrast microscope look for the ratio of spheroplasts to unbroken yeast cells.
5. Repeat the vortexing cycles until ~80% of cells are spheroplasts (look like "ghost" cells).
6. Carefully transfer the cells from glass beads to a sterile 50-mL round bottom centrifugation tube using Pasteur pipettes.
7. Wash beads at least twice with 4 mL NIB solution and transfer into the round bottom centrifugation tube.
8. Spin cells suspended in NIB at 10,000 rpm for 25 min at 4°C.
9. Save pellet and discard the supernatant.
10. Gently resuspend the pellet using 1.5 mL of TEN solution. Gently stir pellet with pipette tip.
11. Add 250 μL of 10% Sarkosyl to the suspended cells.
12. Add a scoop of proteinase K (~0.005 g/sample).
13. Gently swirl the mixture to ensure Sarkosyl and proteinase K are evenly mixed.
14. Incubate in 37°C incubator for 1 h (without shaking).
15. Centrifuge at 4°C 5000 rpm for 5 min.

16. Decant the supernatant into a 15-mL conical tube and make the volume up to 2.5 mL using TEN solution.
17. For 3.3 mL opti-seal tubes, weigh out 3.7 g of cesium chloride and gently rock tube back and forth until cesium chloride is dissolved.
18. Add 70 μL of Hoechst 33,258 dye 5 mg/mL and gently rock conical tube once.

5.3.4 Notes

1. Steps (1–7) should be kept on ice.
2. After step 6 do not vortex any samples. Vortexing can shear the DNA.
3. After second step of bead wash the solution should be clear. If it is not wash one more time with 4 mL of NIB.
4. Resuspension of pellet into TEN should be done carefully by gently stirring so as not to break the spheroplasts. Do not pipette up and down or vortex.
5. For opti-seal tube of size 4.5 mL make up volume of TEN in step 16 to 4 mL and add 4.3 g cesium chloride at step 17. Finally, add 125 μL of Hoechst stain 33258 dye in step 18.

5.4 DNA Separation and Collection

5.4.1 Equipment

- Analytical balance
- 1.5-mL microcentrifuge tubes
- 5.75 in. disposable Pasteur pipettes (Fisher, #13-678-20A)
- 3.5 mL opti-seal tubes (Beckman Coulter, #361627)
- 100,000 rpm ultracentrifuge at 20°C (Beckman Coulter Optima Max-E tabletop Ultracentrifuge 100,000 rpm) with TLN100 rotor
- Benchtop UV imager
- Ring stand tube clamp
- 18-gauge needle (BD, #305185)
- 3 mL syringe (BD, #309657)
- 4°C refrigerator

5.4.2 Solutions and Reagents

- MilliQ water
- Isopropanol and MilliQ water mixture: 5:1 isopropanol to water mixture
- 70% ethanol in MilliQ water

5.4.3 Procedure

1. Transfer the TEN and cesium chloride solution to a 3.3-mL opti-seal tube and seal the tube with the plug.
2. Place balanced tubes in rotor.
3. Place gold cap spacer on top of tubes in rotor.
4. Grease the threads of the rotor screws and tighten the rotor screws to 120 lb–ft using a torque wrench.
5. Spin samples at 100,000 rpm at 20°C for 12 h.
6. Carefully remove the opti-seal tube and carry the tube to the dark room as smoothly as possible. (Too much shaking will mix DNA layers).
7. Remove opti-seal plug and place the opti-seal tube in a ring stand tube holder and set up the UV imager to visualize DNA stained by Hoechst dye.
8. White gooey polysaccharide layer (probably trehalose) should be visible without UV. Mark the location of this layer with a permanent marker.
9. While using face shield for protection, turn on UV imager where three bands should fluoresce. The bottom should be the visible gooey polysaccharide band, the middle is genomic DNA, and the top band is mitochondrial DNA.
10. Using an 18-gauge needle and syringe with the needle hole facing upward, pierce through the side of the tube at the base of the chromosomal DNA layer.
11. Carefully remove the chromosomal DNA layer using the needle (do this quickly and smoothly).
12. Remove needle from the syringe and transfer the DNA to a 15-mL conical tube.
13. Chill mixture containing 5:1 isopropanol to H_2O and 70% ethanol in ice bucket.
14. Add equal volume of 5:1 isopropanol mixture to the collected DNA sample. For instance, add 100 μL of 5:1 isopropanol mixture to 100 μL of extracted DNA.
15. Swirl the tube with a flick of the wrist then incubate on ice for 2 min.
16. Remove and discard the top phase of the mixture.
17. Repeat steps 14 through 16 two more times. (On the last wash remove as much of top layer as possible).
18. Slowly add 2× the volume of chilled 70% ethanol to the tube and mix carefully by a single quick swirling motion. (DNA will come out of solution as a white precipitate in a middle phase).
19. Melt the tip of a Pasteur pipette to seal the end into a little ball.

20. DNA can be removed from the solution by touching the cooled melted tip to the DNA clot (the DNA will stick to the melted tip), and then DNA can be spooled on the stick.
21. Dip the DNA clot in 70% ethanol several times and air dry DNA pellet by placing the Pasteur pipette in a rack DNA end up.
22. Place the tip of Pasteur pipette with the DNA clot attached into a microcentrifuge tube filled with 100 μL MilliQ water.
23. Keep the tip of Pasteur pipette in the tube for half an hour to allow DNA to dislodge from the tip.
24. Allow DNA clot to dissolve at 4°C in the microcentrifuge tube overnight.

5.4.4 Notes

1. Opti-seal tube must be balanced to each other before the centrifugation as accurately as possible.
2. Before inserting needle, be sure to move plunger up and down for smoother pulling of the plunger.
3. Needle should be inserted at the base of the chromosomal DNA band with the needle hole up to prevent the clogging of the needle.
4. Needle is removed, and DNA is transferred into 15-mL conical tube. Removal of needle is critical to avoid breaking of DNA structures.
5. Residual DNA can be removed from the needle by one full push of the syringe through the needle.
6. 5:1 isopropanol to MilliQ water is added to DNA to remove the Hoechst dye.
7. Care should be taken throughout all the steps so as to not break DNA structures that are formed.

5.5 DNA Digestion and Gel Separation in the First Dimension

5.5.1 Equipment

- 37°C water bath
- −20°C freezer
- 1.5-mL microcentrifuge tubes
- Tabletop microcentrifuge (LabNet International, Inc. #Spectrafuge 24D or similar model)
- Electrophoresis chamber (Thermo Scientific Owl A1 large gel system #09528100 or similar model)
- Electrophoresis power pack (Bio-Rad PowerPac or similar model)

5.5.2 Solutions and Reagents

- MilliQ water
- NEBuffer 3.1 (NEB, #B7203S)
- *Pst*I 10 U/μL (NEB, #R0140S)
- Agarose (Sigma-Aldrich, #A0169)
- 70% ethanol in water
- 200 proof ethanol (Deacon, #64-17-5)
- 3 *M* Sodium acetate pH 5.5 (Fisher, #S210-500): Add 24.6 g sodium acetate in 100 mL of MilliQ water, and pH using glacial acetic acid (Fisher, #64-19-7)
- 5× TBE: 0.45 *M* Tris base, 0.45 *M* boric acid (RPI, #10043-35-3), 0.02 *M* EDTA
- Promega blue orange 6× loading dye (Promega, #G1881)

5.5.3 Procedure

1. Prepare enzyme digestion mix using the correct restriction enzyme. The choice of enzyme depends on specific chromosomal location where BIR is analyzed. *Pst*I restriction enzyme is used here.
   a. 100 μL dissolved DNA
   b. 256 μL ddH_2O
   c. 40 μL NEBuffer 3.1
   d. 4 μL *Pst*I
2. Incubate 16 h at 37°C.
3. Add 40 μL of 3 *M* sodium acetate followed by 1 mL of cold 200 proof ethanol to the digestion mix and invert once.
4. Incubate on ice for 30 min.
5. Spin in tabletop microcentrifuge at 13,000 rpm for 15 min.
6. Discard supernatant.
7. Wash DNA pellet with 70% EtOH.
8. Spin at 13,000 rpm for 5 min.
9. Carefully discard supernatant.
10. Dry the pellet on the bench with the tubes open until dry.
11. Add 6 μL 1× TBE and 6 μL 6× loading dye to the sample (do not vortex).
12. Allow the sample to dissolve for at least an hour.
13. In 300 mL of 1× TBE make a 0.4% agarose gel (no ethidium bromide) and pour with the tiny well comb (<2 mm wide comb).
14. Load samples carefully so as to not overflow the well and run at 55 V for 17 h (for optimal separation of 5 and 10 kb DNA fragments).

5.5.4 Notes

1. Restriction enzymes were chosen based on the restriction map of the area where BIR intermediates are expected to form. The fragment size is usually chosen to be between 5 and 6 kb in size.
2. The well for the first dimension is to be as small as possible. This will allow the DNA to migrate through the gel in as thin a band as possible. The well has to be large enough so that the sample will not overflow.
3. The first dimension is run without ethidium bromide. Running with ethidium bromide may lead to issues in DNA migration.

5.6 Two-Dimensional Gel Electrophoresis: Second Dimension

5.6.1 Equipment

- Large electrophoresis box with buffer pump (Thermo Scientific Owl A3 electrophoresis chamber with Bio-Rad variable speed pump or similar model)
- Benchtop UV transilluminator
- Razor blades

5.6.2 Solutions and Reagents

- 5× TBE: 0.45 *M* Tris base, 0.45 *M* boric acid, 0.02 *M* EDTA
- Agarose (Sigma-Aldrich, #A0169)
- Ethidium bromide 10 mg/mL (Fisher, #15-585-011)

5.6.3 Procedure

1. After running the first dimension, stain gel with ethidium bromide (3 μL per 100 mL of buffer).
2. Using a transilluminator to excise the separated DNA from the gel in a thin strip. Excise the gel containing all DNA molecules larger than 5 kb.
3. Place gel strip with DNA horizontally in the large 2D gel tray.
4. Make 5 L of 1× TBE.
5. Using 1× TBE make 1 L of 1.2% agarose gel.
6. Cool remaining 4 L of 1× TBE to 4°C and add 120 μL of ethidium bromide.
7. When agarose cools to 65°C, add 30 μL of ethidium bromide, mix, and pour into gel box around the first-dimension strips of gel.
8. When gel solidifies, place gel into gel box with prechilled 1× TBE.
9. Run gel for 12 h at 340 V while in the cold room while pumping buffer through the gel box.

5.6.4 Notes

1. Fragments of size 5 kb (the region flanked by two *Pst*I restriction sites) and larger fragments (until the well) are excised. Other areas of chromosomes can be studied using different restriction enzymes. Excision of this band would have to be adjusted accordingly for different restriction fragment sizes.
2. Two gel slices will fit at the top of the gel box and two more will fit at halfway down the gel box for a total of four samples.
3. Do not pour the 1.2% agarose gel until it is 65°C so as to not denature the DNA structures within the first-dimension gel slice.

5.7 Capillary Transfer and Hybridization

5.7.1 Equipment

- Glass baking dish
- Glass plate
- Whatman paper 45 × 45 cm
- Scott C-fold paper towels (Fisher, #06-666-32B)
- DNA cross-linker wavelength 254 nm (Stratagene Stratalinker UV 2400 or similar model)
- 20 × 25 cm GE Phosphor screen (#GE28-9564-75)
- Phosphorimager (GE Typhoon FLA 7000 cat #28983618 or similar model)
- Geiger counter (Ludlum 5000 or similar model)

5.7.2 Solutions and Reagents

- 0.2 *M* HCl: 21 mL of concentrated HCl (Fisher, #A144-500) into 979 mL MilliQ water.
- Alkaline solution: 20 g NaOH (RPI, #1310-73-2), 87.7 g NaCl (RPI, #7647-14-5) in 1 L of MilliQ water.
- Neutralization solution: 87.7 g NaCl (RPI, #7647-14-5), 121.4 g Tris base (RPI, #77-86-1) in 1 L of MilliQ water, pH 7.
- 20× SSC pH 7: 175.3 g NaCl, 88.2 g sodium citrate, 900 mL MilliQ water.
- MilliQ water.
- 1× TBE: Dilute 5× TBE by adding 200 mL 5× TBE in 1 L of MilliQ water.
- 1% agarose gel: 1 g of agarose (Sigma-Aldrich, #A0169) in 100 mL of 1× TBE.

- QIAGEN gel extraction kit: (QIAGEN, #28704).
- Amersham Hybond-N+ charged transfer membrane (GE Healthcare, #RPN303B).
- PerfectHyb Plus hybridization buffer (Sigma-Aldrich, #H7033).
- Prime-it RmT Random primer kit (Agilent, #300392).
- P^{32} radiolabeled dCTP (Perkin Elmer, #BLU513Z250).
- Low stringency wash buffer: 80 mL 20× SSC, 4 mL 20% SDS (Fisher, #BP166-5) made up to 800 mL MilliQ water.

5.7.3 Procedure

1. Cut the gel in half (separating top from bottom) and image using UV transilluminator with camera.
2. Submerge gel in 0.25 *M* HCl for 30 min and keep with constant gentle shaking.
3. Submerge gel in alkaline solution and keep for 30 min with constant gentle shaking.
4. Submerge gel in neutralization solution and keep for 30 min with constant gentle shaking.
5. Transfer via capillary transfer methods described in Sambrook and Russell (2006). Refer to Section 3.4.3 for further details.
6. Cross-link membrane in UV cross-linker at 254 nm wavelength light.
7. DNA probe specific to the chromosome III position located 24 kb centromere distal from *MAT* is made using AM1003 genomic DNA as a template and the following primers:
   FP: 5′-AAGGGAGCAAACTGTCTGTC-3′
   RP: 5′-GGTACGGTCTACAGGCAAGG-3′
8. Run the PCR probe fragment on a 1% agarose 1× TBE gel and excise the fragment.
9. Using the QIAGEN Gel extraction kit extract the DNA for making the probe.
10. Resuspend and dissolve the contents of one tube of the RmT Random primer kit in 37 μL of MilliQ water and transfer to a 1.5-mL microcentrifuge tube.
11. Add 5 μL (~70 ng) of the gel-extracted PCR probe fragment.
12. Incubate RmT Random primer kit and probe fragment mixture at 100°C for 5 min. Be sure to add a microcentrifuge tube locker to keep the lid closed.
13. After incubation period, place reaction mixture on ice for 1 min.

14. Add 5 μL of P^{32}-dCTP to the cooled reaction mixture, and add 3 μL of magenta polymerase from the RmT Random primer kit, and incubate at 37°C dry bath for 10 min.
15. Add 2 μL of stop mix from the RmT Random primer kit to stop the reaction.
16. Add 200 μL of MilliQ water to the reaction mixture and denature at 100°C dry bath for 10 min and snap cool on ice for 1 min.
17. Prehybridize the cross-linked membrane from step 6 in 25 mL of PerfectHyb Plus hybridization buffer at 65°C for at least 10 min.
18. Hybridize the membrane at 65°C using 25 mL of the PerfectHyb Plus hybridization buffer for 24 h.
19. Wash the membrane using 65°C low stringency wash buffer in an open tray. Gently rock the wash buffer and membrane for 10 min. Discard waste under the guidelines of your radiation safety officer.
20. Repeat wash two more times or until background is washed away, as being checked using a Geiger counter.
21. Using 20 × 25 cm GE Phosphor screen, expose membrane for 5 days and image on a Typhoon Phosphorimager FLA7000.

5.7.4 Notes

1. Gel is cut in half to fit in the glass baking dish. If there is a glass baking dish large enough then cutting may not be necessary.
2. Machine pressure transfer could be an alternative to capillary DNA transfer to membranes.

6. ALTERNATIVE METHOD OF GENOMIC DNA PREPARATION FOR 2D ANALYSIS BY SPHEROPLASTING CELLS

6.1 Plug Preparation

6.1.1 Equipment

- Hemocytometer (Hausser scientific, #3100)
- Plug mold: (Bio-Rad, #1703713)
- Phase-contrast microscope (Leica DMRBE trinocular research microscope or similar model)
- 30°C incubator
- Tube mixer (Fisher, #120529060)

6.1.2 Solutions and Reagents

- 1× TE: 1 *M* Tris (RPI, #77-86-1) and 0.1 *M* EDTA (Fisher Bioreagents, #BP120-1), pH 7.5
- Digestion buffer: 1 *M* sorbitol (RPI, #50-70-4) and 0.1 *M* EDTA pH 7.5
- Lyticase 5 mg/mL (Sigma-Aldrich, #L2524)
- TEN solution pH 8: 50 m*M* Tris base, 50 m*M* EDTA, 100 m*M* NaCl (RPI, #7647-14-5)
- Low melting agarose (Fisher, #9012-36-6)
- Incubation buffer: 10 m*M* Tris, 50 m*M* EDTA, 0.5% Sarkosyl (Fisher Bioreagents, #137-16-6), 0.005 g/mL proteinase K (RPI, #39450-01-6) added fresh

6.1.3 Procedure

1. Spin down cells at 3000 rpm for 5 min at 4°C following UV cross-linking by Psoralen (see Section 5.2 for details) and keep on ice.
2. Resuspend cells in 1× TE and using a hemocytometer count the number of cells.
3. Determine cell concentration and add ~2.5×10^9 cells per plug.
4. Take cells and spin down to remove excess 1× TE buffer. Resuspend cells in 500 μL digestion buffer.
5. Add 10 μL of lyticase (5 mg/mL) to the cells and incubate at 37°C for 1 h. Note that lyticase stock is made in digestion buffer.
6. Use phase-contrast microscope to check for spheroplasts. Cells must be approximately 80% ghosts. If cells remain undisrupted, add another 15 μL of lyticase and incubate for another hour.
7. Following lyticase digestion, spin down the cells at 13,000 rpm for 1 min.
8. Wash cells in 750 μL of TEN solution.
9. Repeat the wash with TEN solution two times.
10. Place cells in a tube mixer to loosen the cells (do not vortex).
11. Take ~50 μL of loosened cells and place it in a 50°C water bath to equilibrate (do not leave it for too long or it will degrade the spheroplasts).
12. Add 30 μL of 2% molten low melting agarose to the cells and mix the cells using the pipette approximately three times.
13. Add this mixture (cell + agarose) to the mold and allow it to solidify for 30 min.

14. After plugs solidify, transfer the plugs into a 50-mL conical tube containing 0.005 mg Proteinase K per 10 mL of incubation buffer. Incubation buffer and incubate at 30°C overnight.

6.2 1D Gel Electrophoresis

6.2.1 Equipment

- 50-mL conical tubes
- Electrophoresis chamber (Thermo Scientific Owl A1 large gel system #09528100 or similar model)
- Electrophoresis power pack (Bio-Rad PowerPac or similar model)

6.2.2 Solutions and Reagents

- 1× TE: 1 *M* Tris (RPI, #77-86-1) and 0.1 *M* EDTA (Fisher Bioreagents, #BP120-1), pH 7.5
- Isopropanol (Fisher, #67-63-0)
- 100 μ*M* Phenylmethane sulfonyl fluoride (PMSF): 17.4 g of PMSF (RPI, #329-98-6) in 100 mL isopropanol
- MilliQ water
- Buffer O restriction buffer: (Thermo Fisher Scientific, #B05)

6.2.3 Procedure

1. Cut the plugs to desired size (approximately 2–3 mm in width).
2. Place the plug in 1× TE (30 mL) and leave it at 25°C for 16 h.
3. Remove 1× TE and replace 30 mL of fresh 1 × TE and incubate for 1 h at 25°C.
4. Discard 1× TE and replace with fresh 1 × TE containing100m*M* PMSF and incubate for 1 h at 25°C.
5. Wash for 1 h with MilliQ water.
6. Transfer plug from 50-mL conical tube to 2 mL sterile microcentrifuge tubes.
7. Submerge plugs in 2× restriction buffer (buffer O) and incubate for 1 h at 25°C.
8. Remove 2× restriction buffer and add 1× restriction buffer with 10 μL of restriction enzyme. Incubate at 37°C for 16 h.
9. After restriction digestion, equilibrate the plugs in 1× TBE for 1 h at 25°C.
10. Insert the plug into the well of a set 1% agarose gel and seal the well with the molten agarose to prevent plugs from shifting during electrophoresis.

11. Run the 1D and 2D as previously described earlier in Section 5.5.3 through Section 5.6.3.

6.2.4 Notes

- Enzymes using the Orange buffer from Thermo Fisher Scientific are the most compatible with in-plug digestion.

7. CONCLUSIONS

The experimental system and methods described in this review allow to study BIR initiated by a site-specific HO endonuclease in yeast. This system is convenient for identification of proteins that are involved in BIR and for investigation of their role in BIR. This system can be used to determine the frequency and efficiency of BIR by using genetic and physical (molecular biology) methods, for exploring the mechanism of BIR and its kinetics. In addition, this system allows to investigate mutagenesis and chromosomal rearrangements associated with BIR.

ACKNOWLEDGMENT

The research in Malkova lab is supported by R01GM084242 grant from NIGMS.

REFERENCES

Bhowmick, R., Minocherhomji, S., & Hickson, I. D. (2016). *RAD52* facilitates mitotic DNA synthesis following replication stress. *Molecular Cell*, *64*(6), 1117–1126. https://doi.org/10.1016/j.molcel.2016.10.037.

Brewer, B. J., & Fangman, W. L. (1987). The localization of replication origins on ARS plasmids in *S. cerevisiae*. *Cell*, *51*(3), 463–471.

Cho, N. W., Dilley, R. L., Lampson, M. A., & Greenberg, R. A. (2014). Interchromosomal homology searches drive directional ALT telomere movement and synapsis. *Cell*, *159*(1), 108–121. https://doi.org/10.1016/j.cell.2014.08.030.

Church, G. M., & Gilbert, W. (1984). Genomic sequencing. *Proceedings of the National Academy of Sciences of the United States of America*, *81*(7), 1991–1995.

Costantino, L., Sotiriou, S. K., Rantala, J. K., Magin, S., Mladenov, E., Helleday, T., et al. (2014). Break-induced replication repair of damaged forks induces genomic duplications in human cells. *Science*, *343*(6166), 88–91. https://doi.org/10.1126/science.1243211.

Davis, A. P., & Symington, L. S. (2004). *RAD51*-dependent break-induced replication in yeast. *Molecular and Cellular Biology*, *24*(6), 2344–2351.

Deem, A., Barker, K., Vanhulle, K., Downing, B., Vayl, A., & Malkova, A. (2008). Defective break-induced replication leads to half-crossovers in *Saccharomyces cerevisiae*. *Genetics*, *179*(4), 1845–1860. https://doi.org/10.1534/genetics.108.087940.

Deem, A., Keszthelyi, A., Blackgrove, T., Vayl, A., Coffey, B., Mathur, R., et al. (2011). Break-induced replication is highly inaccurate. *PLoS Biology*, *9*(2), e1000594. https://doi.org/10.1371/journal.pbio.1000594.

Dilley, R. L., Verma, P., Cho, N. W., Winters, H. D., Wondisford, A. R., & Greenberg, R. A. (2016). Break-induced telomere synthesis underlies alternative telomere maintenance. *Nature*, *539*(7627), 54–58. https://doi.org/10.1038/nature20099.

Hashimoto, Y., & Costanzo, V. (2011). Studying DNA replication fork stability in *Xenopus* egg extract. *Methods in Molecular Biology*, *745*, 437–445. https://doi.org/10.1007/978-1-61779-129-1_25.

Lydeard, J. R., Jain, S., Yamaguchi, M., & Haber, J. E. (2007). Break-induced replication and telomerase-independent telomere maintenance require Pol32. *Nature*, *448*(7155), 820–823. https://doi.org/10.1038/nature06047.

Malkova, A., Ivanov, E. L., & Haber, J. E. (1996). Double-strand break repair in the absence of *RAD51* in yeast: A possible role for break-induced DNA replication. *Proceedings of the National Academy of Sciences of the United States of America*, *93*(14), 7131–7136.

Malkova, A., Naylor, M. L., Yamaguchi, M., Ira, G., & Haber, J. E. (2005). *RAD51*-dependent break-induced replication differs in kinetics and checkpoint responses from *RAD51*-mediated gene conversion. *Molecular and Cellular Biology*, *25*(3), 933–944. https://doi.org/10.1128/MCB.25.3.933-944.2005.

Mehta, A., & Haber, J. E. (2014). Sources of DNA double-strand breaks and models of recombinational DNA repair. *Cold Spring Harbor Perspectives in Biology*, *6*(9). . a016428https://doi.org/10.1101/cshperspect.a016428.

Minocherhomji, S., Ying, S., Bjerregaard, V. A., Bursomanno, S., Aleliunaite, A., Wu, W., et al. (2015). Replication stress activates DNA repair synthesis in mitosis. *Nature*, *528*(7581), 286–290. https://doi.org/10.1038/nature16139.

Morrow, D. M., Connelly, C., & Hieter, P. (1997). "Break copy" duplication: A model for chromosome fragment formation in *Saccharomyces cerevisiae*. *Genetics*, *147*(2), 371–382.

Rodgers, K., & McVey, M. (2016). Error-prone repair of DNA double-strand breaks. *Journal of Cellular Physiology*, *231*(1), 15–24. https://doi.org/10.1002/jcp.25053.

Saini, N., Ramakrishnan, S., Elango, R., Ayyar, S., Zhang, Y., Deem, A., et al. (2013). Migrating bubble during break-induced replication drives conservative DNA synthesis. *Nature*, *502*(7471), 389–392. https://doi.org/10.1038/nature12584.

Sakofsky, C. J., Ayyar, S., Deem, A. K., Chung, W. H., Ira, G., & Malkova, A. (2015). Translesion polymerases drive microhomology-mediated break-induced replication leading to complex chromosomal rearrangements. *Molecular Cell*, *60*(6), 860–872. https://doi.org/10.1016/j.molcel.2015.10.041.

Sakofsky, C. J., & Malkova, A. (2017). Break induced replication in eukaryotes: Mechanisms, functions, and consequences. *Critical Reviews in Biochemistry and Molecular Biology*, *52*(4), 395–413. https://doi.org/10.1080/10409238.2017.1314444.

Sakofsky, C. J., Roberts, S. A., Malc, E., Mieczkowski, P. A., Resnick, M. A., Gordenin, D. A., et al. (2014). Break-induced replication is a source of mutation clusters underlying kataegis. *Cell Reports*, 7(5), 1640–1648. https://doi.org/10.1016/j.celrep.2014.04.053.

Sambrook, J., & Russell, D. W. (2006). Southern blotting: Capillary transfer of DNA to membranes. *CSH Protocols*, *2006*(1). https://doi.org/10.1101/pdb.prot4040.

Shcherbakova, P. V., & Pavlov, Y. I. (1996). $3' \rightarrow 5'$ Exonucleases of DNA polymerases epsilon and delta correct base analog induced DNA replication errors on opposite DNA strands in *Saccharomyces cerevisiae*. *Genetics*, *142*(3), 717–726.

Smith, C. E., Lam, A. F., & Symington, L. S. (2009). Aberrant double-strand break repair resulting in half crossovers in mutants defective for Rad51 or the DNA polymerase delta complex. *Molecular and Cellular Biology*, *29*(6), 1432–1441. https://doi.org/10.1128/MCB.01469-08.

Sotiriou, S. K., Kamileri, I., Lugli, N., Evangelou, K., Da-Re, C., Huber, F., et al. (2016). Mammalian RAD52 functions in break-induced replication repair of collapsed DNA replication forks. *Molecular Cell, 64*(6), 1127–1134. https://doi.org/10.1016/j.molcel.2016.10.038.

Teng, S. C., Chang, J., McCowan, B., & Zakian, V. A. (2000). Telomerase-independent lengthening of yeast telomeres occurs by an abrupt Rad50p-dependent, Rif-inhibited recombinational process. *Molecular Cell, 6*(4), 947–952.

Teng, S. C., & Zakian, V. A. (1999). Telomere-telomere recombination is an efficient bypass pathway for telomere maintenance in *Saccharomyces cerevisiae*. *Molecular and Cellular Biology, 19*(12), 8083–8093.

Tran, H. T., Keen, J. D., Kricker, M., Resnick, M. A., & Gordenin, D. A. (1997). Hypermutability of homonucleotide runs in mismatch repair and DNA polymerase proofreading yeast mutants. *Molecular and Cellular Biology, 17*(5), 2859–2865.

Vasan, S., Deem, A., Ramakrishnan, S., Argueso, J. L., & Malkova, A. (2014). Cascades of genetic instability resulting from compromised break-induced replication. *PLoS Genetics, 10*(2), e1004119. https://doi.org/10.1371/journal.pgen.1004119.

FURTHER READING

Donnianni, R. A., & Symington, L. S. (2013). Break-induced replication occurs by conservative DNA synthesis. *Proceedings of the National Academy of Sciences of the United States of America, 110*(33), 13475–13480. https://doi.org/10.1073/pnas.1309800110.

Wilson, M. A., Kwon, Y., Xu, Y., Chung, W. H., Chi, P., Niu, H., et al. (2013). Pif1 helicase and Poldelta promote recombination-coupled DNA synthesis via bubble migration. *Nature, 502*(7471), 393–396. https://doi.org/10.1038/nature12585.

CHAPTER TEN

Targeting Allostery with Avatars to Design Inhibitors Assessed by Cell Activity: Dissecting MRE11 Endo- and Exonuclease Activities

Davide Moiani*, Daryl A. Ronato[†,‡], Chris A. Brosey*, Andrew S. Arvai[§], Aleem Syed*, Jean-Yves Masson[†,‡], Elena Petricci[¶,1], John A. Tainer[*,‖,1]

*The University of Texas, M.D. Anderson Cancer Center, Houston, TX, United States
[†]Genome Stability Laboratory, CHU de Québec Research Center, Québec City, QC, Canada
[‡]Laval University Cancer Research Center, Québec City, QC, Canada
[§]The Scripps Research Institute, La Jolla, CA, United States
[¶]University of Siena, Siena, Italy
[‖]Lawrence Berkeley National Laboratory, Berkeley, CA, United States
[1]Corresponding authors: e-mail address: petricci@unisi.it; jatainer@gmail.com

Contents

Abstract

For inhibitor design, as in most research, the best system is question dependent. We suggest structurally defined allostery to design specific inhibitors that target regions beyond active sites. We choose systems allowing efficient quality structures with conformational changes as optimal for structure-based design to optimize inhibitors.

Methods in Enzymology, Volume 601
ISSN 0076-6879
https://doi.org/10.1016/bs.mie.2017.11.030

We maintain that evolutionarily related targets logically provide molecular avatars, where this Sanskrit term for descent includes ideas of functional relationships and of being a physical embodiment of the target's essential features without requiring high sequence identity. Appropriate biochemical and cell assays provide quantitative measurements, and for biomedical impacts, any inhibitor's activity should be validated in human cells. Specificity is effectively shown empirically by testing if mutations blocking target activity remove cellular inhibitor impact. We propose this approach to be superior to experiments testing for lack of cross-reactivity among possible related enzymes, which is a challenging negative experiment. As an exemplary avatar system for protein and DNA allosteric conformational controls, we focus here on developing separation-of-function inhibitors for meiotic recombination 11 nuclease activities. This was achieved not by targeting the active site but rather by geometrically impacting loop motifs analogously to ribosome antibiotics. These loops are neighboring the dimer interface and active site act in sculpting dsDNA and ssDNA into catalytically competent complexes. One of our design constraints is to preserve DNA substrate binding to geometrically block competing enzymes and pathways from the damaged site. We validate our allosteric approach to controlling outcomes in human cells by reversing the radiation sensitivity and genomic instability in BRCA mutant cells.

1. INTRODUCTION

Allostery is much discussed, but very few drug compounds target allosteric sites despite the extremely successful ribosomal antibacterial drugs revealing the tremendous and under-utilized power and specificity of targeting allostery with inhibitors binding outside the active site (Wang et al., 2012). To successfully target allostery, one needs to understand functional conformations. In particular for enzymes, targeting allostery requires a knowledge of the communication between protein conformation and the active site that approaches atomic resolution. Furthermore in developing inhibitors, the optimization of leads is an expensive preclinical effort, so targeting allostery has partly been limited by practicalities. With the above points in mind, we here suggest an approach of crystallography combined with small-angle X-ray scattering (SAXS) on structurally feasible targets: this empirical method allows one to efficiently produce the necessary knowledge aided by recently developed analysis software (Lai et al., 2016; Schneidman-Duhovny, Hammel, Tainer, & Sali, 2016) and then proceed with structurally informed optimization. Here we outline our strategy for efficiently targeting allostery in human cells with atomic level information even when human protein structures of a target enzyme are unavailable. As an exemplary target that forms a biologically critical multifunctional complex, we describe the design and testing of allosteric inhibitors for the DNA repair nuclease termed meiotic recombination 11 (MRE11).

MRE11 is critical for genome stability during DNA replication and DNA repair. It is the fundamental core component of the MRE11, ABC ATPase RAD50, and phosphopeptide-binding Nijmegen breakage syndrome 1 (NBS1) protein Mre11–Rad50–Nbs1 (MRN) complex in humans (also known as MRN/X (Mre11–Rad50–Nbs1/Xrs2) in eukaryotes and MR (Mre11–Rad50) in archaea and SbcCD in bacteria (Fig. 1A) (Chahwan, Nakamura, Sivakumar, Russell, & Rhind, 2003; D'Amours & Jackson, 2002; Hopfner et al., 2000; Lafrance-Vanasse, Williams, & Tainer, 2015; Lamarche, Orazio, & Weitzman, 2010; Seeber et al., 2016; Stracker & Petrini, 2011; Williams, Lees-Miller, & Tainer, 2010; Williams & Tainer, 2007). Through the MRN complex, MRE11 interfaces with multiple DNA damage response pathways, including double-strand break (DSB) repair involving homologous recombination (HR) and nonhomologous end joining (NHEJ) (Acharya et al., 2008; Bennardo, Cheng, Huang, Stark, & Haber, 2008; Biehs et al., 2017; Bierne, Ehrlich, & Michel, 1997; Shibata et al., 2014) and replication fork processing to maintain genome stability (Fig. 1B) (Schlacher et al., 2011; Schlacher, Wu, & Jasin, 2012). In this context, the MRE11 catalytic domain provides structure-specific endo- and exonucleolytic activities to prepare DNA ends for annealing and end-joining repair (Buis et al., 2008; Krogh, Llorente, Lam, & Symington, 2005; Lewis et al., 2004; Limbo, Porter-Goff, Rhind, & Russell, 2011; Majka, Alford, Ausio, Finn, & McMurray, 2012; Paull & Gellert, 1998). In fact, MRN can gain access to occluded DNA ends by removing Ku or other DNA adducts via its Mre11-dependent nucleolytic reaction (Myler et al., 2017; Shibata et al., 2014). In humans, mutations in MRE11 can cause ataxia-telangiectasia-like disorder (ATLD), featuring cerebellar degeneration and variable cancer predisposition due to an alteration in the mechanical activity of the enzyme during DNA processing (Limbo et al., 2012).

The MRE11 complex is furthermore emblematic in terms of forming multiple distinct functional networks with different activities (Williams, Williams, & Tainer, 2007). DNA is susceptible to breaks leading to chromosome abnormalities associated with cancer wherever B-DNA is destabilized by sequence (Bacolla, Tainer, Vasquez, & Cooper, 2016) or opened during DNA processing (Lindahl, 1993). In eukaryotes, MRN/X interfaces with other protein partners to facilitate DNA processing at replication forks (Schlacher et al., 2011, 2012) and repair pathway selection

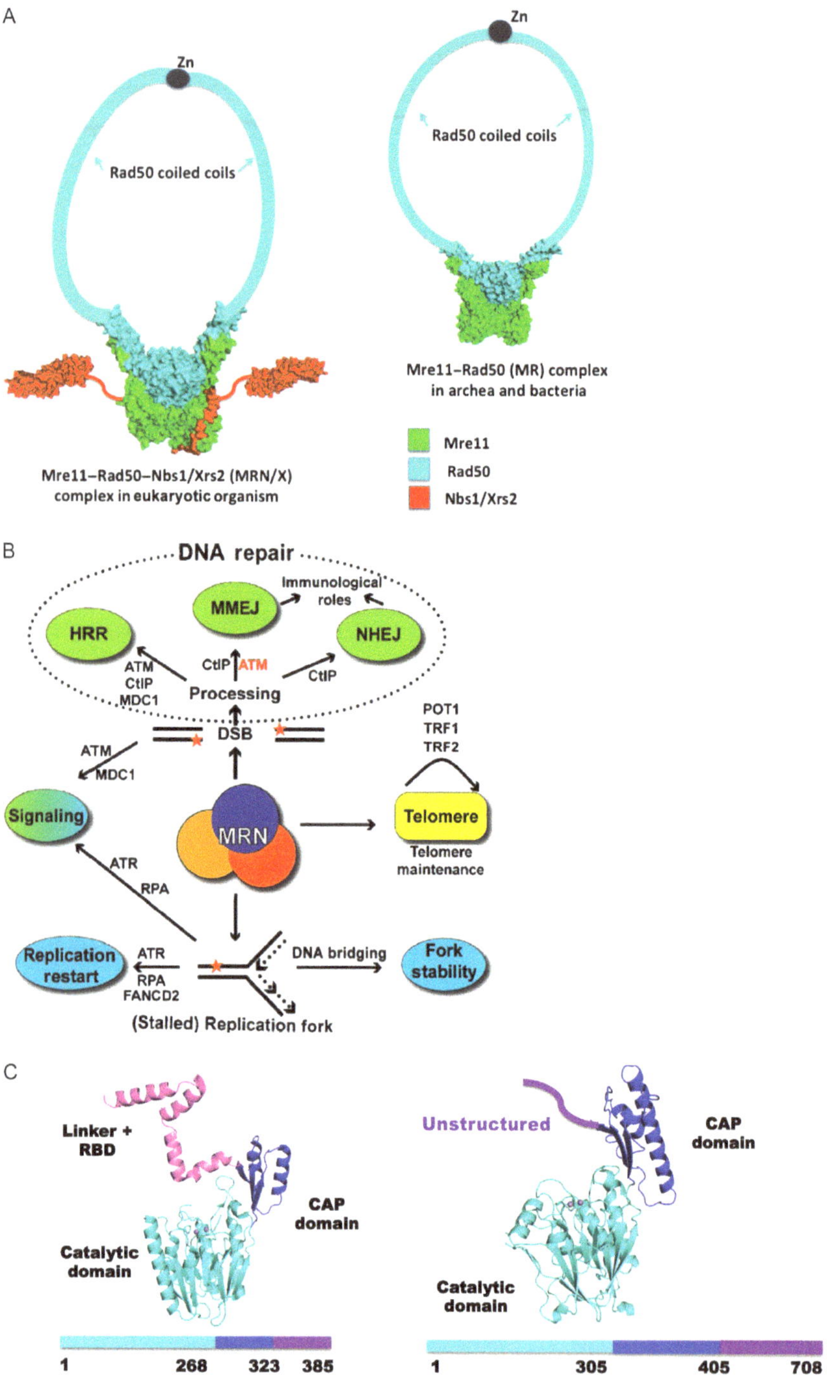

Fig. 1 Mre11–Rad50–Nbs1 (MRN) architecture, pathways, and nuclease core fold. (A) Architectural structures of MRN/X complex from eukaryotes (*left*) and MR vs bacterial and archaeal (*right*). (B) DNA repair pathway schematic for MR and MRN/X functions in

(Shibata et al., 2014). It cooperates with CtIP (Sae2 or ctp1 in budding or fission yeast, respectively) in replication and recombination pathways, where CtIP can promote the endonucleolytic activity of human MRN on protein-blocked ends (Andres et al., 2015; Deshpande, Lee, Arora, & Paull, 2016; Limbo et al., 2011; Makharashvili et al., 2014; Sartori et al., 2007). MRE11 nuclease can either protect or promote the degradation of stalled replication forks when either BRCA2 or RAD51 function is impaired (Schlacher et al., 2011, 2012). Additionally, the MRN complex can activate and signal via the ATM kinase, allowing ATM to regulate MRE11 nuclease to promote DNA microhomology-mediated end joining over degradation (Rahal et al., 2010). In general, nucleases such as MRE11 tend to be points where inhibitors can impact well-regulated pathway elements that may be sculpting the nucleic acid substrate (Tsutakawa, Lafrance-Vanasse, & Tainer, 2014).

The dimeric MRE11 nuclease works synergistically with Rad50 and Nbs1/Xrs2 to govern the multiple conformations of the MRN molecular machine to sense, bind, protect, and process damaged DNA (Crown et al., 2013; de Jager et al., 2001; Deshpande et al., 2014; Hopfner et al., 2002, 2001; Roset et al., 2014; Williams et al., 2008). Structure-based models of MR complexes (from both prokaryotes and eukaryotes) reveal that the Mre11 dimer is the core scaffold that provides the platform for the flexible operations of its partner Rad50 in tethering DNA ends and regulation of endo- and exonuclease processing of DNA through modulation of enzyme activity of Mre11 (Fig. 1A). The Mre11 subunit consists of a catalytic domain, a CAP domain, and a linked Rad50-binding motif in prokaryotes (Fig. 1C). In mammals, MRE11 additionally bears a glycine–arginine-rich (GAR) motif (Yu et al., 2012) that binds poly(ADP) ribose synthetized by PARP1 at DNA DSBs, therefore stimulating DNA damage signaling (Haince et al., 2008). From the back face of the MRE11 catalytic domain, its interactions with NBS1 optimize MRN's allosteric organization to perform the correct function based on the substrate and protein partners, which are dictated by cell cycle status (Fig. 1)

cell biology. (C) The TmMre11 (PDB ID: 3THO) fold (*ribbon diagram*) and sequence (*left*) with domain organization (catalytic domain with metal ion active site, *cyan*), CAP domain (*blue*), and linker plus Rad50-binding domain (RBD, *violet*). Human MRE11 fold and sequence with the domain organization from PDB ID 3T1I (*right*) with catalytic domain with metal ion active site (*cyan*), CAP domain (*blue*), and the unstructured C-terminus (*violet*) containing the RBD. (Human MRE11 has GAR domain, RBD, and C-terminus unstructured yet.)

(Crown et al., 2013; Lammens et al., 2011; Mockel, Lammens, Schele, & Hopfner, 2012; Rojowska et al., 2014; Schiller et al., 2012; Williams et al., 2009). Structures of prokaryotic MR further validate Mre11 as a blueprint for orienting the core MR complex (Fig. 1A).

In humans, the MRN complex binds DNA DSBs and initiates damage-induced signaling cascades via activation of the ataxia-telangiectasia-mutated (ATM) (Lee & Paull, 2005; You, Chahwan, Bailis, Hunter, & Russell, 2005) and ataxia-telangiectasia- and rad3-related (ATR) kinases (Falck, Coates, & Jackson, 2005; Regal, Festerling, Buis, & Ferguson, 2013; Shiotani et al., 2013; Williams et al., 2010). Yet, detailed structural information from microbial Mre11 allowed the development of human MRE11 inhibitors that can control the ability of a cell to switch among DSB repair pathways (Shibata et al., 2014). These MRE11 inhibitors therefore have the potential as radioprotectants (Koukourakis, 2012) and are tools for testing MRE11 activity in the clinic in both combined or single-therapy generation for diseases of aging (McPherson, Shen, & Ford, 2014; Sun et al., 2017), autoimmunity, and cancer (Kelley, Logsdon, & Fishel, 2014; Kuroda, Urata, & Fujiwara, 2012; Nicholson et al., 2017; Sun et al., 2017).

While these observations reveal how this complex may function similarly and differently between prokaryotes and eukaryotes, the core cell biology surrounding Mre11 complex activity in detection and processing of damaged DNA ends is similar throughout the domains of life (Williams et al., 2007). Thus, useful knowledge obtained from high-resolution analyses of microbial MR frequently applies to physical and mechanistic hypotheses for eukaryotic MRN homologs.

In fact, MRE11 is under active investigation as a target for enhancing DNA damage-induced tumor cell killing (Dynlacht, Batuello, Lopez, Kim, & Turchi, 2011; Hengel, Spies, & Spies, 2017; Hosoya & Miyagawa, 2014; Thompson, Montano, & Eastman, 2012; Velic et al., 2015; Vilar et al., 2011). As MRE11 knockouts are embryonic lethal (Stracker & Petrini, 2011), the development of inhibitor tools to evaluate MRE11 functions in cells has a biomedical value. In contrast, current strategies for MRE11 inhibition based on RNAi have limitations on specificity and efficiency, as MRE11 is multifunctional.

Importantly, small molecule approaches promise targeting of specific MRE11 nuclease activities that govern repair pathway choice (Shibata et al., 2014). Small molecules also have the advantage of yielding highly penetrant effects across whole-cell populations compared to other targeting therapies, such as RNA interference. Moreover, they can be efficiently tested across cell lines and therapeutic conditions. Early high-throughput screening for inhibition of double-strand break repair in *Xenopus laevis* egg extracts identified Mirin as the Mre11 nuclease inhibitor (Dupré et al., 2008; Garner, Pletnev, & Eastman, 2009; Roques et al., 2009). Recent elaboration of the original Mirin scaffold to produce the PFM library (representing the names Petricci–Forli–Moiani) identified derivatives, which selectively target either endo- or exonuclease activity of Mre11 in vitro. Since inhibition of Mre11 endonuclease activity inhibits repair by HR and promotes NHEJ-mediated repair, these next-generation inhibitors provide a means to guide pathway switching between NHEJ and HR (see Fig. 1B) (Shibata et al., 2014).

Structure-based inhibitor analyses and design were critical to understanding and refining the molecular mechanisms underlying functional selection of Mre11 endo- or exonuclease activity by the next-generation PFM inhibitors. As many mammalian DNA repair factors, MRE11 included, remain challenging for structural study, utilization of a microbial molecular avatar of the Mre11 nuclease from eubacterial *Thermotoga maritima* was critical for understanding structure–activity relationships (SARs) between Mre11 and the next-generation selective inhibitors PFM01, PFM03, PFM04, and PFM39. We employ the term avatar for our approach, rather than homolog or other term for sequence similarity or evolutionary relationship, because our strategy depends upon the physical embodiment of the target's essential features including allostery without necessarily having high sequence identity. Here, we describe our avatarian approach to develop chemical inhibitors that target allostery to dissect multifunctionality and control biological outcomes with the MRE11 nuclease as a prototypic system. The methods discussed provide the basis to employ structural allostery and avatars for complex mammalian targets in order to develop chemical inhibitors that are increasingly relevant for accurate understanding and prediction of mechanisms in cancer cell biology and biomedicine. By focusing on avatars and allosteric switches the inhibitor optimization cycle may be focused more tightly than often otherwise done which may improve efficient progress and decrease preclinical costs for the design of inhibitors for possible clinical trials.

2. METHODS FOR AVATAR INHIBITORS TARGETING ALLOSTERY

In this section, we describe the design and generation of the PFM-derivative library from the Mirin scaffold. We furthermore note strategies for selecting the *T. maritima* Mre11 as a molecular avatar along with characterizations of PFM library interactions with the Mre11 avatar, which tested endo- and exonucleolytic selectivity among the PFM-derivative library. We outline strategies for cocrystallization of Mre11/PFM inhibitor complexes and for the analysis of allosteric effects within the resulting crystal structures of the complex driving endo- or exonucleolytic selectivity. As the validation of SARs derived from microbial avatars in a mammalian system is critical, we highlight in vitro and in cell approaches for testing the effects of PFM inhibitors on human MRE11 and DSB repair in human cells.

2.1 Chemical Design and Synthesis of the PFM Library

Chemical analysis of the early Mre11 inhibitor Mirin (Dupré et al., 2008) reveals a pseudothiohydantoin core containing a Z *p*-hydroxybenzylidene moiety at the 5-C position. Functional group modifications of Mirin, in both the aromatic and the pseudothiohydantoinic ring, were identified as logical targets for testing SARs. A good candidate probe should be efficiently synthesized in high yields and be achiral. Initial substitution of the core pseudothiohydantoinic moiety with a rhodanine and/or *N*-alkyl-rhodanine ring generated a compound series with variable modulation of Mre11 nuclease activity. From these pilot compounds, a targeted library of PFM derivatives was designed and synthesized by using the general synthetic approaches described in Fig. 2 for a detailed characterization of Mre11 nuclease modulation.

The protocol developed represents a straightforward pragmatic approach for the preparation of decorated rhodanine and pseudothiohydantoin derivatives. These compounds provided interesting probes to elucidate selective inhibition of Mre11 endo- or exoonuclease activity. From the screening of our PFM derivatives, we infer that the aromatic portion impacts the activity and the OH or NH_2 group in the 4-C position (on the benzene ring) is crucial for inhibition (Fig. 2). The pseudothiohydantoin core is important for the selective inhibition of the Mre11 exonuclease activity. Importantly, we furthermore found it possible to switch the exo inhibitor into an endo one by the simple replacement of the core with a rhodanine ring demonstrating the power of a focused library for this system.

2.2 Identifying Nonhuman Mre11 Avatars

Human MRE11 has remained intractable to the robust high-resolution crystallography required for efficiently elucidating protein/inhibitor

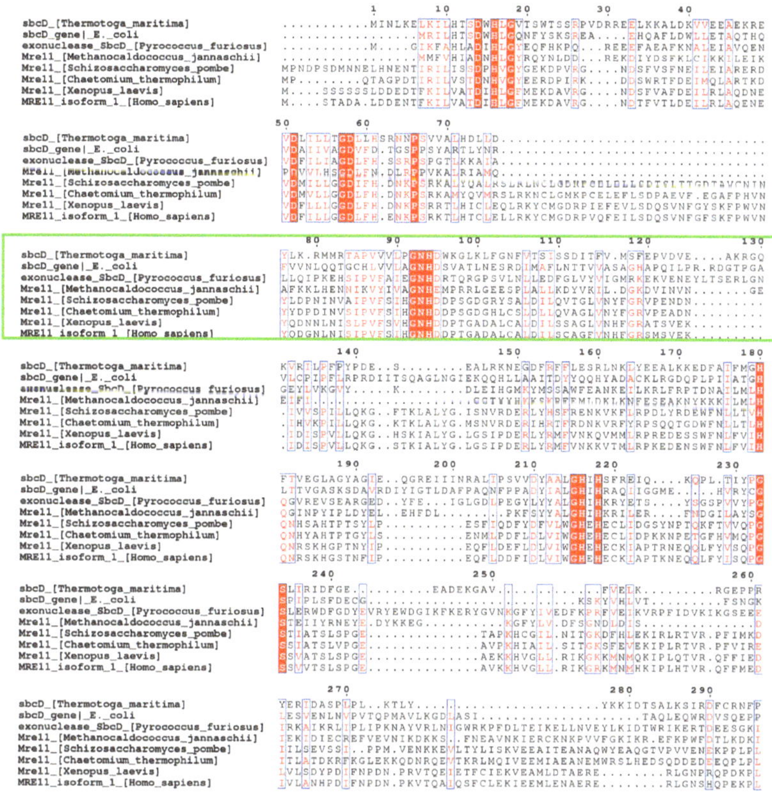

Fig. 2 Sequence conservation from an alignment performed with Cobalt and Espript 3.0 showing the Mre11 catalytic domain among organisms having a structure deposited into the PDB plus the *X. laevis* Mre11 used for in vivo identification of Mirin activity. Conserved regions (*red fill*) and the region showing an allosteric conformational change in *T. maritima* centered at N93–H94 is highlighted (*green box*). Organisms from the *top* are shown like bacterial, through archeal, yeast, fungi, vertebrate, and mammals.

complexes: the sole X-ray MRE11 structure, containing the CAP and catalytic nuclease domains, only reaches to 3.0 Å resolution (Park, Chae, Kim, & Cho, 2011). To achieve the higher resolution desired for detailed structures of MRE11-PFM inhibitor complexes, we chose to employ a microbial molecular avatar of the Mre11 nuclease to test and improve inhibitors. Our starting point for identifying nonhuman avatar systems is to examine existing crystal structures and to identify critical residues conserved among all species implying fundamental mechanistic features. The full sequence alignment of Mre11 obtained with Cobalt (Constraint-based Multiple Alignment Tool), as displayed in Fig. 3, highlights residues corresponding

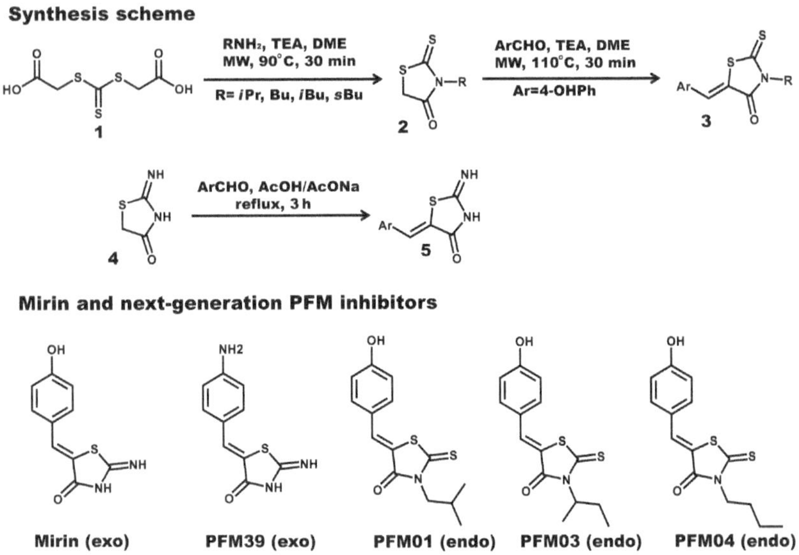

Fig. 3 Chemical synthesis scheme and inhibitors analyzed. Schematic representation of the two-step synthetic chemical method to produce *N*-alkyl rhodanines (*top*). A one-step protocol to produce mirin-like pseudothiohydantoin derivatives (*middle*). The 2D representation of all inhibitors discussed (*bottom*) starting from left mirin (exo), PFM39 (exo), PFM01 (endo), PFM03 (endo), and PFM04 (endo). Procedures: In **1** properly functionalized rhodanines (Mendgen, Steuer, & Klein, 2012) moieties were synthesized by cyclization with primary amines in the presence of TEA (triethanolamine) in DME, at 90°C under microwave (MW) irradiation for 30 min. Compound **2** was obtained in different yields depending on the starting amine used; the worst yields were obtained for the sterically hindered and problematic isopropylamine. A Knoevenagel's condensation in the presence of an aromatic aldehyde and TEA furnished **3** compounds by irradiation at 110°C for 30 min. When the pseudothiohydantoin ring is maintained, MW as the heating source generated a mixture of *Z/E* isomers. Complete region selectivities toward the *Z* isomers were possible refluxing the pseudothiohydantoin with traditional heating conditions in the presence of arylaldehydes in a mixture of AcONa/AcOH for 3 h.

to the GDL motif (57–59 in *Tm*Mre11), GNHD motif (92–95 in *Tm*Mre11), GHxH motif (215–218 in *Tm*Mre11), and all residues that are part of the active site, as exhibiting the greatest conservation across all domains of life.

Informed by conserved residues identified by the alignment, we examined the best available MRE11 structures, using a resolution cutoff of 2.7 Å, to map the location of these key residues and identify components to improve structure-based design of small molecule inhibitors. This combined sequence and structural analysis identified the bacterial Mre11 homolog from *T. maritima* (*Tm*Mre11, Fig. 1C) as the best avatar candidate for human

MRE11. Following structural superimposition between the human and bacterial nucleases, we had a suitable scaffold for performing in silico docking experiments for rational chemical design. In these efforts we also employed the Mre11 dimer structure validated in solution by SAXS (Rambo & Tainer, 2011) and in our two DNA–Mre11 complex crystal structures and SAXS shapes for Mre11 dimers (Williams et al., 2008). Mutations impacting dimer formation in solution lost DNA binding validating the dimer interface as contributing the DNA-binding channel.

2.3 Cocrystallization and Analysis of *Tm*Mre11/Inhibitor Complexes

Although structures of protein–ligand complexes can be obtained by soaking, we recommend cocrystallizations. We find that the best technique for obtaining high inhibitor occupancy when targeting allostery is cocrystallization which may allow visualization of altered conformations.

2.3.1 *Cocrystallization of* Tm*Mre11/Inhibitor Complex Structures*

In this section we outline the steps employed to generate X-ray crystal structures of *Tm*Mre11 inhibitor complexes, which are exemplary of the methods we have found useful for determining structures of enzyme–inhibitor complexes.

After expression and purification (Das et al., 2010), *Tm*Mre11 is exchanged into its target crystallization buffer and concentrated to approximately 10 mg/mL. Target compounds are prepared at 100 m*M* in DMSO and added directly to the protein solution to a final concentration of 5 m*M*. If precipitation occurs, the solution should be centrifuged at $15{,}000 \times g$ for 10 min to remove particulates prior to preparing crystallization trays, either hanging drop or sitting drop setups, which are incubated at 18°C for 5 days. Once crystals have matured, a suitable cryo-harvest condition is identified by testing crystal resilience and diffraction upon transfer to well solution supplemented with 5%–20% glycerol and/or low-molecular-weight polyethylene glycols (PEGs). Benefits of maintaining 5 m*M* compound within the cryo-harvest condition should also be explored. Once crystals have been frozen without damage, X-ray diffraction datasets may be collected, preferably from a synchrotron source, such as the SIBYLS beamline at the Advanced Light Source (Classen et al., 2013), which will provide the highest resolution data. After data processing, initial phasing is accomplished by molecular replacement, followed by model refinement. In the best-case scenario, compound electron density can be unambiguously identified within $F_o - F_c$ difference maps and subsequently refined using PDB and *cif* library parameter files created with the PRODRG server (http://davapc1.bioch.dundee.ac.uk/prodrg/).

Once an inhibitor has been successfully cocrystallized with its target, one must analyze the resulting target/inhibitor complex model to develop hypotheses linking compound orientation at the binding site to allosteric movement of the target for testing in future in vitro and in vivo experiments.

2.3.2 Performing In Silico Docking Against TmMre11

X-ray crystallography allows the use of relatively small fragments and the growth of better binders from structures defined with initial fragments (Chan et al., 2017). We employed a general strategy to test and improve the activities of the small molecules consisting of testing a focused library of compounds, solving structures of selected binders, and obtaining the SDF (structure-data file) files of them. The best way to obtain SDF files is from chemical SMILE using a translator (https://cactus.nci.nih.gov/translate/).

However, our strategy further involves targeting allosteric conformational changes that are required for the enzyme to form a productive complex with the DNA substrate. Lessons from multiple repair enzyme structures have taught us the advantages of focusing on allosteric interactions instead of primarily targeting the active site. First, allosteric interactions can enable multiple potential sites for inhibitors and thereby allow dissection of related functions, as we found for MRE11. Second, allosteric interactions tend to be system specific, whereas compounds targeting active sites may show cross-reactivity to related enzyme classes. Third, we can interfere with conformational changes required for catalysis without blocking substrate binding, so the inactivated enzyme can itself provide protection against alternative reactions that might otherwise cause unwanted or unpredictable outcomes in cells.

The use of a docking software, such as the Schrodinger suite for drug discovery (Schrodinger, 2017), makes it possible to run a calculation for multiple ligand series per job to permit a detailed analysis of key features and improve the molecular design. In this approach, the in silico characterization is completed with a computational energy evaluation for implicit or explicit solvation models that can be calculated and efficiently compared with calorimetric experimental data (Moiani, Cavallotti, Famulari, & Schmuck, 2008). The best candidate is thereby identified for additional in vitro experiments, such as biophysical characterizations, and for testing in several cell lines.

2.3.3 Mre11 Endo- and Exonuclease Inhibitor Complex Structures

We crystallized and solved informative Mre11 structures with PFM01, PFM03 (endo inhibitors), and PFM39 (exo inhibitor) to 2.3, 2.4, and 2.3 Å, respectively. The Mre11 microbial avatar pinpointed two specific sites where small molecules can bind and modulate nuclease activity. Specifically, exonuclease inhibitors bind with variable orientation near the active site, blocking His61 (His63 in human MRE11), which mediates the 3′-to-5′ exonuclease activity by opening the ends of dsDNA substrates. In contrast, binding of endonuclease inhibitors is shifted toward the dimer interface with an orientation that places their alkyl chain emerging from a rhodanine ring toward the hydrophobic residues of helix α3.

To examine conformational changes in Mre11 due to inhibitor binding, *Tm*Mre11 endo and exo inhibitor structures were superimposed onto the *Tm*Mre11 apo structure (Fig. 4). Viewed from two different angles, the

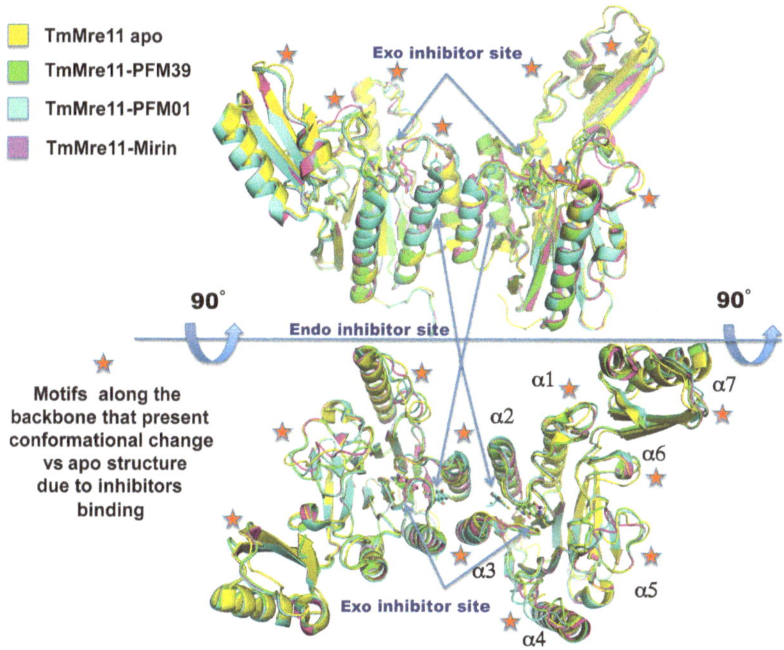

Fig. 4 Structural variation revealed by superimposition of TmMre11 apo structure (*yellow*, PDB ID: 4NZV), TmMre11-PFM39 exo inhibitor structure (*green*, PDB ID: 4O5G), TmMre11-PFM01 endo inhibitor structure (*cyan*, PDB ID: 4O24), and TmMre11-Mirin exonuclease inhibitor structure (*magenta*, PDB ID: 4O4K). The *top* and *bottom* show different orientations (90° tilted) of the Mre11 dimers. *Arrows* point to the two different binding sites for endo and exo inhibitors. *Red stars* highlight backbone regions shifted vs the apo structure. These reorganizations revealed a mechanism for dissecting the endo- and exonuclease activities by chemical inhibitors.

alpha helices in the catalytic domain and in the CAP domain appear to restrict inhibitor access to catalytic area of the active site. The PFM inhibitors, however, are more distant from the active site, localizing near the Mre11 dimerization interface and contacting the N93–H94 motif.

The Mre11 α4 and α5 helices show a conformational change with different patterns in the endo- and exonuclease inhibitor structures. The helices α1, α3, α6, and α7 show a small conformational change vs the apo structure, but the endo and exo inhibitor-bound structures have a similar structural pattern. Only helix α2 shows no conformational change in the presence of bound inhibitor, implying that it is part of the solid structural core of the catalytic domain in *Tm*Mre11 between residues Asn65 and Ala85. Conformational changes in a similar region were reported in the characterization of ATLD mutant W243R, showing that the mutation close to the inhibitor-binding region causes conformational change of these regions with clastogenic activity in fission yeast (Limbo et al., 2012).

2.3.4 High-Resolution Structure of PFM04, Endo Inhibitor, Bound to TmMre11 and Its in vitro Activity

Structures provide a means of visualizing and understanding the impact of inhibitor binding. As a test of our methods, we introduce in this section the novel high-resolution structure of *Tm*Mre11 with bound PFM04. This small molecule completes our initial investigational series for structural and activity characterization as presented in Fig. 5A. Compound PFM04 is a

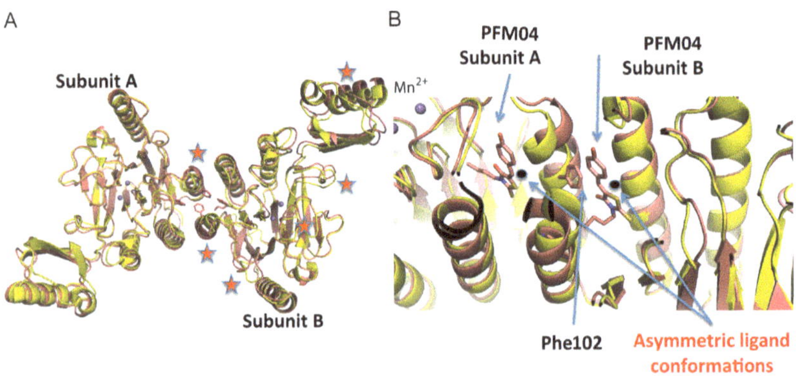

Fig. 5 The apo and novel PFM04 inhibitor-bound structures. (A) The superimposition of TmMre11 apo (*yellow*) and the novel TmMre11–PFM04 inhibitor complex structure (*pink*) showing the different organization of the two domains due to the asymmetric binding of PFM04. *Red stars* highlight main chain shifts in the inhibited vs the apo structure. (B) A zoom-in on the inhibitor-binding site shows the different conformation of ligands between subunit A and subunit B, along with the interaction of Phe102 in subunit A and PFM04 in subunit B.

member of the *N*-alkyl-rhodanine derivative family with an endonucleolytic inhibitory effect: it has a linear *N*-alkylic chain instead of the branched *N*-alkyl chain of PFM01 and PFM03. The compound PFM04 was cocrystallized as described in Section 2.3.1, with crystals being grown by using sitting drop vapor diffusion in 0.1 *M* MES pH 6.5, 10% PEG3350, 20 m*M* $CaCl_2$, then soaked in 20% ethylene glycol as cryoprotectant. A dataset to 2.15 Å resolution was obtained at the SIBYLS beamline (ALS BL 12.3.1) (Classen et al., 2013), and the structure was refined and submitted to the Protein Data Bank (PDB ID: 6ASC). X-ray diffraction data collection and refinement statistics are shown in Table 1. In Fig. 5B and C, we present the superimposition of the new *Tm*Mre11–PFM04 (color pink) with the apo structure PDB ID: 4NZV (color yellow).

Differing from the previous PFM structures, this novel structure shows the compound PFM04 oriented in a new orientation between the two Mre11 subunits, as shown in Fig. 5C. The alkyl chain is also organized differently and shows evidence of flexibility. The effect of asymmetric orientations of the two small molecules in the dimer is translated into different organizations of Mre11 subunit A and subunit B, as shown in Fig. 5B where red stars identify regions of conformational change in the backbone. In fact, one domain resembles the organization in apo Mre11 with a small difference in the CAP domain. However, the other domain shows notable differences from the apo structure with two crucial regions differently organized. This effect is also correlated with the direct interaction of PFM04 in one domain with Phe102 that is an important residue governing *Tm*Mre11 dimerization.

2.4 Targeting Mre11 Dimer Allostery With a Molecular Avatar

We chose eubacteria *Tm*Mre11 to represent essential features of human MRE11 catalytic domain for inhibitor development. The size and complexity of eukaryotic MRN/X assembly vs the bacterial MR complex (Lafrance-Vanasse et al., 2015; Yu et al., 2012), and the absence of posttranslational modifications support its practical advantages. Compared to human MRN, *Tm*Mre11 provides a more efficient and accurate structural characterization of inhibitor complexes with implications for design of potential therapeutic compounds. Defining the conserved features of the catalytic domain including both substrate recognition and sculpting unveils similarities whereby small molecules designed to act on bacterial Mre11 can efficiently provide testable inhibitors for human MRE11. Such inhibitors have general interest as possible chemical tools to control DNA repair pathway selection in human cells.

Table 1 X-Ray Diffraction Data Collection and Refinement Statistics (Molecular Replacement)

	*Tm*Mre11–PFM04
Data	
Space group	*P*21
Cell *a*, *b*, *c* (Å)	48.053, 112.682, 81.162
α, β, γ (°)	90.00, 100.26, 90.00
Resolution (Å)	2.15
R_{sym} or R_{merge}	4.9
I/σ	46.7
Completeness	94.2
Redundancy	4.8
Refinement	
Resolution (Å)	2.15
No. reflections/ observed	72,439
R_{work}/R_{free}	19.69/23.18
No. atoms	5375
Protein	5187
Ligand/ion	$(Mn)_4$
	(PFM04) 38
Waters	142
Protein	67.3
Ligand/ion	85.6
Water	55.4
Bond	0.004
Angle	0.789

We therefore used bacterial Mre11 as a molecular avatar that embodies the key features of human MRE11 that we wished to target. The essential structural biology for avatars from microbial cells may provide many of the features needed to experimentally test the means to inhibit the human enzyme, as seen for DNA replication and repair enzymes. In Fig. 3, the green highlighted rectangle shows the conserved region that in the *T. maritima* homolog shows the allosteric motion of the NH loop in the superimposition of PDB files 2Q8U, 3QG5 (Fig. 6A–C), in the asymmetric conformation

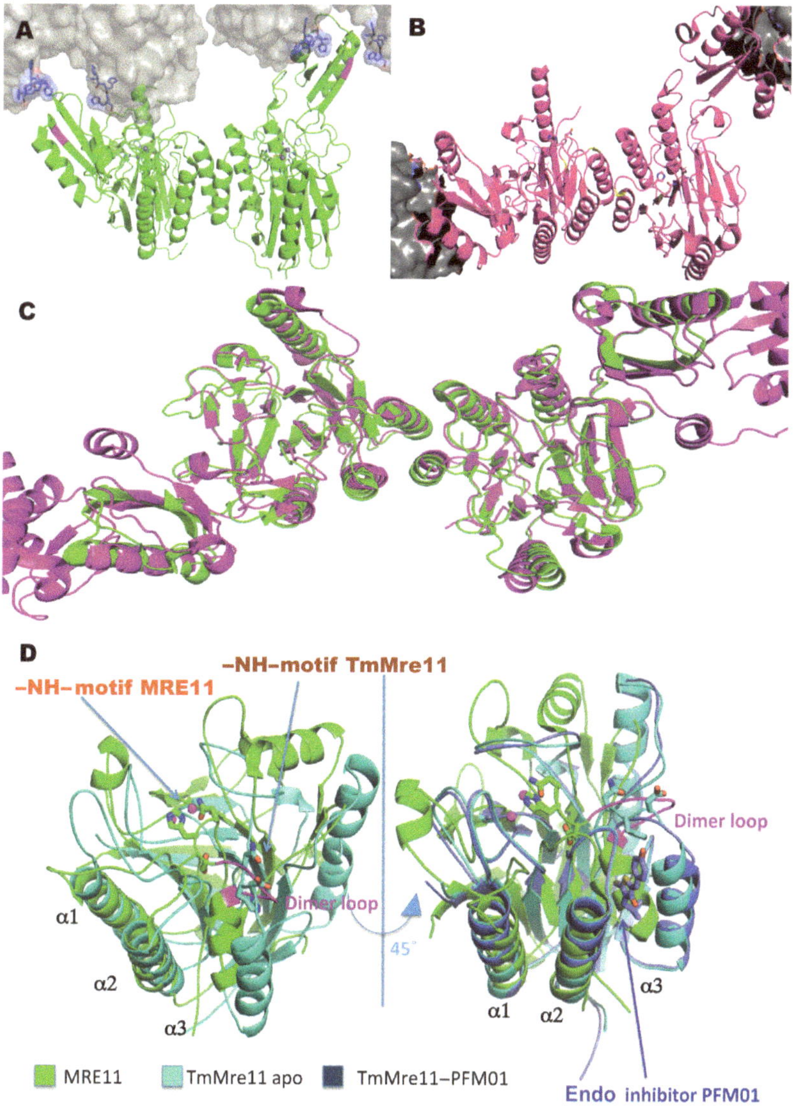

Fig. 6 Human MRE11 compared to TmMre11 apo and endonuclease PFM01-bound structures. (A) PDB ID: 4NZV TmMre11 apo dimer (*green* with crystal mates in the *gray surface* with N-terminal His-tags in sticks) shows how the allosteric opening of a small *(Continued)*

induced by the N-terminal 6His-tag (Das et al., 2010), and in the open asymmetric conformation induced by Rad50 torsional rotation (Lammens et al., 2011). Fig. 6A and B shows the bacterial assembly with the electrostatic surface and the asymmetric orientation of the N93–H94 loop. His94 is conserved among all Mre11 homologs and acts in sculpting the DNA for the exonuclease process (Williams et al., 2008). Thus, His94 mutants show the inhibition of the exonuclease activity and a mutant mouse for the equivalent histidine mutation (H129N) has specific sensibility DSB repair by HR but not in NHEJ or other processes that do not involve the Mre11 exonuclease activity (Buis et al., 2008). However, the structural analyses based on the sequence of Mre11 from higher eukaryotes uncover different features in the assembly of the nuclease domain that contain that NH loop motif, as well as differences in the dimerization interface and its placement relative to the NH loop motif.

The asymmetric organization of MR from *T. maritima* (Lammens et al., 2011) in the PDB 3QG5 does not show the pockets that can bind small molecules to modulate exonuclease vs endonuclease activities and thereby consequent Mre11 functions in regulating outcomes at DSBs and stalled replication forks. The data obtained with the *T. maritima* Mre11 construct codeveloped with JCSG (Joint Center for Structural Genomics) (Das et al., 2010) were first crystallized and deposited as PDB: 2Q8U, and then improved by the authors and deposited as PDB 4NZV. This structure reveals that the allosteric opening and closing of the loop contain His94

Fig. 6—Cont'd molecule accessible cavity was revealed in the switch of the N93–H94. The crystal packing uncovered the proposed allosteric conformation in the middle of its hypothesized cycle. (B) Representation of structure PDB ID: 3QG5 open dimer TmMre11 (*magenta*) where the gray cloud represents the Rad50 surface providing a torsional opening that reorganizes the asymmetric Mre11 allosteric dimerization. (C) Two levels of allosteric conformational reorganizations are visible in superimposition of PDB files uncovering differences in the organization of the NH loop and the opened cavity adjacent to the dimerization site. (D) The TmMre11 catalytic domain structure (*cyan*, PDB ID: 4NZV) and human MRE11 catalytic domain (*green*, PDB ID: 3T1I), superimposed on alpha helix α1 and α2 showing the human α3 tilted vs the TmMre11 as driven by the dimerization site represent in the human loop (*magenta loop*). The Mn^{2+} ions of both catalytic sites (*spheres*) and NH regions of both constructs are shown (*sticks*) with one engaged and one disengaged. A view tilted 45 degree with an added structure of the complex TmMre11–PFM01 catalytic domain superimposed (*dark blue*, PDB ID: 4O24). PFM01 (*sticks*) at the dimerization site shows an overlap with the human MRE11 α3. The engagement of NH domain in human MRE11 in this structure closes the site where small molecules can act as endonuclease inhibitors.

and the hidden pocket close to the dimerization interface and able to contain small molecules. The crystal-packing arrangement suitable to open that loop was created by leaving the N-terminus His-tag uncleaved; in fact, the first histidine stacks with the residues Y276–Y277 inducing a conformational change of the CAP domain relative to the catalytic domain, as shown in Fig. 6A. We proposed that the crystal packing simulates the action of Rad50 that is shown in the Mre11–Rad50 complex structure (Lammens et al., 2011). This Mre11 conformation allowed cocrystallizing the first endonuclease inhibitor PFM01 (Fig. 6D) in its vertical orientation close to dimerization site (PDB ID: 4024) and was successively employed for other endo- and exonuclease inhibitors. The exonuclease inhibitors, e.g., Mirin and PFM39, do not show unambiguous occupancy in electron density maps, but their general binding position appears to be displaced toward the active site relative to the endonuclease inhibitors, which are shifted toward the dimer interface.

The Mre11 dimer forms the core of the MRN complex as shown in Fig. 1A (Williams et al., 2010). Mre11 dimerization is necessary to bind and process dsDNA sites and to connect Mre11 nuclease activities to Rad50 conformational changes (Williams et al., 2008). These Rad50 conformational changes are predicted to regulate the Mre11 endo- and exonuclease activity and thus the pathway choice in vivo. Thus, we viewed the dimer interface region and its connections to the active site as a logical target for allosteric control of this system including for DNA repair pathway selection based on the DNA-substrate binding impacting cell cycle-dependent activities and protein partners involved in functional networks and biological outcomes. The dimerization interface of human MRE11 captured by its crystal structure shows unique features relative to other homologs: the human dimer assembly (Park et al., 2011) is maintained by a loop–loop interaction (Fig. 6D).

This sequence (DPTGADALC spanning Asp131–Cys139 from subunit A to subunit B) is conserved in *X. laevis*, the organism used in the identification of the original Mirin inhibitor (Dupré et al., 2008). The dimerization loop is immediately adjacent to the Asn128–His129 (NH loop) loop region near the N-terminal Cap domain, suggesting that dimerization may reorganize the N-terminal domain at the active site based on the macromolecular organization of the full MRN complex in the presence of substrates. Nbs1 may act naturally to induce this macromolecular reorganization, and this change may also be promoted by its interactions with partners such as CtIP (Deshpande et al., 2016; Kim et al., 2017). Importantly, small molecule compounds may provide chemical master keys to lock this conformational change into a single position. In other organisms with structurally characterized Mre11 assemblies, Mre11 retains an analogous dimer with coupled alpha-helical hairpins to maintain a specific interaction.

Our strategy for the dissection of MRE11 nuclease provides a prototypic exemplary system to describe methods to develop chemical inhibitors that dissect multifunctionality and control biological outcomes. We observed that DNA repair and replication enzymes typically sculpt their DNA substrates, as we found for nucleotide flipping by DNA repair glycosylases (Guan et al., 1998; Slupphaug et al., 1996; Thayer, Ahern, Xing, Cunningham, & Tainer, 1995), AP endonucleases (Tsutakawa et al., 2013), and endonuclease V (Dalhus et al., 2009), as well as for the flap exo- and endonuclease FEN1 (Rashid et al., 2017; Tsutakawa et al., 2017). Thus, part of our strategy to obtain specificity is to retain DNA binding but to block protein-mediated DNA conformational changes needed to position the substrate complex for catalysis. The methods discussed for this Mre11 system thus provide the basis to employ structural allostery and avatars to develop chemical inhibitors. Such inhibitors are increasingly relevant for accurate understanding and prediction of mechanisms in cancer cell biology and biomedicine.

Our goal was therefore to develop small molecules that bind near the dimer interface and specifically interact where they can impact the motion of the NH loop. We therefore aligned the structures of apo-*Tm*Mre11, *Tm*Mre11 in complex with the endo inhibitor PFM01 (Shibata et al., 2014), and human MRE11 catalytic domain (Fig. 6D). After pruning *Tm*Mre11 and human MRE11 catalytic domains, the superimposition was optimized on helices α1 and α2. This analysis shows that human MRE11 α3 is shifted toward the active site due to the engagement of the NH loop (Fig. 6D). The *Tm*Mre11 α3 helix is more open than the human homolog due to the disengagement of the NH loop enabling the fit of the endo inhibitors (Fig. 6D). In human MRE11 the dimerization loop (Fig. 6D, magenta) is directly connected to α3 and its allosteric movement is connected to reorganization of the heterocomplex based on the substrate bound. Small molecules that interfere with Mre11 sculpting of DNA into the active site make it feasible to control nuclease activity and consequently turn off nuclease functions to induce a predetermined DSB repair pathway (Fig. 1B).

2.5 In Vitro Validation of PFM Inhibitors Against Human MRE11

PFM compounds were first evaluated for their activity as nuclease inhibitors of MRE11. For efficiency, they were tested as a group of five at time for endo- or exonuclease-specific inhibition in parallel with a control for DMSO and a specific Mre11 mutant that inhibits the nuclease activity tested.

2.5.1 *In Vitro Characterization of PFM Library Using* TmMre11

Exonuclease assays were done by using a phospho-labeled radioactive substrate (Paull & Gellert, 1998) or aminofluoropurine substrate and then using a fluorimeter to measure the activity (Williams et al., 2008). The endonuclease assay was performed by using a circular DNA substrate (Das et al., 2010; Lee et al., 2012). Where feasible additional biochemistry tests are recommended for more specific characterizations. The in vitro activities of PFM04 against *Tm*Mre11 in an endo assay and in a resection assay against human MRN are shown in Fig. 7A and B and described later.

Using an exonuclease assay with a double-stranded oligonucleotide substrate, we found that the newly characterized MRE11 endonuclease inhibitor only has a minor effect on the exonuclease activity of the MRN complex. The assay was performed along with a few other previously characterized MRE11 inhibitors and the reaction was allowed to proceed at 37°C for 60 min. As described by our group, PFM39, an exonuclease inhibitor, blocks the exonuclease activity of MRN in the assay (Shibata et al., 2014). On the other hand, the endonuclease inhibitors (PFM01, PFM03, and PFM04) in comparison slightly inhibit the exonuclease activity of MRN.

For the endonuclease assay, ΦX174 circular ssDNA was incubated with purified *Tm*Mre11 catalytic domain and inhibitors Mirin, PFM01, PFM03, and PFM04 at 500 m*M* final concentration. The inhibition assay was performed at 37°C for 15 min. Fig. 7B shows the percent of circular ssDNA degraded relative to the positive control (DMSO treated) and negative control *Tm*Mre11 H94S that present null endonucleolytic activity (Das et al., 2010; Shibata et al., 2014).

2.5.2 *In Vitro Characterization of PFM Library Using Human MRE11*

The in vitro validation process of PFM inhibitors was performed with human MRE11 and human MRN complex following the protocol described in our publication (Shibata et al., 2014). The exonuclease assay with human MRN is performed in an exo buffer consisting of 25 m*M* MOPS pH 7.0, 60 m*M* KCl, 0.2% Tween-20, 2 m*M* DTT, 2 m*M* ATP, 5 m*M* $MnCl_2$, 5 n*M* MRN, and 0.8 m*M* PFM inhibitors. Labeled DNA (100 n*M*) was incubated in exo buffer for 30 min at 37°C, followed by deproteinization in one-fifth volume of stop buffer (20 m*M* Tris–Cl pH 7.5 and 2 mg/mL

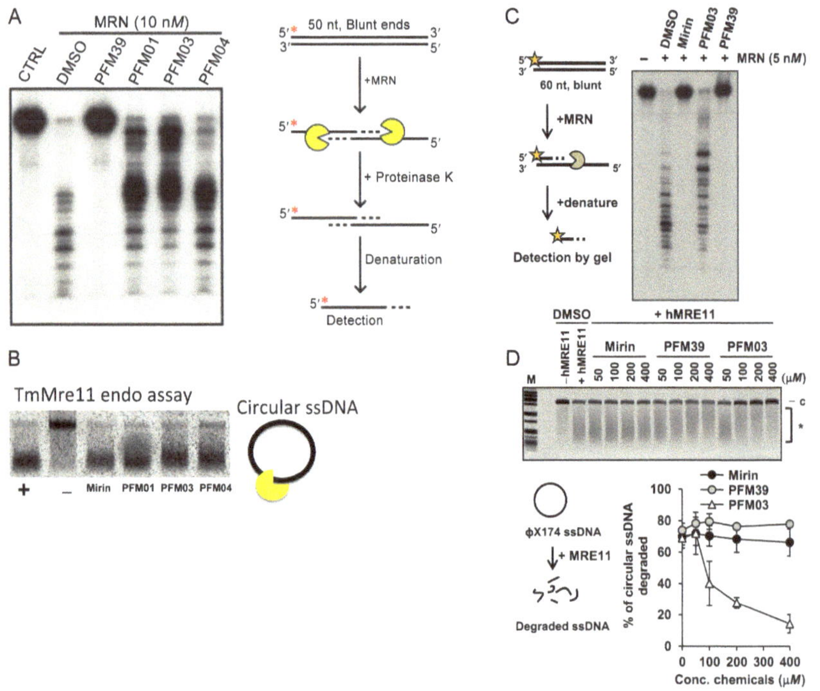

Fig. 7 Nuclease activities and inhibitor impact. (A) MRE11 inhibitor PFM39 blocks MRN exonuclease activity (10 n*M*). In contrast, the MRE11 endonuclease inhibitors (PFM01, PFM03, and PFM04) have little effect on the MRN exonuclease activity. Radiolabeled DNA (100 n*M*) was incubated with MRN at 37°C for 60 min. Reactions contained a final concentration of 0.5 m*M* of the inhibitors. *Red asterisk* represents the 5′ radioactive label on the DNA. (B) Endo inhibitors PFM01, PFM03, and PFM04 primarily reduce the endo activity of TmMre11 vs the exo inhibitor Mirin. The nuclease assay is done at constant temperature and stopped at specific times. (C, D) The strong exo inhibition effect of PFM39 and Mirin vs endo inhibitor PFM03 for human MRN. (D) The strong endo inhibition activity of PFM03 vs the exo inhibitor mirin and PFM39 to human MRE11. The figure shows the percentage of circular ssDNA degraded relative to the control. *Panels (C) and (D): Reproduced from Shibata, A., Moiani, D., Arvai, A. S., Perry, J., Harding, S. M., Genois, M. M., et al. (2014). DNA double-strand break repair pathway choice is directed by distinct MRE11 nuclease activities.* Molecular Cell, 53*(1), 7–18. https://doi.org/10.1016/j.molcel.2013.11.003.*

proteinase K) for 15 min at 37°C. Reactions were loaded on a 8% acrylamide/urea gel, run at 75 W for 60 min, dried onto DE81 filter paper, and autoradiographed with typical results shown in Fig. 7C.

Endonuclease assays were performed with ΦX174 circular ssDNA virion DNA (New England Biolabs), and 100 ng substrate was incubated with 300 ng purified WT hMRE11 in a 20 μL reaction (30 m*M* Tris–HCl pH

7.5, 1 m*M* dithiothreitol, 25 m*M* KCl, 200 ng acetylated bovine serum albumin (BSA), 0.4% DMSO, and 5 m*M* $MnCl_2$) at 37°C for 30 min with inhibitors. The assay with human MRN complex was performed with 100 ng of substrate incubated with 400 ng of protein for 45 min in a 20 μL reaction (30 m*M* Tris–HCl pH 7.5, 1 m*M* dithiothreitol, 25 m*M* KCl, 200 ng acetylated BSA, 0.5% DMSO, and 5 m*M* $MnCl_2$). Reactions were terminated with the addition of 1/10 volume of stop solution (3% SDS, 50 m*M* EDTA) and proteinase K to a final concentration of 0.1 mg/mL and incubation for 10 min at 37°C. The reaction products were run in a 0.8% agarose gel (1xTAE) for 90 min at 100 mA. DNA was stained with ethidium bromide and visualized using a Typhoon 9200 scanner with typical results shown in Fig. 7D for the specific PFM inhibitor tested.

2.6 From Molecular Avatar to Selective Human MRE11 Inhibitors Controlling Repair Pathway Selection in Human Cells

While in vitro cross-validation between human targets and their microbial avatars supports essential molecular mechanisms for inhibitor action, it is nevertheless critical to validate in vitro observations in the context of the cellular environment. To investigate the effects of MRE11 inhibitors on HR or NHEJ repair pathways, we chose to irradiate untreated or treated cells with X-rays (typical doses of 7–10 Gy) to induce DNA DSBs and monitor the outcomes and kinetics of repair by using a variety of assays. To establish whether PFM MRE11 inhibitors effectively prevent DNA end resection at sites of breaks, the first biochemical step mediated by MRE11, inhibitor-treated cells were monitored for the appearance of chromatin-bound RPA using FACS detection and the formation of Rad51 foci using immunofluorescence (Fig. 8A–C) (Shibata et al., 2014). For these studies, G2 phase cells were specifically examined (isolated via FACS sorting) to ensure observation of two-ended DSBs and avoid confounding observations from MRE11 action at replication fork structures (Roy et al., 2018). The selected PFM inhibitors successfully exhibited a reduction of chromatin-bound RPA and Rad51 foci, but still allowed MRE11 to localize to DSBs, highlighting their ability to impair MRE11 DNA end resection in vivo.

Monitoring the kinetics of DSB repair via immunofluorescence monitoring of γH2AX foci was used to probe repair differences between the MRE11 exonuclease and endonuclease inhibitors. Most DSB repair events (~80%) in irradiated cells follow the fast kinetics of NHEJ (~2 h); the remaining 20% are restored more slowly by HR (~8 h). Neither

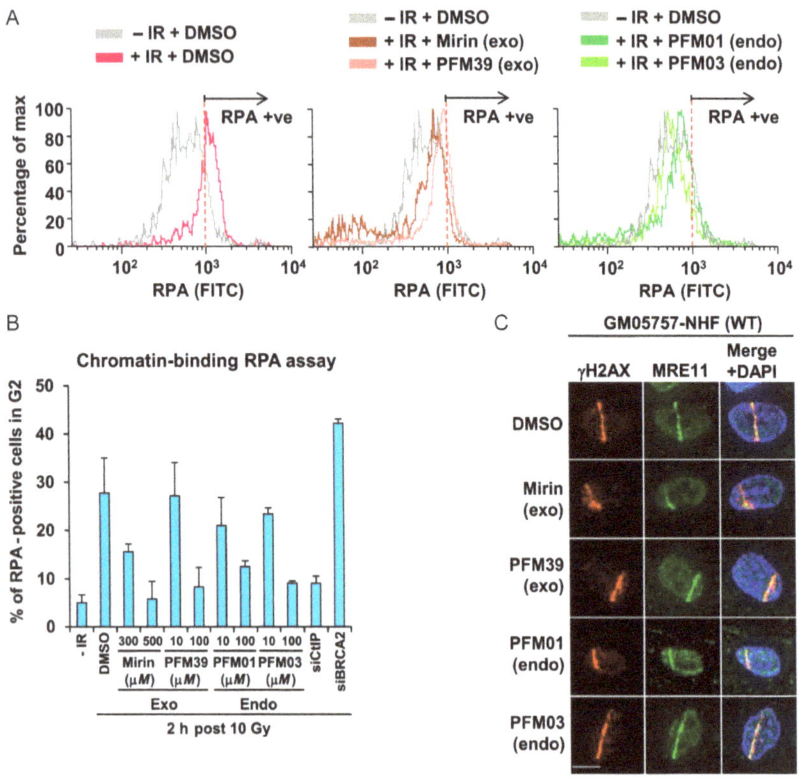

Fig. 8 In cell validation assays from our publication (Shibata et al., 2014). (A) The effect that MRE11 inhibitors have in reducing chromatin-bound RPA level after IR treatment. IR-induced RPA retention in A549 cells was monitored by FACS analysis at 2 h post 10 Gy IR. (B) The quantification of RPA–FACS analysis at two different inhibitor concentrations. (C) Treatment with inhibitor does not affect the recruitment of MRE11 to the site of DNA damage. The cell lines used are primary human fibroblasts GM05757 subjected to UV-microbeam. The analysis of recruitment was done 1 h after inhibitor treatment. To identify the damage site, the histone γH2AX was labeled. (D, E) Treatment with exo inhibitor (but not endo inhibitors) causes a DSB repair defect in G2. DSB repair in G2 (CENPF+) cells was investigated by γH2AX foci analysis. Inhibitors were added 30 min before ionizing radiation (IR) treatment. 1BR3(WT) hTERT was fixed and stained at 8 h posttreatment with 3 Gy IR. (F) MRE11 endo inhibitors rescue the repair defect in HR-defective cells. γH2AX foci were enumerated in 48BR (WT) and HSC62 (BRCA2-defective) primary cells at 8 h posttreatment with 3 Gy IR. (G) The chromosome break defect in BRCA2 siRNA-treated 1BR3 (WT) hTERT was reversed by treatment with an MRE11 endonuclease inhibitor.

endo- nor exonuclease inhibitors impaired the fast kinetics of NHEJ repair (γH2AX foci number was similar between mock and inhibitor-treated cells 2 h posttreatment) (Fig. 8D and E) (Shibata et al., 2014). For extended HR repair, however, exonuclease inhibitors caused a repair defect (resulting

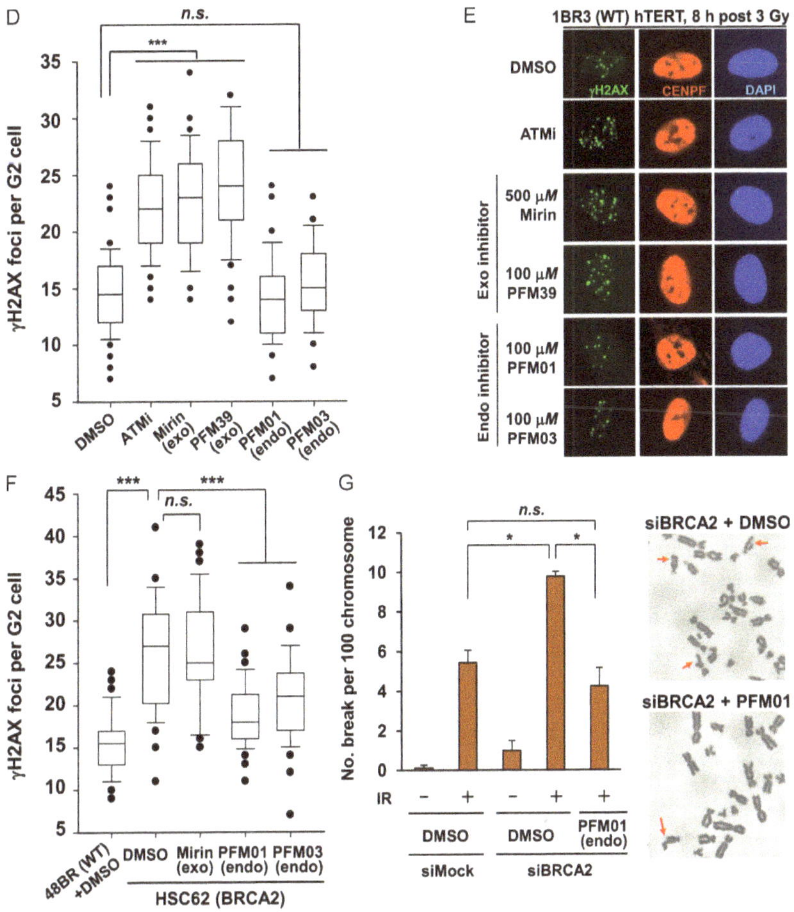

Fig. 8—Cont'd

in elevated γH2AX foci). These results show the ability of different MRE11 inhibitor classes to generate unique DNA repair phenotypes in a cellular context.

Since MRE11 endonuclease inhibitors do not result in a DNA repair defect and are capable of preventing DNA end resection of DSBs, we postulated that inhibiting MRE11 endonuclease activity might reduce HR initiation and promote NHEJ-driven repair in cells defective for HR. In fact, applying MRE11 endonuclease inhibitors to BRCA2-defective or knockdown cells rescued repair in a manner consistent with non-HR-deficient cells (Fig. 8F and G) (Shibata et al., 2014), indicating that MRE11 endonuclease activity is key for initial licensing of HR vs NHEJ although the exonuclease is required for excision and HR repair commitment. Thus, not only were we able to demonstrate the in vivo efficacy and biological impact of the avatar-derived human MRE11 inhibitors, but we were also to extend knowledge of MRE11 function in cells using these chemical tools.

3. FUTURE CONSIDERATIONS AND PROSPECTS

For our research we developed and describe here an avatarian approach to targeting allostery with chemical inhibitors in order to test specific functions in cell biology. Human proteins and their structures are not always efficiently obtainable, despite the development of advanced methods such as MacroBac with Hi-5 insect cells and baculovirus expression vectors with polypromotors (Gradia et al., 2017) for difficult eukaryotic multiprotein complex expression. There is moreover often a timely need for chemical knockdowns to examine protein functions in multiple cell types and under multiple conditions, such as in testing possible resistance factors for radiation or chemotherapeutic agents in cancer cells. Avatars can provide the critical tools for inhibitor development by structural biology. Thus, for example, the high-resolution structure of gammaH2A bound to Brc1 provides a reasonable avatar for inhibitor design against the human PTIP complex (Williams et al., 2010).

Proteins may be part of multiple networks and of complexes with different functions, so knockdowns or knockouts may not accurately inform specific activities and functions. Furthermore, active site inhibitors may not answer the most relevant questions. In contrast, allosteric inhibitors can offer advantages of specificity and of dissection of function that may otherwise be difficult or impossible because conformations often control networks and assemblies. As noted here for Mre11, developing inhibitor tools that target allosteric conformations can allow the dissection of function for multifunctional complexes to inform cell biology.

3.1 Targeting Allostery

We propose general advantages to targeting allostery rather than the traditional approach of focusing primarily on active sites. Allostery, which can involve conformations (that control orientations and shape) and assemblies (that alter local concentrations and orientations), can be as much a part of protein biological function as its catalytic activity. Furthermore, allostery can endow proteins with the multifunctionality that is the mechanistic underpinning of how many proteins are able to act in multiple steps and even in multiple pathways. Thus, proteins often have their interactions and activity allosterically regulated such that particular conformational states correlate to distinct functions. Allosteric events typically extend beyond active sites and include plastic deformations, order-to-disorder transitions, and large scale domain rotations (Goldsmith, 1996) that may be the rate-limiting and regulated steps. Such conformational changes may also provide the basis for sculpting DNA and RNA substrates, as seen for the nuclease activities of Mre11–Rad50 complex.

Targeting allostery expands inhibitor target sites from local active site pockets to include larger interfaces and conformational switches that are often unique to a given functional assembly. In contrast, an active site may have chemistry and shape similarities among multiple proteins. In an analogous approach to our loop movements, active site specificity can be achieved despite similar binding chemistry by directing allowed shifts in a binding pocket (Garcin et al., 2008), which was in fact the inspiration for our approach for MRE11. For the Mre11 nuclease activities, we were able to target adjacent loop and interface regions that open dsDNA ends and those that sculpt ssDNA to enter the active site. This type of dissection-of-function inhibitor can sometimes be mimicked by mutations, as seen for the Mre11 partner Rad50 (Deshpande et al., 2014; Williams et al., 2011). Notably, the Rad50 allosteric conformational change is further influenced by DNA binding and can span the entire molecule and extend into cross-molecular complexes to alter protein interactions, joining DNA ends across 1000 Å (Hopfner et al., 2002; Roset et al., 2014). However for Mre11 nuclease, we were unable to dissect the endo- and exonuclease functions with mutations. Furthermore, an active site inhibitor would logically most likely block both activities. All these observation support the value of the allosteric inhibitor approach as dissection-of-function tools for cell biology. Advanced imaging allowing time and localization measurements provides a means to measure changes in assembly and disassembly of MRE11 DNA repair complexes (Brosey, Ahmed, Lees-Miller, & Tainer, 2017) to test the protein interactions and conformations controlling rate-limiting steps that are often biologically regulated.

3.2 An Avatarian Approach

In the avatarian approach, the tremendous power of comparative genomics and structural similarity can be effectively applied to the design of inhibitors by targeting accessible structures that embody the features critical for control of the system by small molecules. We maintain that there is no compelling need to do structures of human complexes in order to develop and optimize inhibitors through structural analyses. There is of course an absolute need to test resulting inhibitors in human cells and in an appropriate biological system validating their activities for situations where they will be applied. We suggest that the best practical test of specificity is to empirically test if the inhibitor phenotype is removed by mutating the target protein in human cells. This general avatarian approach is valuable as it allows efficient progress in the absence of target human protein structures, and it leverages optimal systems and informative tools for structural biology. In an example supporting the methods and strategies in this report, we used the first archaeal full-length Rad51 structure and assembly to develop and test the BRCA2 assembly in human cells by redesigning the microbial Rad51 at the site of the human RAD51 interface for BRCA2. This work showed that the redesigned microbial Rad51 went to regions of damaged DNA and bound BRCA2 in human cells: the specific targeted binding was validated by its being competed for by high-level expression of BRCA2 BRC3/4 domains (Shin et al., 2003).

3.3 Inhibitor Tools as Biological Probes and Scaffolds for Drug Design

To examine the possible value of specific protein targets and before developing possible drugs, it is generally useful to have chemical tools to probe the biological importance of specific activities. Structures provide the means to see where inhibitors are binding and what they may do to the target protein structure. Combined computational and experimental X-ray crystallographic structures can efficiently and accurately identify small molecule-binding fragments important for inhibition and inform the design of new inhibitors based upon the observed ligand-bound structure (Blundell et al., 2006; Erlanson, Fesik, Hubbard, Jahnke, & Jhoti, 2016; Moiani et al., 2009; Murray & Blundell, 2010; Perry, Harris, Moiani, Olson, & Tainer, 2009; Thomas et al., 2017; Winter et al., 2012). Starting from small fragments bound to a protein, further optimization including merging adjacent fragments with a linker can lead to the generation of better inhibitors. In general, we choose to grow onto fragments for our projects rather than link them to avoid possible strain between binding pockets and linker design issues; nevertheless, linked approaches are also successful and may be optimal for a given target site (Patrone et al., 2016; Waterson et al., 2015).

We find that the experimental X-ray electron density maps for protein-inhibitor complexes will frequently aid comprehension of how functional moieties interact with specific residues of the target enzyme. These interactions may impact proteins' conformational states, and as seen for Mre11 these changes may also perturb possible conformational changes to its substrates. A detailed understanding of the mechanism-of-action from combining structures and measured cell activities usually makes it possible to efficiently improve binding of identified fragments and also to make directed chemical libraries on its small molecular scaffold. Structure-based improvements can include inhibitor optimizations for therapeutic use and with better properties in terms of solubility, toxicity, and specificity. An emerging method will be to examine inhibitor binding by X-ray scattering in solution, which has become high throughput (Hura et al., 2009), and able to detect allosteric assembly changes (Rambo & Tainer, 2013a), changes in flexibility (Rambo & Tainer, 2013b), and relatively high-resolution changes in fold and compactness due to inhibitor binding (Rambo & Tainer, 2013b), as seen for abscisic acid-binding protein (Nishimura et al., 2009). Unlike other structural techniques, SAXS enables comprehensive assessment of conformation and assembly under near physiological conditions (Hura et al., 2013). So we can expect it to be increasingly used for ligand-binding assessments and their impact on assembly, as seen for the apoptosis-inducing factor that links mitochondrial metabolism, DNA damage responses, and programmed cell death (Brosey et al., 2016).

Whereas drug discovery involves many steps beyond the generation of a chemical inhibitor, an allosteric inhibitor with validated cellular activity provides a valuable tool for multiple scientific and biomedical applications. For example, the use of both classes of the MRE11 inhibitors showed that MRE11 endonuclease and exonuclease activities are both required for microhomology-mediated end joining, an error-prone pathway for DSB repair implicated in genomic rearrangement and oncogenic transformation that involves other repair enzymes including CtIP, PARP1, FEN1, plus DNA ligase I and III (Dutta et al., 2017). This MRE11 nuclease requirement was previously masked without the inhibitors.

By describing our strategies and approaches for the development of DNA repair inhibitors exemplified by Mre11, we hope that these methods may benefit the efficient discovery of chemical tools to knockdown- or knockout-specific activities to separate the complex contributions of multifunctional proteins and networks. We view the potential of applying structure-based inhibitor design to impact human disease and patient care as truly great. We therefore propose the avatarian allosteric approach as a possible enabling technology for the design of inhibitor probes and scaffolds for those human systems that are intractable to robust structure determinations and traditional active-site targeting.

ACKNOWLEDGMENTS

We thank the many researchers who have made collaborative and intellectual contributions to our studies of the MRE11 complex, especially Caroline Austin, Yunje Cho, David Ferguson, Jean Gautier, Martin Gellert, Karl-Peter Hopfner, Steve Jackson, Penny Jeggo, Maria Jasin, Cynthia McMurray, Sankar Mitra, Tanya Paull, John Petrini, Guy Poirier, Michael Resnick, Paul Russell, Katharina Schlacher, Lorraine Symington, John Turchi, Robert Scott Williams, Gareth Williams, Claire Wyman, and Miriam Zolan. John Tainer's work on the Mre11–Rad50–Nbs1 complex is aided by Elizabeth D. Getzoff and Tammy Woo and supported in part by the National Institutes of Health (NIH) Grants NIH R01CA117638 and P01CA92584. Added support comes from the Robert A. Welch Distinguished Chair in Chemistry (JAT), the Cancer Prevention and Research Institute of Texas, and the University of Texas STARs program. SAXS at the Advanced Light Source SIBYLS beamline is supported in part by the US Department of Energy program Integrated Diffraction Analysis Technologies (IDAT). This work was also supported by research grants from the Canadian Institutes of Health Research to J.Y.M. (PJ/375408). J.Y.M. is a FRQS Chair in Genome stability. This work was supported in part by the Italian Ministero dell'Istruzione, dell'Università e della Ricerca (MIUR) to E.P. (PRIN 2015 project 2015LZE994).

REFERENCES

Acharya, S., Many, A., Schroeder, A., Kennedy, F., Savytskyy, O., Grubb, J., et al. (2008). Coprinus cinereus rad50 mutants reveal an essential structural role for Rad50 in axial element and synaptonemal complex formation, homolog pairing and meiotic recombination. *Genetics*, *180*(4), 1889–1907. https://doi.org/10.1534/genetics.108.092775.

Andres, S. N., Appel, C. D., Westmoreland, J. W., Williams, J. S., Nguyen, Y., Robertson, P. D., et al. (2015). Tetrameric Ctp1 coordinates DNA binding and DNA bridging in DNA double-strand-break repair. *Nature Structural & Molecular Biology*, *22*(2), 158–166. https://doi.org/10.1038/nsmb.2945.

Bacolla, A., Tainer, J. A., Vasquez, K. M., & Cooper, D. N. (2016). Translocation and deletion breakpoints in cancer genomes are associated with potential non-B DNA-forming sequences. *Nucleic Acids Research*, *44*(12), 5673–5688. https://doi.org/10.1093/nar/gkw261.

Bennardo, N., Cheng, A., Huang, N., Stark, J., & Haber, J. (2008). Alternative-NHEJ is a mechanistically distinct pathway of mammalian chromosome break repair. *PLoS Genetics*, *4*(6), e1000110. https://doi.org/10.1371/journal.pgen.1000110.g005.

Biehs, R., Steinlage, M., Barton, O., Juhasz, S., Kunzel, J., Spies, J., et al. (2017). DNA double-strand break resection occurs during non-homologous end joining in G1 but is distinct from resection during homologous recombination. *Molecular Cell*, *65*(4), 671–684, e675. https://doi.org/10.1016/j.molcel.2016.12.016.

Bierne, H., Ehrlich, S. D., & Michel, B. (1997). Deletions at stalled replication forks occur by two different pathways. *The EMBO Journal*, *16*(11), 3332–3340.

Blundell, T. L., Sibanda, B. L., Montalvao, R. W., Brewerton, S., Chelliah, V., Worth, C. L., et al. (2006). Structural biology and bioinformatics in drug design: Opportunities and challenges for target identification and lead discovery. *Philosophical Transactions of the Royal Society of London. Series B, Biological Sciences*, *361*(1467), 413–423. https://doi.org/10.1098/rstb.2005.1800.

Brosey, C. A., Ahmed, Z., Lees-Miller, S. P., & Tainer, J. A. (2017). What combined measurements from structures and imaging tell us about DNA damage responses. *Methods in Enzymology*, *592*, 417–455. https://doi.org/10.1016/bs.mie.2017.04.005.

Brosey, C. A., Ho, C., Long, W. Z., Singh, S., Burnett, K., Hura, G. L., et al. (2016). Defining NADH-driven allostery regulating apoptosis-inducing factor. *Structure*, *24*(12), 2067–2079. https://doi.org/10.1016/j.str.2016.09.012.

Buis, J., Wu, Y., Deng, Y., Leddon, J., Westfield, G., Eckersdorff, M., et al. (2008). Mre11 nuclease activity has essential roles in DNA repair and genomic stability distinct from ATM activation. *Cell*, *135*(1), 85–96.

Chahwan, C., Nakamura, T. M., Sivakumar, S., Russell, P., & Rhind, N. (2003). The fission yeast Rad32 (Mre11)-Rad50-Nbs1 complex is required for the S-phase DNA damage checkpoint. *Molecular and Cellular Biology*, *23*(18), 6564–6573.

Chan, D. S., Mendes, V., Thomas, S. E., McConnell, B. N., Matak-Vinkovic, D., Coyne, A. G., et al. (2017). Fragment screening against the EthR-DNA interaction by native mass spectrometry. *Angewandte Chemie (International Ed. in English)*, *56*(26), 7488–7491. https://doi.org/10.1002/anie.201702888.

Classen, S., Hura, G. L., Holton, J. M., Rambo, R. P., Rodic, I., McGuire, P. J., et al. (2013). Implementation and performance of SIBYLS: A dual endstation small-angle X-ray scattering and macromolecular crystallography beamline at the advanced light source. *Journal of Applied Crystallography*, *46*(Pt. 1), 1–13. https://doi.org/10.1107/S0021889812048698.

Crown, K. N., Savytskyy, O. P., Malik, S. B., Logsdon, J., Williams, R. S., Tainer, J. A., et al. (2013). A mutation in the FHA domain of Coprinus cinereus Nbs1 leads to Spo11-independent meiotic recombination and chromosome segregation. *G3 (Bethesda, Md)*, *3*(11), 1927–1943. https://doi.org/10.1534/g3.113.007906.

Dalhus, B., Arvai, A. S., Rosnes, I., Olsen, O. E., Backe, P. H., Alseth, I., et al. (2009). Structures of endonuclease V with DNA reveal initiation of deaminated adenine repair. *Nature Structural & Molecular Biology*, *16*(2), 138–143. https://doi.org/10.1038/nsmb.1538.

D'Amours, D., & Jackson, S. P. (2002). The Mre11 complex: At the crossroads of dna repair and checkpoint signalling. *Nature Reviews. Molecular Cell Biology*, *3*(5), 317–327.

Das, D., Moiani, D., Axelrod, H. L., Miller, M. D., Mcmullan, D., Jin, K. K., et al. (2010). Crystal structure of the first eubacterial Mre11 nuclease reveals novel features that may discriminate substrates during DNA repair. *Journal of Molecular Biology*, *397*(3), 647–663. https://doi.org/10.1016/j.jmb.2010.01.049.

de Jager, M., Dronkert, M. L., Modesti, M., Beerens, C. E., Kanaar, R., & van Gent, D. C. (2001). DNA-binding and strand-annealing activities of human Mre11: Implications for its roles in DNA double-strand break repair pathways. *Nucleic Acids Research*, *29*(6), 1317–1325.

Deshpande, R. A., Lee, J. H., Arora, S., & Paull, T. T. (2016). Nbs1 converts the human Mre11/Rad50 nuclease complex into an endo/exonuclease machine specific for protein-DNA adducts. *Molecular Cell*, *64*(3), 593–606. https://doi.org/10.1016/j.molcel.2016.10.010.

Deshpande, R. A., Williams, G. J., Limbo, O., Williams, R. S., Kuhnlein, J., Lee, J. H., et al. (2014). ATP-driven Rad50 conformations regulate DNA tethering, end resection, and ATM checkpoint signaling. *The EMBO Journal*, *33*(5), 482–500. https://doi.org/10.1002/embj.201386100.

Dupré, A., Boyer-Chatenet, L., Sattler, R., Modi, A., Lee, J., Nicolette, M., et al. (2008). A forward chemical genetic screen reveals an inhibitor of the Mre11-Rad50-Nbs1 complex. *Nature Chemical Biology*, *4*(2), 119–125. https://doi.org/10.1038/nchembio.63.

Dutta, A., Eckelmann, B., Adhikari, S., Ahmed, K. M., Sengupta, S., Pandey, A., et al. (2017). Microhomology-mediated end joining is activated in irradiated human cells due to phosphorylation-dependent formation of the XRCC1 repair complex. *Nucleic Acids Research*, *45*(5), 2585–2599. https://doi.org/10.1093/nar/gkw1262.

Dynlacht, J. R., Batuello, C. N., Lopez, J. T., Kim, K. K., & Turchi, J. J. (2011). Identification of Mre11 as a target for heat radiosensitization. *Radiation Research*, *176*(3), 323–332.

Erlanson, D. A., Fesik, S. W., Hubbard, R. E., Jahnke, W., & Jhoti, H. (2016). Twenty years on: The impact of fragments on drug discovery. *Nature Reviews. Drug Discovery*, *15*(9), 605–619. https://doi.org/10.1038/nrd.2016.109.

Falck, J., Coates, J., & Jackson, S. P. (2005). Conserved modes of recruitment of ATM, ATR and DNA-PKcs to sites of DNA damage. *Nature*, *434*(7033), 605–611.

Garcin, E. D., Arvai, A. S., Rosenfeld, R. J., Kroeger, M. D., Crane, B. R., Andersson, G., et al. (2008). Anchored plasticity opens doors for selective inhibitor design in nitric oxide synthase. *Nature Chemical Biology*, *4*(11), 700–707. https://doi.org/10.1038/nchembio.115.

Garner, K. M., Pletnev, A. A., & Eastman, A. (2009). Corrected structure of mirin, a small-molecule inhibitor of the Mre11-Rad50-Nbs1 complex. *Nature Chemical Biology*, *5*(3), 129–130. author reply 130. https://doi.org/10.1038/nchembio0309-129.

Goldsmith, E. J. (1996). Allosteric enzymes as models for chemomechanical energy transducing assemblies. *The FASEB Journal*, *10*(7), 702–708.

Gradia, S. D., Ishida, J. P., Tsai, M. S., Jeans, C., Tainer, J. A., & Fuss, J. O. (2017). MacroBac: New technologies for robust and efficient large-scale production of recombinant multiprotein complexes. *Methods in Enzymology*, *592*, 1–26. https://doi.org/10.1016/bs.mie.2017.03.008.

Guan, Y., Manuel, R. C., Arvai, A. S., Parikh, S. S., Mol, C. D., Miller, J. H., et al. (1998). MutY catalytic core, mutant and bound adenine structures define specificity for DNA repair enzyme superfamily. *Nature Structural Biology*, *5*(12), 1058–1064. https://doi.org/10.1038/4168.

Haince, J. F., McDonald, D., Rodrigue, A., Dery, U., Masson, J. Y., Hendzel, M. J., et al. (2008). PARP1-dependent kinetics of recruitment of MRE11 and NBS1 proteins to multiple DNA damage sites. *The Journal of Biological Chemistry*, *283*(2), 1197–1208. https://doi.org/10.1074/jbc.M706734200.

Hengel, S. R., Spies, M. A., & Spies, M. (2017). Small-molecule inhibitors targeting DNA repair and DNA repair deficiency in research and cancer therapy. *Cell Chemical Biology*, *24*(9), 1101–1119. https://doi.org/10.1016/j.chembiol.2017.08.027.

Hopfner, K., Craig, L., Moncalian, G., Zinkel, R., Usui, T., Owen, B., et al. (2002). The Rad50 zinc-hook is a structure joining Mre11 complexes in DNA recombination and repair. *Nature*, *418*(6897), 562–566. https://doi.org/10.1038/nature00922.

Hopfner, K. P., Karcher, A., Craig, L., Woo, T. T., Carney, J. P., & Tainer, J. A. (2001). Structural biochemistry and interaction architecture of the DNA double-strand break repair Mre11 nuclease and Rad50-ATPase. *Cell*, *105*(4), 473–485.

Hopfner, K. P., Karcher, A., Shin, D., Fairley, C., Tainer, J. A., & Carney, J. P. (2000). Mre11 and Rad50 from Pyrococcus furiosus: Cloning and biochemical characterization reveal an evolutionarily conserved multiprotein machine. *Journal of Bacteriology*, *182*(21), 6036–6041.

Hosoya, N., & Miyagawa, K. (2014). Targeting DNA damage response in cancer therapy. *Cancer Science*, *105*(4), 370–388. https://doi.org/10.1111/cas.12366.

Hura, G. L., Budworth, H., Dyer, K. N., Rambo, R. P., Hammel, M., McMurray, C. T., et al. (2013). Comprehensive macromolecular conformations mapped by quantitative SAXS analyses. *Nature Methods*, *10*(6), 453–454. https://doi.org/10.1038/nmeth.2453.

Hura, G. L., Menon, A. L., Hammel, M., Rambo, R., Ii, F. L. P., Tsutakawa, S. E., et al. (2009). Robust high-throughput solution structural analyses by small angle X-ray scattering (SAXS). *Nature Methods*, *6*(8), 606–612. https://doi.org/10.1038/nmeth.1353.

Kelley, M. R., Logsdon, D., & Fishel, M. L. (2014). Targeting DNA repair pathways for cancer treatment: What's new? *Future Oncology*, *10*(7), 1215–1237. https://doi.org/10.2217/fon.14.60.

Kim, J. H., Grosbart, M., Anand, R., Wyman, C., Cejka, P., & Petrini, J. H. (2017). The Mre11-Nbs1 interface is essential for viability and tumor suppression. *Cell Reports*, *18*(2), 496–507. https://doi.org/10.1016/j.celrep.2016.12.035.

Koukourakis, M. I. (2012). Radiation damage and radioprotectants: New concepts in the era of molecular medicine. *The British Journal of Radiology*, *85*(1012), 313–330. https://doi.org/10.1259/bjr/16386034.

Krogh, B. O., Llorente, B., Lam, A., & Symington, L. S. (2005). Mutations in Mre11 phosphoesterase motif I that impair Saccharomyces cerevisiae Mre11-Rad50-Xrs2 complex stability in addition to nuclease activity. *Genetics*, *171*(4), 1561–1570. https://doi.org/10.1534/genetics.105.049478.

Kuroda, S., Urata, Y., & Fujiwara, T. (2012). Ataxia-telangiectasia mutated and the Mre11-Rad50-NBS1 complex: Promising targets for radiosensitization. *Acta Medica Okayama*, *66*(2), 83–92. https://doi.org/10.18926/AMO/48258.

Lafrance-Vanasse, J., Williams, G. J., & Tainer, J. A. (2015). Envisioning the dynamics and flexibility of Mre11-Rad50-Nbs1 complex to decipher its roles in DNA replication and repair. *Progress in Biophysics and Molecular Biology*, *117*(2–3), 182–193. https://doi.org/10.1016/j.pbiomolbio.2014.12.004.

Lai, Y. T., Hura, G. L., Dyer, K. N., Tang, H. Y., Tainer, J. A., & Yeates, T. O. (2016). Designing and defining dynamic protein cage nanoassemblies in solution. *Science Advances*, *2*(12). e1501855. https://doi.org/10.1126/sciadv.1501855.

Lamarche, B. J., Orazio, N. I., & Weitzman, M. D. (2010). The MRN complex in double-strand break repair and telomere maintenance. *FEBS Letters*, *584*(17), 3682–3695. https://doi.org/10.1016/j.febslet.2010.07.029.

Lammens, K., Bemeleit, D. J., Möckel, C., Clausing, E., Schele, A., Hartung, S., et al. (2011). The Mre11:Rad50 structure shows an ATP-dependent molecular clamp in DNA double-strand break repair. *Cell*, *145*(1), 54–66. https://doi.org/10.1016/j.cell.2011.02.038.

Lee, K. C., Padget, K., Curtis, H., Cowell, I. G., Moiani, D., Sondka, Z., et al. (2012). MRE11 facilitates the removal of human topoisomerase II complexes from genomic DNA. *Biology Open*, *1*(9), 863–873. https://doi.org/10.1242/bio.20121834.

Lee, J. H., & Paull, T. T. (2005). ATM activation by DNA double-strand breaks through the Mre11-Rad50-Nbs1 complex. *Science*, *308*(5721), 551–554.

Lewis, L. K., Storici, F., Van Komen, S., Calero, S., Sung, P., & Resnick, M. A. (2004). Role of the nuclease activity of Saccharomyces cerevisiae Mre11 in repair of DNA double-strand breaks in mitotic cells. *Genetics*, *166*(4), 1701–1713. https://doi.org/10.1534/genetics.166.4.1701.

Limbo, O., Moiani, D., Kertokalio, A., Wyman, C., Tainer, J. A., & Russell, P. (2012). Mre11 ATLD17/18 mutation retains Tel1/ATM activity but blocks DNA double-strand break repair. *Nucleic Acids Research*, *40*(22), 11435–11449. https://doi.org/10.1093/nar/gks954.

Limbo, O., Porter-Goff, M. E., Rhind, N., & Russell, P. (2011). Mre11 nuclease activity and Ctp1 regulate Chk1 activation by Rad3ATR and Tel1ATM checkpoint kinases at double-strand breaks. *Molecular and Cellular Biology*, *31*(3), 573–583. https://doi.org/10.1128/MCB.00994-10.

Lindahl, T. (1993). Instability and decay of the primary structure of DNA. *Nature*, *362*(6422), 709–715. https://doi.org/10.1038/362709a0.

Majka, J., Alford, B., Ausio, J., Finn, R. M., & McMurray, C. T. (2012). ATP hydrolysis by RAD50 protein switches MRE11 enzyme from endonuclease to exonuclease. *The Journal of Biological Chemistry*, *287*(4), 2328–2341. https://doi.org/10.1074/jbc.M111.307041.

Makharashvili, N., Tubbs, A. T., Yang, S. H., Wang, H., Barton, O., Zhou, Y., et al. (2014). Catalytic and noncatalytic roles of the CtIP endonuclease in double-strand break end resection. *Molecular Cell*, *54*(6), 1022–1033. https://doi.org/10.1016/j.molcel.2014.04.011.

McPherson, L. A., Shen, Y., & Ford, J. M. (2014). Poly (ADP-ribose) polymerase inhibitor LT-626: Sensitivity correlates with MRE11 mutations and synergizes with platinums and irinotecan in colorectal cancer cells. *Cancer Letters*, *343*(2), 217–223. https://doi.org/10.1016/j.canlet.2013.10.034.

Mendgen, T., Steuer, C., & Klein, C. D. (2012). Privileged scaffolds or promiscuous binders: A comparative study on rhodanines and related heterocycles in medicinal chemistry. *Journal of Medicinal Chemistry*, *55*(2), 743–753. https://doi.org/10.1021/jm201243p.

Mockel, C., Lammens, K., Schele, A., & Hopfner, K. P. (2012). ATP driven structural changes of the bacterial Mre11:Rad50 catalytic head complex. *Nucleic Acids Research*, *40*(2), 914–927. https://doi.org/10.1093/nar/gkr749.

Moiani, D., Cavallotti, C., Famulari, A., & Schmuck, C. (2008). Oxoanion binding by guanidiniocarbonylpyrrole cations in water: A combined DFT and MD investigation. *Chemistry*, *14*(17), 5207–5219. https://doi.org/10.1002/chem.200701745.

Moiani, D., Salvalaglio, M., Cavallotti, C., Bujacz, A., Redzynia, I., Bujacz, G., et al. (2009). Structural characterization of a protein A mimetic peptide dendrimer bound to human IgG. *The Journal of Physical Chemistry. B*, *113*(50), 16268–16275. https://doi.org/10.1021/jp909405b.

Murray, C. W., & Blundell, T. L. (2010). Structural biology in fragment-based drug design. *Current Opinion in Structural Biology*, *20*(4), 497–507. https://doi.org/10.1016/j.sbi.2010.04.003.

Myler, L. R., Gallardo, I. F., Soniat, M. M., Deshpande, R. A., Gonzalez, X. B., Kim, Y., et al. (2017). Single-molecule imaging reveals how Mre11-Rad50-Nbs1 initiates DNA break repair. *Molecular Cell*, *67*(5), 891–898. e894. https://doi.org/10.1016/j.molcel.2017.08.002.

Nicholson, J., Jevons, S. J., Groselj, B., Ellermann, S., Konietzny, R., Kerr, M., et al. (2017). E3 ligase cIAP2 mediates downregulation of MRE11 and radiosensitization in response to HDAC inhibition in bladder cancer. *Cancer Research*, 77(11), 3027–3039. https://doi.org/10.1158/0008-5472.CAN-16-3232.

Nishimura, N., Hitomi, K., Arvai, A. S., Rambo, R. P., Hitomi, C., Cutler, S. R., et al. (2009). Structural mechanism of abscisic acid binding and signaling by dimeric PYR1. *Science*, *326*(5958), 1373–1379. https://doi.org/10.1126/science.1181829.

Park, Y. B., Chae, J., Kim, Y. C., & Cho, Y. (2011). Crystal structure of human Mre11: Understanding tumorigenic mutations. *Structure*, *19*(11), 1591–1602. https://doi.org/10.1016/j.str.2011.09.010.

Patrone, J. D., Pelz, N. F., Bates, B. S., Souza-Fagundes, E. M., Vangamudi, B., Camper, D. V., et al. (2016). Identification and optimization of anthranilic acid based inhibitors of replication protein A. *ChemMedChem*, *11*(8), 893–899. https://doi.org/10.1002/cmdc.201500479.

Paull, T. T., & Gellert, M. (1998). The 3′ to 5′ exonuclease activity of Mre 11 facilitates repair of DNA double-strand breaks. *Molecular Cell, 1*(7), 969–979.

Perry, J. J., Harris, R. M., Moiani, D., Olson, A. J., & Tainer, J. A. (2009). p38alpha MAP kinase C-terminal domain binding pocket characterized by crystallographic and computational analyses. *Journal of Molecular Biology, 391*(1), 1–11. https://doi.org/10.1016/j.jmb.2009.06.005.

Rahal, E. A., Henricksen, L. A., Li, Y., Williams, R. S., Tainer, J., & Dixon, K. (2010). ATM regulates Mre11-dependent DNA end-degradation and microhomology-mediated end joining. *Cell Cycle, 9*(14), 2866–2877.

Rambo, R. P., & Tainer, J. A. (2011). Characterizing flexible and intrinsically unstructured biological macromolecules by SAS using the Porod-Debye law. *Biopolymers, 95*(8), 559–571. https://doi.org/10.1002/bip.21638.

Rambo, R. P., & Tainer, J. A. (2013a). Accurate assessment of mass, models and resolution by small-angle scattering. *Nature, 496*(7446), 477–481. https://doi.org/10.1038/nature12070.

Rambo, R. P., & Tainer, J. A. (2013b). Super-resolution in solution X-ray scattering and its applications to structural systems biology. *Annual Review of Biophysics, 42*, 415–441. https://doi.org/10.1146/annurev-biophys-083012-130301.

Rashid, F., Harris, P. D., Zaher, M. S., Sobhy, M. A., Joudeh, L. I., Yan, C., et al. (2017). Single-molecule FRET unveils induced-fit mechanism for substrate selectivity in flap endonuclease 1. *eLife, 6*. https://doi.org/10.7554/eLife.21884.

Regal, J. A., Festerling, T. A., Buis, J. M., & Ferguson, DO. (2013). Disease-associated MRE11 mutants impact ATM/ATR DNA damage signaling by distinct mechanisms. *Human Molecular Genetics, 22*(25), 5146–5159. https://doi.org/10.1093/hmg/ddt368.

Rojowska, A., Lammens, K., Seifert, F. U., Direnberger, C., Feldmann, H., & Hopfner, K. P. (2014). Structure of the Rad50 DNA double-strand break repair protein in complex with DNA. *The EMBO Journal, 33*(23), 2847–2859. https://doi.org/10.15252/embj.201488889.

Roques, C., Coulombe, Y., Delannoy, M., Vignard, J., Grossi, S., Brodeur, I., et al. (2009). MRE11-RAD50-NBS1 is a critical regulator of FANCD2 stability and function during DNA double-strand break repair. *The EMBO Journal, 28*(16), 2400–2413. https://doi.org/10.1038/emboj.2009.193.

Roset, R., Inagaki, A., Hohl, M., Brenet, F., Lafrance-Vanasse, J., Lange, J., et al. (2014). The Rad50 hook domain regulates DNA damage signaling and tumorigenesis. *Genes & Development, 28*(5), 451–462. https://doi.org/10.1101/gad.236745.113.

Roy, S., Tomaszowski, K. H., Luzwick, J. W., Park, S., Li, J., Murphy, M., et al. (2018). p53 suppresses mutagenic RAD52 and POLtheta pathways by orchestrating DNA replication restart homeostasis. *Elife* 7. https://doi.org/10.7554/eLife.31723.

Sartori, A. A., Lukas, C., Coates, J., Mistrik, M., Fu, S., Bartek, J., et al. (2007). Human CtIP promotes DNA end resection. *Nature, 450*(7169), 509–514. https://doi.org/10.1038/nature06337.

Schiller, C. B., Lammens, K., Guerini, I., Coordes, B., Feldmann, H., Schlauderer, F., et al. (2012). Structure of Mre11-Nbs1 complex yields insights into ataxia-telangiectasia-like disease mutations and DNA damage signaling. *Nature Structural & Molecular Biology, 19*(7), 693–700. https://doi.org/10.1038/nsmb.2323.

Schlacher, K., Christ, N., Siaud, N., Egashira, A., Wu, H., & Jasin, M. (2011). Double-strand break repair-independent role for BRCA2 in blocking stalled replication fork degradation by MRE11. *Cell, 145*(4), 529–542. https://doi.org/10.1016/j.cell.2011.03.041.

Schlacher, K., Wu, H., & Jasin, M. (2012). A distinct replication fork protection pathway connects Fanconi anemia tumor suppressors to RAD51-BRCA1/2. *Cancer Cell, 22*(1), 106–116. https://doi.org/10.1016/j.ccr.2012.05.015.

Schneidman-Duhovny, D., Hammel, M., Tainer, J. A., & Sali, A. (2016). FoXS, FoXSDock and MultiFoXS: Single-state and multi-state structural modeling of proteins and their complexes based on SAXS profiles. *Nucleic Acids Research*, *44*(W1), W424–429. https://doi.org/10.1093/nar/gkw389.

Schrodinger. (2017). *Small-molecule drug discovery suite 2017-2*.

Seeber, A., Hegnauer, A. M., Hustedt, N., Deshpande, I., Poli, J., Eglinger, J., et al. (2016). RPA mediates recruitment of MRX to forks and double-strand breaks to hold sister chromatids together. *Molecular Cell*, *64*(5), 951–966. https://doi.org/10.1016/j.molcel.2016.10.032.

Shibata, A., Moiani, D., Arvai, A. S., Perry, J., Harding, S. M., Genois, M. M., et al. (2014). DNA double-strand break repair pathway choice is directed by distinct MRE11 nuclease activities. *Molecular Cell*, *53*(1), 7–18. https://doi.org/10.1016/j.molcel.2013.11.003.

Shin, D. S., Pellegrini, L., Daniels, D. S., Yelent, B., Craig, L., Bates, D., et al. (2003). Full-length archaeal Rad51 structure and mutants: Mechanisms for RAD51 assembly and control by BRCA2. *The EMBO Journal*, *22*(17), 4566–4576. https://doi.org/10.1093/emboj/cdg429.

Shiotani, B., Nguyen, H. D., Hakansson, P., Marechal, A., Tse, A., Tahara, H., et al. (2013). Two distinct modes of ATR activation orchestrated by Rad17 and Nbs1. *Cell Reports*, *3*(5), 1651–1662. https://doi.org/10.1016/j.celrep.2013.04.018.

Slupphaug, G., Mol, C. D., Kavli, B., Arvai, A. S., Krokan, H. E., & Tainer, J. A. (1996). A nucleotide-flipping mechanism from the structure of human uracil-DNA glycosylase bound to DNA. *Nature*, *384*(6604), 87–92. https://doi.org/10.1038/384087a0.

Stracker, T. H., & Petrini, J. H. (2011). The MRE11 complex: Starting from the ends. *Nature Reviews Molecular Cell Biology*, *12*(2), 90–103. https://doi.org/10.1038/nrm3047.

Sun, C., Fang, Y., Yin, J., Chen, J., Ju, Z., Zhang, D., et al. (2017). Rational combination therapy with PARP and MEK inhibitors capitalizes on therapeutic liabilities in RAS mutant cancers. *Science Translational Medicine*, *9*(392), 1–18. https://doi.org/10.1126/scitranslmed.aal5148.

Thayer, M. M., Ahern, H., Xing, D., Cunningham, R. P., & Tainer, J. A. (1995). Novel DNA binding motifs in the DNA repair enzyme endonuclease III crystal structure. *The EMBO Journal*, *14*(16), 4108–4120.

Thomas, S. E., Mendes, V., Kim, S. Y., Malhotra, S., Ochoa-Montano, B., Blaszczyk, M., et al. (2017). Structural biology and the design of new therapeutics: From HIV and cancer to mycobacterial infections: A paper dedicated to John Kendrew. *Journal of Molecular Biology*, *429*(17), 2677–2693. https://doi.org/10.1016/j.jmb.2017.06.014.

Thompson, R., Montano, R., & Eastman, A. (2012). The Mre11 nuclease is critical for the sensitivity of cells to Chk1 inhibition. *PLoS One*, 7(8). e44021https://doi.org/10.1371/journal.pone.0044021.

Tsutakawa, S. E., Lafrance-Vanasse, J., & Tainer, J. A. (2014). The cutting edges in DNA repair, licensing, and fidelity: DNA and RNA repair nucleases sculpt DNA to measure twice, cut once. *DNA Repair (Amst)*, *19*, 95–107. https://doi.org/10.1016/j.dnarep.2014.03.022.

Tsutakawa, S. E., Shin, D. S., Mol, C. D., Izumi, T., Arvai, A. S., Mantha, A. K., et al. (2013). Conserved structural chemistry for incision activity in structurally non-homologous apurinic/apyrimidinic endonuclease APE1 and endonuclease IV DNA repair enzymes. *The Journal of Biological Chemistry*, *288*(12), 8445–8455. https://doi.org/10.1074/jbc.M112.422774.

Tsutakawa, S. E., Thompson, M. J., Arvai, A. S., Neil, A. J., Shaw, S. J., Algasaier, S. I., et al. (2017). Phosphate steering by flap endonuclease 1 promotes 5′-flap specificity and incision to prevent genome instability. *Nature Communications*, *8*, 15855. https://doi.org/10.1038/ncomms15855.

Velic, D., Couturier, A. M., Ferreira, M. T., Rodrigue, A., Poirier, G. G., Fleury, F., et al. (2015). DNA damage signalling and repair inhibitors: The long-sought-after Achilles' heel of cancer. *Biomolecules*, *5*(4), 3204–3259. https://doi.org/10.3390/biom5043204.

Vilar, E., Bartnik, C. M., Stenzel, S. L., Raskin, L., Ahn, J., Moreno, V., et al. (2011). MRE11 deficiency increases sensitivity to poly(ADP-ribose) polymerase inhibition in microsatellite unstable colorectal cancers. *Cancer Research*, *71*(7), 2632–2642. https://doi.org/10.1158/0008-5472.CAN-10-1120.

Wang, L., Pulk, A., Wasserman, M. R., Feldman, M. B., Altman, R. B., Cate, J. H., et al. (2012). Allosteric control of the ribosome by small-molecule antibiotics. *Nature Structural & Molecular Biology*, *19*(9), 957–963. https://doi.org/10.1038/nsmb.2360.

Waterson, A. G., Kennedy, J. P., Patrone, J. D., Pelz, N. F., Feldkamp, M. D., Frank, A. O., et al. (2015). Diphenylpyrazoles as replication protein a inhibitors. *ACS Medicinal Chemistry Letters*, *6*(2), 140–145. https://doi.org/10.1021/ml5003629.

Williams, R. S., Dodson, G. E., Limbo, O., Yamada, Y., Williams, J. S., Guenther, G., et al. (2009). Nbs1 flexibly tethers Ctp1 and Mre11-Rad50 to coordinate DNA double-strand break processing and repair. *Cell*, *139*(1), 87–99. https://doi.org/10.1016/j.cell.2009.07.033.

Williams, G. J., Lees-Miller, S. P., & Tainer, J. A. (2010). Mre11-Rad50-Nbs1 conformations and the control of sensing, signaling, and effector responses at DNA double-strand breaks. *DNA Repair*, *9*(12), 1299–1306. https://doi.org/10.1016/j.dnarep.2010.10.001.

Williams, R. S., Moncalian, G., Williams, J. S., Yamada, Y., Limbo, O., Shin, D. S., et al. (2008). Mre11 dimers coordinate DNA end bridging and nuclease processing in double-strand-break repair. *Cell*, *135*(1), 97–109. https://doi.org/10.1016/j.cell.2008.08.017.

Williams, R. S., & Tainer, J. (2007). Learning our ABCs: Rad50 directs MRN repair functions via adenylate kinase activity from the conserved ATP binding cassette. *Molecular Cell*, *25*(6), 789–791. https://doi.org/10.1016/j.molcel.2007.03.004.

Williams, J., Williams, R. S., Dovey, C. L., Guenther, G., Tainer, J., & Russell, P. (2010). gammaH2A binds Brc1 to maintain genome integrity during S-phase. *The EMBO Journal*, *29*(6), 1136–1148. https://doi.org/10.1038/emboj.2009.413.

Williams, R., Williams, J., & Tainer, J. (2007). Mre11-Rad50-Nbs1 is a keystone complex connecting DNA repair machinery, double-strand break signaling, and the chromatin template. *Biochemistry and Cell Biology*, *85*(4), 509–520. https://doi.org/10.1139/O07-069.

Williams, G. J., Williams, R. S., Williams, J. S., Moncalian, G., Arvai, A. S., Limbo, O., et al. (2011). ABC ATPase signature helices in Rad50 link nucleotide state to Mre11 interface for DNA repair. *Nature Structural & Molecular Biology*, *18*(4), 423–431. https://doi.org/10.1038/nsmb.2038.

Winter, A., Higueruelo, A. P., Marsh, M., Sigurdardottir, A., Pitt, W. R., & Blundell, T. L. (2012). Biophysical and computational fragment-based approaches to targeting protein-protein interactions: Applications in structure-guided drug discovery. *Quarterly Reviews of Biophysics*, *45*(4), 383–426. https://doi.org/10.1017/S0033583512000108.

You, Z., Chahwan, C., Bailis, J., Hunter, T., & Russell, P. (2005). ATM activation and its recruitment to damaged DNA require binding to the C terminus of Nbs1. *Molecular and Cellular Biology*, *25*(13), 5363–5379.

Yu, Z., Vogel, G., Coulombe, Y., Dubeau, D., Spehalski, E., Hebert, J., et al. (2012). The MRE11 GAR motif regulates DNA double-strand break processing and ATR activation. *Cell Research*, *22*(2), 305–320. https://doi.org/10.1038/cr.2011.128.

CHAPTER ELEVEN

Integrating Experimental and *In Silico* HTS in the Discovery of Inhibitors of Protein–Nucleic Acid Interactions

Quinn Li, Laura Folly da Silva Constantino, M. Ashley Spies[1]
The University of Iowa, Iowa City, IA, United States
[1]Corresponding author: e-mail address: m-ashley-spies@uiowa.edu

Contents

Abstract

Discovery of novel tool compounds and drug leads against a range of unorthodox protein targets has pushed both experimental screening methodologies as well as the field of structure-based design to the limit in recent years. Increasingly, it has been recognized that some of the most desirable targets for the development of small-molecule effectors are actually protein–protein and protein–nucleic acid interactions. There are numerous nontrivial challenges to pursuing small-molecule lead compounds directed toward PPIs and PNIs: relatively shallow cavities, large surface areas that are natively complexed to macromolecules, complex patterns of interstitial waters, a paucity of "hot spots," large conformational changes upon ligand binding, etc. Although there

Methods in Enzymology, Volume 601
ISSN 0076-6879
https://doi.org/10.1016/bs.mie.2017.11.036

have been some notable successes targeting PPIs in the last decade, there has been distinctly less success in the realm of targeting PNIs. This chapter focuses on an approach, successfully applied by our group to address the challenge of gaining traction on the PPI target RAD52, which is a protein that binds both single-stranded and double-stranded DNA, and is an anticancer target for certain types of cancer. There are many approaches to tackling the difficult problems of finding effective small molecules that disrupt PPIs and PNIs, but the methods presented here offer a series of elegant solutions, which integrate experimental HTS, biophysical methods, docking, and molecular dynamics in a powerful way. Additionally, the structural knowledge gained from these studies provides a means for rationally understanding what features lead to ligand affinity in these fascinating and highly unorthodox pockets.

1. THE CHALLENGE OF LIGAND DISCOVERY AND DESIGN TARGETING PROTEIN–NUCLEIC ACID (AND PROTEIN–PROTEIN) INTERACTIONS

It has become increasingly desirable to develop drug lead compounds against protein–macromolecular complexes. This is not surprising, as these unorthodox drug targets make up the bulk of the human proteome and are involved in an array of pharmaceutically attractive phenomena, including cellular movement, assembly and disassembly of protein complexes, chaperon activity, subcellular transport, nucleic acid binding, signal transduction, transcription, and other functions (Makley & Gestwicki, 2013; Venter, Adams, Fau-Myers, & Al, 2001). However, regardless of their medical importance, these unorthodox targets have been very recalcitrant to the facile identification of small-molecule inhibitors of their native activities. Unlike classic drug targets, such as enzymes, the native ligand in PPIs or PNIs, even when known (which may often not be the case) does not usually provide a clear picture about how to proceed with small-molecule scaffold discovery and/or development. Additionally, PPIs and PNIs often lack clearly distinguishable binding pockets or have highly diffuse binding interfaces. The nature of these interfaces is complicated by the presence of "hot spots" that largely control the magnitude of the ligand-binding energy (Makley & Gestwicki, 2013; Scott, Bayly, Abell, & Skidmore, 2016; Yap et al., 2012). Furthermore, additional complications arise when considering protein flexibility and the ligand-associated changes pocket distribution (Dean, Whalen, & Spies, 2015; Eyrisch & Helms, 2007; Sperandio, Miteva, Segers, Nicolaes, & Villoutreix, 2008).

Most of what we know about the ability to successfully identify drug lead molecules against protein–macromolecular targets comes from the field of PPIs. There are certain types of PPIs that have been more amenable to drug discovery than others (Scott et al., 2016). Particular success has emerged from strategies that attempt to mimic a peptide or protein ligand, such that small molecule leads directly compete with the native macromolecular ligand (De Vega, Martin-Martinez, & Gonzalez-Muniz, 2007). In the field of PPI research, structural and biophysical approaches play a key role, often in iterative cycles of ligand development (Scott et al., 2016). These discovery campaigns have often revealed the fascinating, but complicated role that water plays in the shape, accessibility, and binding activity of the PPI protein target. A number of key lessons should be noted from PPIs, which are likely to be at least partially relevant to PNIs. The buried regions in PPIs that have been amenable to inhibitor development are usually between 1000 and 6000 $Å^2$ (Scott et al., 2016), with regions larger than 2000 $Å^2$ often consisting of a patchwork of buried and solvent accessible groups. Often there are increases in Trp and Tyr residues around the PPI-binding pocket, as well as polar residues (e.g., Arg, Asp, and His). The classical PPI possesses some hydrophobic core region flanked by a patchwork of polar regions, often utilizing interstitial waters. Generally, PPI inhibition strategies are more successful when the binding hot spots are relatively contiguous. Experimental approaches that have yielded results include classical HTS, fragment-based drug discovery, design of targeted peptides and peptidomimetics, and computational approaches (i.e., structure-based discovery and design). Traditional computational approaches to identification of PPI inhibitors are extremely challenging, due to the often ambiguous role of water and the inherent ligand-associated protein conformational changes in most PPIs (Huggins, Marsh, & Payne, 2011; Scott et al., 2016).

Although there are a vast array of difficulties in executing drug discovery campaigns against PPIs, it is potentially more challenging to attempt to disrupt PNIs, as much less is known about them (Yap et al., 2012). Nevertheless, our research group was interested in designing a campaign targeting the DNA repair protein RAD52, by specifically disrupting its ssDNA-binding activity. The essence of our approach was to first carry out an experimental (in vitro) HTS campaign targeting the RAD52–ssDNA PNI, and to then employ computational and structural methods to learn about how certain chemotypes were disrupting the PNIs. A statistical approach was used to test the strength of our structural hypotheses about PNI disruption. Finally, this knowledge was then put to the test in an independent in silico HTS campaign against a database of interesting natural product compounds. Amazingly, a single natural product was selected for detailed biophysical study and found to be in the high n*M* to low μ*M* range (depending on the assay employed). A rational strategy for targeting the RAD52 PNI is, ostensibly, a nightmarish challenge for structure-based design, in that only an unliganded crystal structure of RAD52 was available (PDB 1KN0), which suggests an exotic circular continuous ssDNA-binding epitope (Hengel et al., 2016; Kagawa et al., 2002), with a patchwork of alternating hydrophobic and hydrophilic regions and complex water networks in the biding groove. Given the uncertainty of PDB 1KN0 representing the liganded conformation of the protein in solution, the lack of any known ligands that bind in this groove and the complex role of water, we sought to build an integrated workflow for PNI inhibitor discovery, which capitalized on in vitro HTS initially, followed by computational studies built on a rigorous statistical framework for understanding why certain chemotypes disrupt the RAD52 PNI. This validated computational approach is then used to identify completely novel scaffolds in an in silico screening. Finally, we take great care in experimentally validating these PNI inhibitors via biophysical methods that confirm the projected activity of preventing ssDNA binding to the RAD52 target.

This chapter focuses largely on how to parlay the experimental HTS data into models that can be meaningfully tested for accuracy, and how to extrapolate these validated models into in silico screens against the challenging RAD52 target. Not only are these approaches relevant to discovering inhibitors against other DNA repair proteins, such as RAD51 recombinase and RPA ss-DNA-binding proteins, whose binding may be assayed in a similar fashion (Grimme & Spies, 2011; Subramanyam, Jones, Spies, & Spies, 2013), but to many types of PPI targets as well. The sections described below focus on (1) how experimental HTS data was obtained for disrupting the RAD52 PPI, and an analysis of the most promising inhibitors; (2) employing docking and molecular dynamics simulations to obtain information about the locus of action of the new inhibitors obtained from HTS; (3) utilizing the receiver operating characteristic (ROC) (Metz, 1978; Yap et al., 2012) statistical method to determine the accuracy of various computational workflow models; (4) application of the ROC-validated model in an in silico screening campaign against a database of natural products and a discussion of the role of biophysical validation using NMR WaterLOGSY.

2. EMPLOYING EXPERIMENTAL HTS AS A TOOL FOR LEARNING ABOUT THE NATURE OF PNI LEAD COMPOUNDS THAT DISRUPT THE RAD52 PNI

To identify compounds that disrupt the RAD52–ssDNA interactions, a previously established Förster resonance energy transfer (FRET)-based assay was adapted to the HTS format (Grimme et al., 2010; Grimme & Spies, 2011). The assay took advantage of certain known characteristics of RAD52; the ssDNA-binding region of RAD52 is continuous around the circumference of the ring with a shallow nucleic acid interactive groove that repeats in each monomer. Our FRET-based assay relied on the ability of ssDNA to bind and wrap around the narrow groove spanning the protein ring (Grimme et al., 2010; Grimme & Spies, 2011). FRET donor (Cy3) and acceptor (Cy5) were incorporated at the opposite ends of a 30-mer ssDNA to form Cy3-dT_{30}-Cy5 (which can be synthesized commercially from IDT (Integrated DNA Technologies, Inc.)). An increase in the FRET signal occurs when the two fluorophores Cy3 and Cy5 were brought into proximity to one another during the process in which Cy3-dT_{30}-Cy5 forms a 1:1 stoichiometric complex with RAD52. The positive control for disrupting the RAD52–ssDNA interaction was achieved using a stoichiometric complex of RAD52 with Cy3-dT_{30}-Cy5 substrate competing with an excess amount of unlabeled ssDNA (Poly dT100); while the negative control consisted of an unperturbed stoichiometric complex of RAD52 with Cy3-dT_{30}-Cy5 (Fig. 1A). A Multiflo dispenser (Biotek) was used for dispensing control mixtures into the plate, and compounds of interest were dispensed and mixed three times using a Microlab Star liquid handling robot (Hamilton). It is noted that from our experiment, the appropriate reaction volume along with the right plate type are valuable in producing an acceptable single/noise ratio. The quality of the HTS assay was quantified with the *Z*-factor, which accesses the assay's performance in terms of separation between the positive and negative controls. A more detailed derivation of the *Z*-factor is described in Zhang, Chung, and Oldenburg (1999). For the evaluation of any assays, a *Z*-factor of 1 is ideal. The *Z*-factor from our RAD52–ssDNA FRET HTS assay was calculated to be 0.94, which fell between the range of 1 and 0.5, indicating an excellent assay with a large separation band for clear signal detection and superb reliability. Using this assay, the MicroSource SPECTRUM collection library (Microsource,

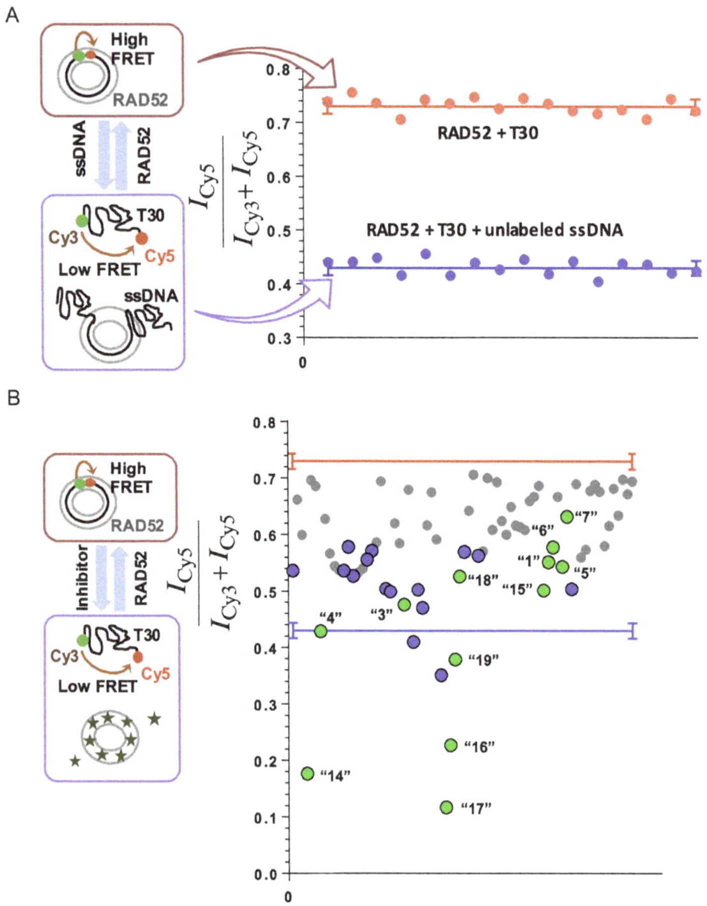

Fig. 1 FRET-based high-throughput screening of the MicroSource Spectrum library. (A) Control lanes from a 384-well plate. The positive control (*blue*) containing stoichiometric RAD52 with Cy3-dT_{30}-Cy5 substrate challenged with an excess amount of unlabeled ssDNA (Poly dT100); while the negative control (*red*) consisted of unperturbed stoichiometric complex of RAD52 with Cy3-dT_{30}-Cy5. *Red* and *blue solid lines* represent the average, and the standard deviation is shown with error bars, respectively. (B) 12 identified hits (*green*) from cherry-picked rescreening of the initial screen of the MicroSource SPECTRUM collection, with several false positives shown in *blue*. *Originally published as a part of fig. 1 (https://doi.org/10.7554/eLife.14740) in Hengel, S. R., Malacaria, E., Folly da Silva Constantino, L., Bain, F. E., Diaz, A., Koch, B. G., et al. (2016). Small-molecule inhibitors identify the RAD52-ssDNA interaction as critical for recovery from replication stress and for survival of BRCA2 deficient cells.* eLife, 5; *used under a CC-BY license.*

Gaylordsville, CT), which contains a wide structural diversity, including 2320 drug, drug-like synthetic compounds, and natural products with a range of known biological activities. Of the 2320 examined compounds, 96 preliminary hits were screened further for signal reproducibility and compound concentration-dependent signal detection (Fig. 1B). Based on the reported FRET signal and biochemical validation, 12 compounds were purchased with chemical structure confirmed by 1D NMR (Table 1).

In addition, similar FRET-based biochemical studies investigated the specificity of inhibition achieved by compounds "**1**" and "**6**" to show that, compound "**6**" interfered with the RAD52–dsDNA interaction while compound "**1**" had no effect on the interaction between RAD52 and dsDNA. Furthermore, unlike the known RAD52 inhibitor 6-hydroxy-DL-dopa (Chandramouly et al., 2015), the inhibition exhibited by our hit compounds "**1**" and "**6**" did not disrupt the ring structural assembly of RAD52 as no tertiary structural changes were observed from our dynamic light scattering experiments. Specifically, through performing WaterLOGSY experiments (see Section 3 for more detailed descriptions), direct-binding interactions were detected between RAD52 and hit compounds "**1**" and "**6**," respectively (Fig. 2).

3. USING DOCKING AND MOLECULAR DYNAMICS SIMULATIONS TO MAKE HYPOTHESES ABOUT RAD52 PNI PHARMACOPHORES

3.1 Understanding Small-Molecule Placement in the ssDNA-Binding Groove

Our next task was to find, with very high confidence, the features that are necessary for molecular recognition of our known (active) ligands with RAD52 (i.e., the binding orientation, ligand confirmations, and eventually the atomistic level structure of the RAD52–small molecule complexes). This information defines the RAD52 pharmacophore for our initial hits. Our primary interests focused on the compounds "**1**" and "**6**" from the HTS screen. For orthodox pockets, this would be an almost routine task using any number of docking approaches. However, due to the many challenges in targeting PNIs listed earlier, we required a multilayer approach that eventually led to a robust in silico screening. As the overall workflow in Fig. 3 shows, our overarching goal is to confidently use in silico screening to identify novel chemical scaffolds that target RAD52. In order to accomplish this, we must first grapple with several stark problems. First, a direct in

Table 1 The 12 Hits From the FRET-Based HTS Assay Designed for Identifying Inhibitors of the RAD52–sDNA Interaction

#	Small Molecule Name; CAS#	Small-Molecule Structure	IC_{50} (DNA Binding); FRET Value at Saturation	IC_{50} (Annealing Extent)	SAEM ΔG (kcal/mol)
"1"	(−)-Epigallocatechin; 970-74-1	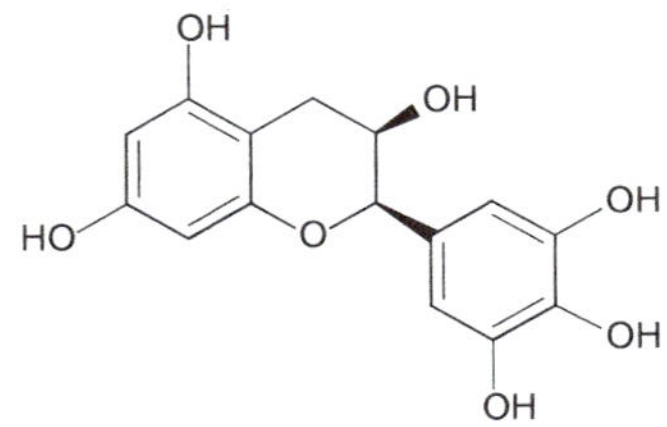	ssDNA: 1.8 ± 0.1 μ*M*; 0.45 ± 0.004 ssDNA–RPA: 1.6 ± 0.1 μ*M*	ssDNA: 4.9 ± 0.4 μ*M* ssDNA–RPA: 4.8 ± 1.8 μ*M*	−8.60
"3"	Methacycline hydrochloride; 3963-95-9	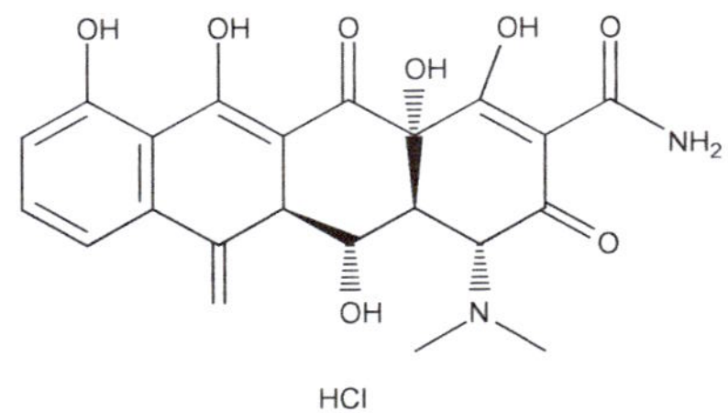	2.0 ± 0.17 μ*M*; 0.47 ± 0.01	3.8 ± 0.2 μ*M*	−4.61
"4"	Rolitetracycline; 751-97-3	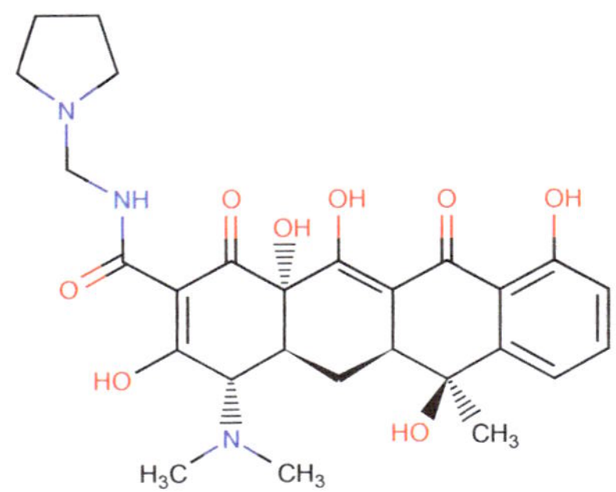	29 ± 8.2 μ*M*; 0.56 ± 0.04	NI[a]	−10.50

“5”	(−)-Epicatechin gallate; 1257-08-5	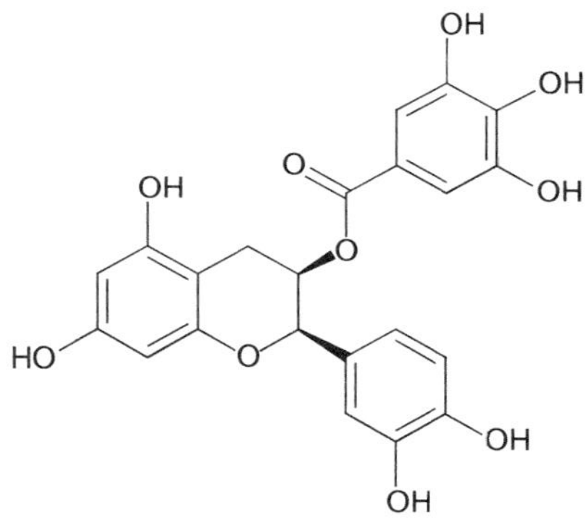	$255 \pm 16\,nM$; 0.41 ± 0.004	$20 \pm 0.7\,\mu M$	−9.87
“6”	Epigallocatechin-3-monogallate; 989-51-5	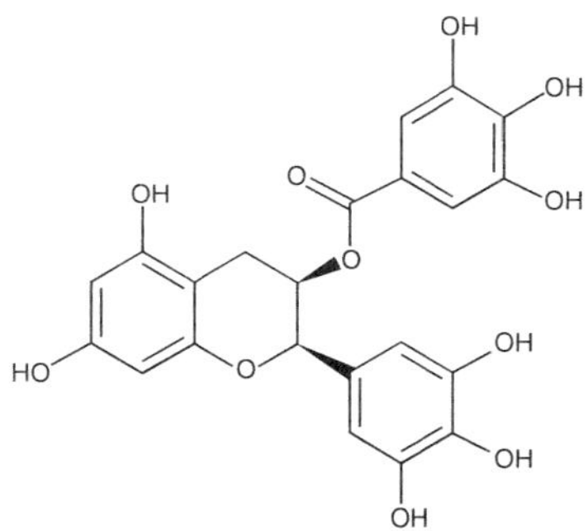	ssDNA: $277 \pm 22\,nM$; 0.46 ± 0.01 ssDNA–RPA: $1.6 \pm 0.5\,\mu M$	ssDNA: $6.7 \pm 2.1\,\mu M$ ssDNA–RPA: $3.7 \pm 0.5\,\mu M$	−10.69

Continued

Table 1 The 12 Hits From the FRET-Based HTS Assay Designed for Identifying Inhibitors of the RAD52–sDNA Interaction—cont'd

#	Small Molecule Name; CAS#	Small-Molecule Structure	IC_{50} (DNA Binding); FRET Value at Saturation	IC_{50} (Annealing Extent)	SAEM ΔG (kcal/mol)
"**7**"	(−)-Epicatechin; 490-46-0	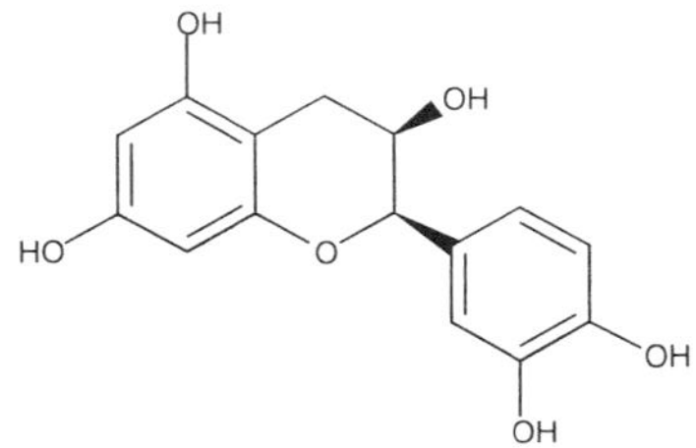	$1.45 \pm 0.11\ \mu M$; 0.51 ± 0.01	NI	−9.03
"**14**"	Oxidopamine; 28094-15-7 1199-18-4	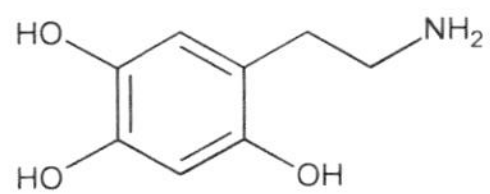	$779 \pm 51\ nM$; 0.50 ± 0.01	NI	−5.71
"**15**"	Quinalizarin; 81-61-8	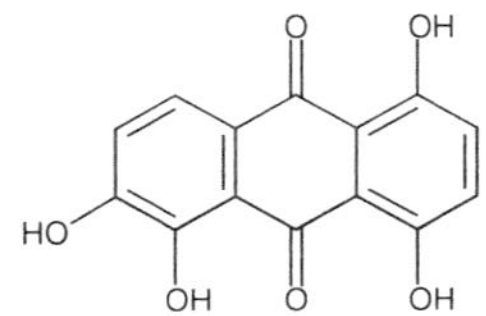	$563 \pm 40\ nM$; 0.51 ± 0.01	$5.6 \pm 0.6\ \mu M$	−9.17

"16"	Cisapride monohydrate; 260779-88-2 81098-60-4	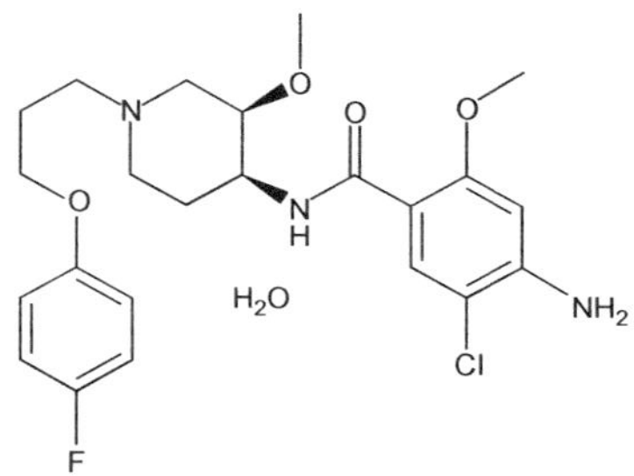	$1.06 \pm 0.05\,\mu M$; 0.50 ± 0.01	NI	−8.39
"17"	Cedrelone; 1254-85-9	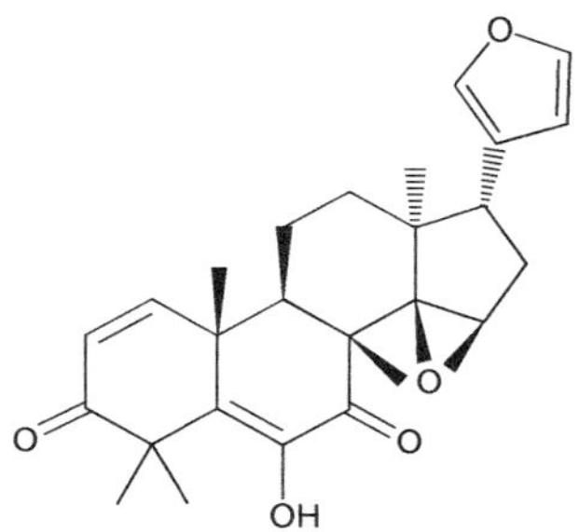	$>300\,\mu M$	NI	−10.0

Continued

Table 1 The 12 Hits From the FRET-Based HTS Assay Designed for Identifying Inhibitors of the RAD52–sDNA Interaction—cont'd

#	Small Molecule Name; CAS#	Small-Molecule Structure	IC_{50} (DNA Binding); FRET Value at Saturation	IC_{50} (Annealing Extent)	SAEM ΔG (kcal/mol)
"**18**"	Asiatic acid; 464-92-6 18449-41-7	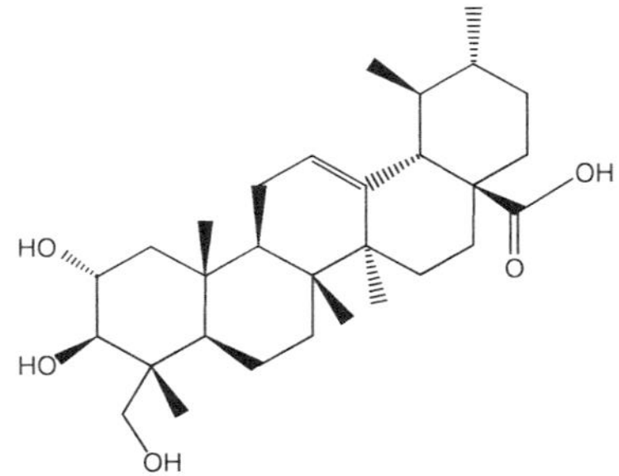	$>800\,\mu M$	$>100\,\mu M$	−11.33
"**19**"	Gossypetin; 489-35-0	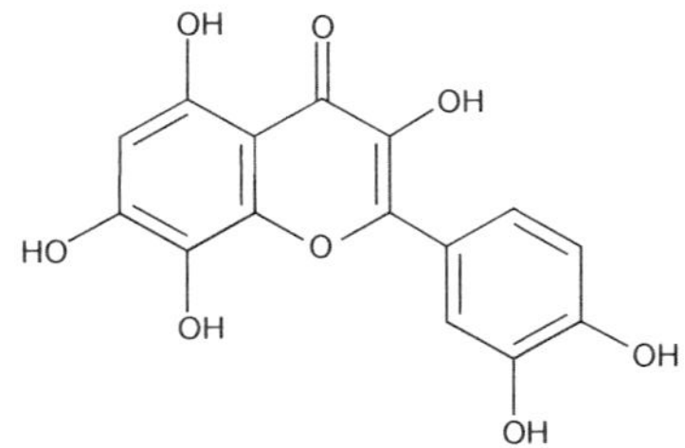	$913 \pm 58\,nM$; 0.49 ± 0.01	$6.0 \pm 2.3\,\mu M$	−9.30

[a]*NI*, no inhibition.

Originally published as a part of table 1 (https://doi.org/10.7554/eLife.14740) in Hengel, S. R., Malacaria, E., Folly da Silva Constantino, L., Bain, F. E., Diaz, A., Koch, B. G., et al. (2016). Small-molecule inhibitors identify the RAD52-ssDNA interaction as critical for recovery from replication stress and for survival of BRCA2 deficient cells. *eLife*, *5*;

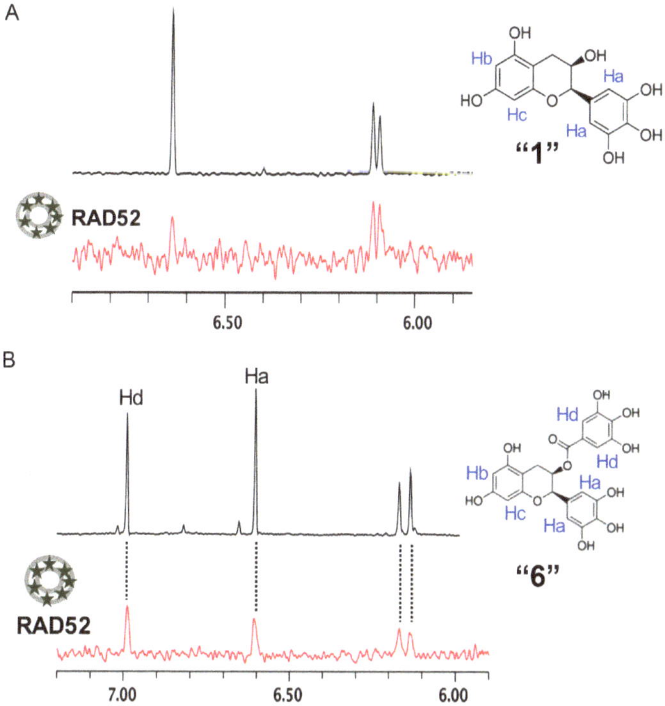

Fig. 2 WaterLOGSY experiments indicate direct interactions between HTS hit compounds and RAD52. (A) Aromatic region of the 1D ^{1}H NMR spectrum of compound "**1**" (*black*) and the WaterLOGSY spectrum of 20 μ*M* compound "**1**" in the presence of 3.3 μ*M* RAD52 (*red*). The nonexchangeable proton peaks are labeled using atom names (*blue*) on the compound structure. (B) Same as (A), except representing the aromatic region of the 1D ^{1}H NMR spectrum of compound "**6**" (*black*) and the WaterLOGSY spectrum of 40 μ*M* compound "**6**" in the presence of 3.3 μ*M* RAD52 (*red*). The nonexchangeable proton peaks are labeled using atom names (*blue*) as indicated on the compound structure. *Originally published as a part of figs. 2A and 3A (doi: 10.7554/eLife.14740) in Hengel, S. R., Malacaria, E., Folly da Silva Constantino, L., Bain, F. E., Diaz, A., Koch, B. G., et al. (2016). Small-molecule inhibitors identify the RAD52-ssDNA interaction as critical for recovery from replication stress and for survival of BRCA2 deficient cells.* eLife, 5; *used under a CC-BY license.*

silico campaign against this particular target is likely to generate too many false positives to be truly useful. Briefly, docking is an excellent tool for identifying the binding loci and orientation of small molecules, but is not good at making meaningful predictions of affinity rank ordering across ligand classes (Head, 2010). Additionally, although the technique can be useful in in silico screening, the large number of false positives is only amplified by increasing the size of the docking library. In addition to these classical problems, our

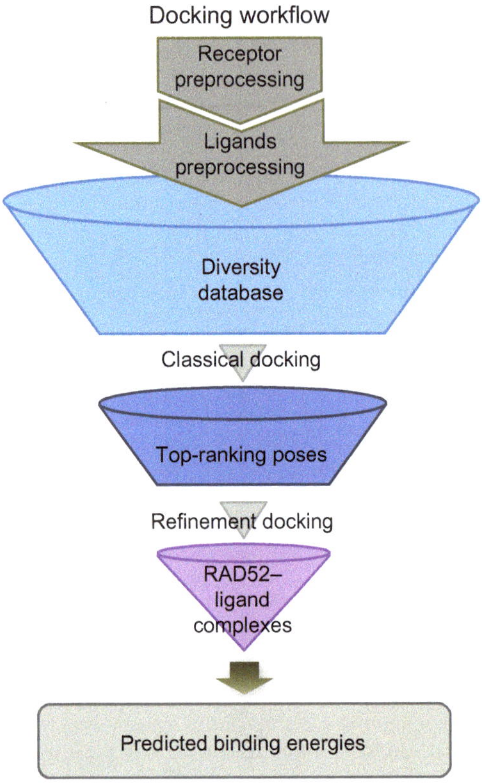

Fig. 3 Docking workflow for screening small-molecule drugs against RAD52. *Originally published as a part of fig. 8A (https://doi.org/10.7554/eLife.14740) in Hengel, S. R., Malacaria, E., Folly da Silva Constantino, L., Bain, F. E., Diaz, A., Koch, B. G., et al. (2016). Small-molecule inhibitors identify the RAD52-ssDNA interaction as critical for recovery from replication stress and for survival of BRCA2 deficient cells.* eLife, 5; *used under a CC-BY license.*

lack of structural knowledge about the native liganded complex of RAD52 and the role of interstitial waters is concerning. Nevertheless, we believed that, given the current apo-structure and our HTS hits, particularly, "**1**" and "**6**," that we could employ various integrated docking and molecular dynamics strategies to inevitably determine the locus and orientation of these RAD52–ssDNA inhibitors. An increasing body of literature has shown that when docking and MD are employed together that meaningful binding affinity rank ordering can be made, and that the many complex roles of water can often be inferred from these simulations (Whalen, Chau, & Spies, 2013; Whalen, Tussey, Blanke, & Spies, 2011).

Generally, the docking of small molecules involves positioning of the ligand with respect to a receptor structure and determining the most likely binding mode, with regard to the geometries and projected affinities using scoring functions (Kitchen, Decornez, Furr, & Bajorath, 2004). The primary goal of a typical in silico screen would be to maximize the detection of known ligand as actives (i.e., true positives), while minimizing the errors of false positives (inactives wrongly assigned as actives) and false negatives (actives wrongly assigned as inactives) (Malo, Hanley, Cerquozzi, Pelletier, & Nadon, 2006). These classic approaches have been successfully applied to numerous receptor types for over three decades now. However, due to the many challenges associated with unorthodox targets, as mentioned earlier, very few in silico campaigns have succeeded in identifying leads against such target classes involved in PPI and PNI, and the prognosis for classical in silico screening approaches is not good (Scott et al., 2016).

The campaign described in this chapter is fundamentally different from the classical approach in that molecular dynamics approaches are used in the early phases to scrutinize both the positioning and the global conformation of the receptor–small molecule complex, such that we are explicitly and globally dealing with protein flexibility and solvation. The utility of performing MD simulations in the early phase of our workflow was to have the highest possible confidence for establishing a set of "true positives" that could be used in a statistical approach for assessing the degree of success in our in silico screening procedure, using the method of ROC (Metz, 1978), which is described in detail later (Fig. 4).

There are an enormous array of potential docking approaches, and there are many factors that can affect the quality (accuracy and selectivity) of in silico screening approaches (Head, 2010). There is no one best method or docking method across target classes (Head, 2010), and certainly in the realm of attempting to disrupt a PPI or PNI, there is no clear indication about which docking algorithms are optimal. It was our contention that it is best to directly empirically assess any in silico screening workflow using quantitative metrics before attempting an actual production screen. Sections 3.2 to 3.4 detail (1) the features used in the initial docking of the HTS lead compounds, (2) how molecular dynamics using a knowledge-based force field approach was used to parse the docking results and to provide a much deeper level of detail in terms dealing with water, and (3) how the ROC statistical approach was used to quantify the accuracy of an in silico screening approach.

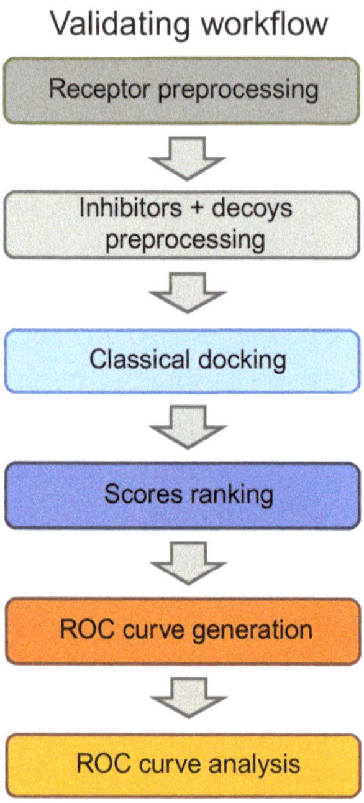

Fig. 4 Implementation of the ROC analysis in our docking studies against the RAD52 target. *Originally published as a part of fig. 8A (https://doi.org/10.7554/eLife.14740) in Hengel, S. R., Malacaria, E., Folly da Silva Constantino, L., Bain, F. E., Diaz, A., Koch, B. G., et al. (2016). Small-molecule inhibitors identify the RAD52-ssDNA interaction as critical for recovery from replication stress and for survival of BRCA2 deficient cells.* eLife, 5; *used under a CC-BY license.*

3.2 Docking of HTS Leads to the ssDNA-Binding Grove of RAD52

As described earlier, RAD52 forms an oligomeric ring (Kagawa et al., 2002; Lloyd, Forget, & Knight, 2002; Singleton, Wentzell, Liu, West, & Wigley, 2002); in which ssDNA-binding site is located in a narrow groove, which is a continuous epitope that spans the ring circumference (Lloyd, McGrew, & Knight, 2005; Mortensen, Erdeniz, Feng, & Rothstein, 2002). There are repeated subpockets within each monomer, with alternating hydrophobic and hydrophilic regions, as well as numerous crystal waters. Our docking workflow employed a combination of sequences, which initially used the Triangle Matcher approach for pose determination, followed by the London dG scoring function implemented in Molecular Operating Environment (MOE) 2013.08 (Chemical Computing Group Inc., 2013). London dG is an empirical scoring function which attempts to approximate the binding energy of a small molecule to a protein (Chemical Computing Group Inc., 2013). We then employed an additional layer of rigor by performing an all atom force field energy minimization of each docked pose, while the receptor was fixed, using the MMFF94x force field in MOE 2013.08, followed by rescoring the low energy pose with a physics-based force field scoring function, GBVI/WSA dG; this approach is a highly parameterized version of the general MM/PBSA (or MM/GBSA) methodologies (Steinbrecher & Labahn, 2010; Wang & Kollman, 2000), in that it employs a force field interaction energy between ligand and receptor, and also calculates the difference in solvation energy using the GB/VI solvation model, and determines the change in the solvent accessible surface area (Naim et al., 2007). Not surprisingly, there were very large discrepancies between the top-ranked poses when comparing the two different approaches. Additionally, there were some dissimilar poses, even within a single metric, which were different from one another as well. We addressed these challenges by using all atom molecular dynamics simulations with a knowledge-based force field (KBFF) to guide the selection process, which is described in detail later.

All of the lead inhibitors identified by HTS were subjected to the two docking protocols described earlier, using MOE 2013.08 to a portion of the ssDNA-binding groove of RAD52 spanning nearly a quarter of its circumference, which consists of three adjacent monomers. The receptor was built using PDB 1KN0 and subjected to preprocessing, including adding hydrogens, optimizing hydrogen bond networks, and assigning charges and protonation states. Additionally, ligands were preprocessed to assign protonation states, assign charges, and energy minimized using the MMFF94x force field in MOE (Chemical Computing Group Inc., 2013). The ligands were then docked into the large (quarter circumference) pocket described earlier, using the systematic Triangle Matcher search algorithm in MOE (Chemical Computing Group Inc., 2013). One then has a database of potential poses with docking scores. We kept the top 30 poses (based on the London dG scoring function) and these were further subjected to the ligand energy minimization with a rigid RAD52 receptor using the MMFF94x force field, followed by a free energy scoring using the GBIV/WSB dG approach described earlier.

3.3 Use of Molecular Dynamics Simulations to Expand on Docking Pose Selection

Following the docking workflow outlined earlier, we subjected all docked poses to a much more physically rigorous treatment, employing an all atom molecular dynamics simulations with simulated annealing, using the Yamber03 KBFF (Krieger, Darden, Nabuurs, Finkelstein, & Vriend, 2004), using fully explicit solvent conditions (Dean et al., 2015). Briefly, each docked complex (i.e., each pose structure) was placed into a simulation cell and solvated, and charge neutralized with physiological salt concentration, followed by optimizations of the solvent and the hydrogen bond network of the receptor–ligand complex, and finally a phased simulated annealing minimization was performed (the full technical details of a very similar MD simulation are described in more detail in Whalen et al. (2011) and Dean et al. (2015). No restraints or fixed atoms were employed (i.e., all atoms in the ligand and the entire RAD52 complex, ions, and solvent were free). When the convergence criteria for the simulated annealing was met, the VINA (Trott & Olson, 2010) docking utility in YASARA (Krieger, Koraimann, & Vriend, 2002) was invoked to perform a "local docking," which means that the VINA simulation cell is placed only in an area just around the ligand, so that the docking score reflects the position via molecular dynamics. Additionally, interstitial water molecules were determined and retained to be used by the VINA scoring function. Our workflow (Hengel et al., 2016), featured a user-specified parameter for defining what constituted an interstitial water. This allowed us to parse all the docking poses from the docking workflow outlined earlier. In these studies, we found that interstitial water structure was a major factor contributing to ligand pose and the quality of the docking scores. In Hengel et al. (2016), we referred to this automated process as Simulated Annealing Energy Minimization (SEAM).

The nature of the many subpockets along this three monomer continuous epitope provided a large number of potential poses for both compounds "**1**" and "**6**." The automated SEAM approach was applied to 34 unique docked poses, respectively, along the large quarter circumference ssDNA-binding groove (Fig. 5A). There was a clear lack of consensus in the docking, which was clarified by the SAEM. Generally, what one finds is that regardless of the docking approach used, the vast majority of docking poses that yield high scores result in poor performance when subjected to SEAM, since introduction of both realistic water structure and full ligand and receptor flexibility expose these poses as weak binders.

Interestingly, the top scoring poses of compounds "**1**" and "**6**" using the SAEM approach yielded complexes with unique binding subpockets along the RAD52-binding groove, suggesting that they may have distinct activities with regard to disrupting complexation with ssDNA (Fig. 5). Both

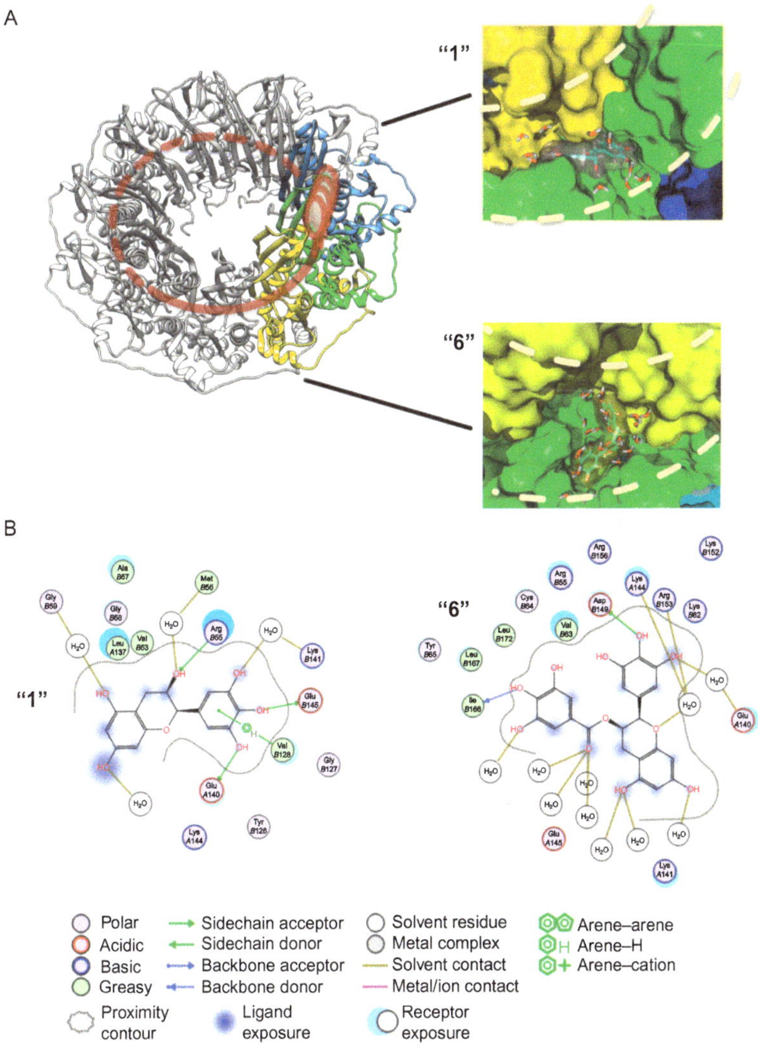

Fig. 5 Spatial arrangement of inhibitors "**1**" and "**6**" with RAD52–ssDNA-binding groove. (A) Binding positions of inhibitors "**1**" and "**6**" along the subpockets of three individual monomers (*yellow*, *green*, and *blue*) of the RAD52 ring (PDB 1KN0). *Dotted lines* indicate the approximate boundaries of ssDNA-binding groove. (B) Ligand interactive maps of inhibitors "**1**" and "**6**" inside the subpockets of RAD52. *Originally published as fig. 4 (https://doi.org/10.7554/eLife.14740) in Hengel, S. R., Malacaria, E., Folly da Silva Constantino, L., Bain, F. E., Diaz, A., Koch, B. G., et al. (2016). Small-molecule inhibitors identify the RAD52-ssDNA interaction as critical for recovery from replication stress and for survival of BRCA2 deficient cells.* eLife, 5; *used under a CC-BY license.*

compounds occupy complex pockets lying at the interface of the two RAD52 monomers. "**1**" mediates its interactions through an array of residues, including R55, V128, E140, as well as through water contacts via G59, M56, and K141. "**6**" binds via hydrogen bonding with D149 and I166, and acts via water-mediated contacts with E140, K144, and R153 (Fig. 5B). It was very interesting and encouraging for our SAEM approach to note that all final compound placements include interactions, directly or through interstitial waters, with key RAD52 residues, which have previously been reported to be involved in ssDNA binding (Lloyd et al., 2005) (Fig. 5B). Particular key residues in the locus of the docked compounds include R55, Y65, K152, R153, and R156, which have been shown to be critical to ssDNA binding to RAD52 (Lloyd et al., 2005) (Fig. 5B). Additionally, K141 and K144, also involved in the SEAM pharmacophore, have been linked to particular RAD52 activities via mutation studies (Mortensen et al., 2002).

Importantly, all of the lead compounds subjected to the SEAM approach yielded complexes in which interstitial water plays a role in the binding of the ligand. In retrospect, it is clear that the unorthodox nature of the RAD52 ssDNA-binding groove demands the use of some type of explicit solvent treatment to accurately capture small-molecule binding. Interestingly, many of the interstitial waters identified in the SEAM method are not involved in high quality hydrogen bonds that one may frequently see in very buried enzyme active sites, but rather, represent a van der Waals binding surface. This suggests that it may be possible to design better ligands, which both displaces these waters, and gain binding energy by filling these potential hydrogen-binding positions with the receptor.

3.4 Use of the ROC Method for Assessing Docking Workflow Accuracy

It should be kept in mind that the overall goal of this project was to build a workflow that would lead to a high accuracy in silico screening method for RAD52 (i.e., a structure-based approach for identification of novel inhibitors). In essence, the SAEM method outlined earlier has provided us with what will be used as "actual positives" in a statistical assessment of docking workflow accuracy. That raises the question about what approach one may use for "actual negatives" in such a docking workflow. An excellent choice in lieu of using experimental negatives is to employ decoy compounds that possess a number of similar properties as true hits, but that are topologically distinct. We employed the Database of Useful Decoys-Enhanced (DUD-E) website to generate a series of 50 "DUDS" for each true positive. The general utility of employing DUDS is discussed in Huang, Shoichet, and Irwin (2006). The idea is to use property matching to compounds of interest, such that compounds with similar molecular weight, estimated water–octanol partitioning coefficient (cLogP), number of rotatable bonds, number of hydrogen bond acceptors, number of hydrogen bond donors, and net charge are selected from among a database.

Having a set of true positives and true negatives allows one to apply quantitative tests of accuracy to docking workflows. We used the statistical method of analyzing ROC curves to optimize the balance of true positives, false positives, true negatives, and false negatives (Alexandre Varnek, 2008; Metz, 1978). The ROC approach allows one to make informed decisions about the type of accuracy desired. For screening purposes, accuracy itself is not the goal. It helps to understand that the performance of a classifier can be defined by the following indices:

$$\text{Sensitivity} = \frac{\text{Number of true positives (TP) classification}}{\text{Number of actual positives}}$$
$$= \text{Fraction of true positives}$$

$$\text{Sensitivity} = \frac{\text{Number of true negative (TN) classifications}}{\text{Number of actual positives}}$$
$$= \text{Fraction of true negatives}$$

One way to view accuracy is that it can be represented by these two different indices:

$$\text{Accuracy} = \frac{\text{Number of correct decisions}}{\text{Number of cases}} = \text{Sensitivity} + \text{Specificity}$$

We can also examine the number of incorrect classifications as well:

$$\text{Fraction of false positives} = \frac{\text{Number of false positive classifications}}{\text{Number of actual negatives}}$$

$$\text{Fraction of false negatives} = \frac{\text{Number of false negative classifications}}{\text{Number of actual negatives}}$$

If we have only two classifications, positive or negative, then the total fraction of correct and incorrect decisions must be unity, thus:

$$\text{Fraction of true positives} + \text{Fraction of false negatives} = 1$$
$$\text{Fraction of true negatives} + \text{Fraction of false positives} = 1$$

Due to these inherent constraints, it is only necessary to express one of the fractions from each of the above equations in order to capture the necessary information for quantifying accuracy. The central point is that accuracy may be achieved in different ways, and that it would be ideal to test how a given workflow performs before actually employing it. Given overlapping populations of true negatives and true positives, a classifier threshold value must be defined, which optimizes these benefits and drawbacks (Fig. 6). Importantly, there is no absolute best classifier threshold; it is a subjective decision that depends on many factors that the researcher must weigh. The best way to make the key decision about what threshold value to employ is to vary its value and see how this affects the fraction of true positives and fraction of false negatives, respectively. ROC analysis is usually performed by changing the classifier threshold and determining how the fraction of true positives and true negatives change (since determination of only these two values reports on all four classifier fractions).

An example of a typical ROC curve is shown in Fig. 7, in which the fraction of true positives is plotted on the ordinate and the fraction of false positives is plotted on the abscissa; the range of values will always be from 0 to 1, since these are limits of the fractions. As we vary the classifier threshold, the form that the function takes in the ROC plot is, of course, dictated by the nature and the shape of the overlap of the two distributions (as in Fig. 6). Importantly, we see that all possible permutations of the fraction of true positives and fraction of true negatives are represented on the ROC curve. This is the real power of the technique, in that it affords the researcher the

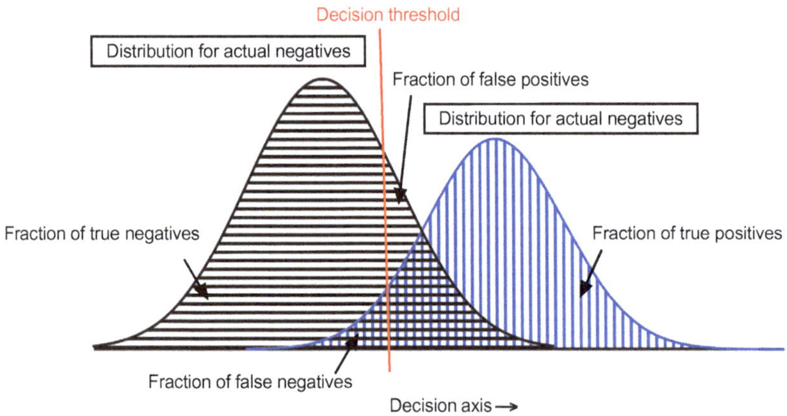

Fig. 6 Decision threshold selection is not absolute. Two simulated distribution (*black* and *blue*) of a quantity with one possible decision threshold (*red*) shown.

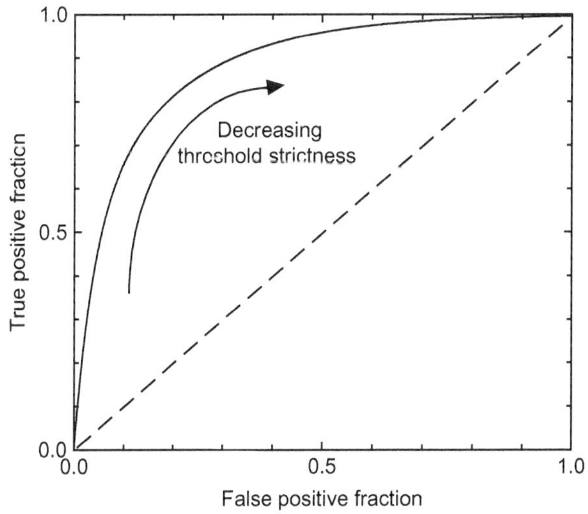

Fig. 7 A typical conventional ROC curve with decreasing threshold strictness.

opportunity to see the performance of the workflow over the complete range of classifier values, and choose what value is most conducive to the current study. Fig. 7 indicates that a very high classifier threshold yields the portion of the ROC curve in the lower left quadrant of the plot, and that very low classifier thresholds would be associated with the upper right quadrant. An ideal ROC curve would reach the upper left corner of the plot, while a classifier that did no better than random would cross the diagonal of the graph (Fig. 7). In other words, the closer to the diagonal an ROC curve is, the worse is the protocol used to distinguish the actives from the decoys (i.e., DUDS in this case), and the closer to the ideal curve, the more the docking protocol gets to fully distinguishing actives from decoys.

The nature of our particular in silico screening campaign is that we would like to focus on screening natural product compounds, since they have a long history of producing excellent lead compounds in challenging ligand discovery campaigns. However, these compounds are usually relatively precious, and significant labor may be expended in their preparation. Thus, in terms of our particular workflow, we were very interested in a high classifier threshold, which would come close to completely separating the populations, even if this means that we get a good deal of false negatives. In other words, we very much wanted to avoid false positives (the usual bane of in silico screening). Fortunately, our particular workflow, employing SAEM, provided a near perfect ROC curve, which allowed us to confidently employ the use of natural product library, as discussed later.

Docking scores were used for determining the ROC optimal classifier threshold values based on the cost–benefit analysis described earlier. The original HTS and its poses were determined to be the only "actual positives," and the DUDS and their respective poses were treated as "actual negatives"; any poses above the classifier threshold value were true positives, and values below were designated as true negatives. The ROC curves were analyzed using the metric of the area under the curve (AUC) (DeLong, DeLong, & Clarke-Pearson, 1988). The AUC reports on the probability that, when we compare the scores for a randomly selected active compound and a randomly selected decoy, the in silico screen will rank the active compound ahead of the decoy (DeLong et al., 1988).

The docking scores for the poses with the most active compounds exhibited bimodal frequency distribution (Fig. 8A), and the docking protocols' ability to distinguish between active compounds and decoys was quantified (Fig. 8B). As mentioned earlier, our goal was to be able to employ a high classifier threshold, in order to reduce the cost of false positives.

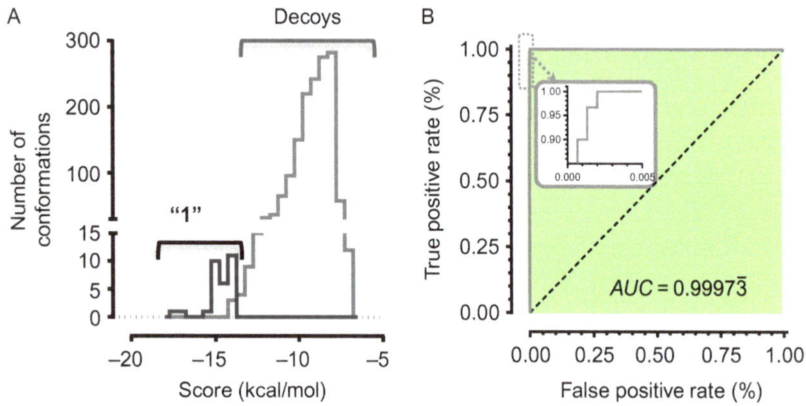

Fig. 8 Application of ROC in combination with binding energy-based docking scores for novel small-molecule inhibitors for RAD52–ssDNA interaction. (A) Calculated docking scores (kcal/mol) for the binding configurations of compound "**1**" in comparison to decoys compounds. As the low docking scores indicate low binding energies, which correlate to favorable interactions, the histogram shows a clear distinction between true positives from compound "**1**" and true negatives from "decoys." (B) Receiving operating characteristics (Metz, 1978) curve used with the threshold shown yielding an AUC value of 0.9973. *Originally published as a part of fig. 8B and C (https://doi.org/10.7554/eLife.14740) in Hengel, S. R., Malacaria, E., Folly da Silva Constantino, L., Bain, F. E., Diaz, A., Koch, B. G., et al. (2016). Small-molecule inhibitors identify the RAD52-ssDNA interaction as critical for recovery from replication stress and for survival of BRCA2 deficient cells.* eLife, *5; used under a CC-BY license.*

The ROC curve for "**1**" illustrates how the optimized in silico classification process is nearly ideal in distinguishing false positives from true positives, since it assigns more negative docking scores (i.e., more favorable predicted binding energies) for active conformations than for those of the decoy compounds. Thus, up to this point we have used retrospective information gathered from HTS and used molecular dynamics and docking to expand our understanding of the complexation of these hits with the ssDNA-binding groove of RAD52. We now pivot our focus to a completely *prospective* study. This represents a much more significant, and risky challenge, since we are no longer retrospectively analyzing HTS results. Indeed, true prospective structure-based discovery of specific low μ*M* to high n*M* hits using strictly in silico screening, especially against challenging or novel targets, is quite rare (Head, 2010). Our philosophy for the campaign described in Hengel et al. (2016) was to use ROC curve analysis in such a way as to be able to focus on screening of interesting and relatively precious natural products. Natural products clearly have an unrivaled history in drug discovery and often represent the primary hits against targets. We employed the AnalytiCon MEGx Natural Products Screen Library (AnalytiCon Discovery GmbH, Postsdam, Germany)—the in silico version of an actual library, consisting of highly purified plant, fungal, and microbial-derived compounds, which are available for purchase.

To summarize, the overall workflow is shown in Fig. 3. A database composed of the natural products and a control selected from the HTS hits were created and preprocessed for virtual screening, as described earlier. The Dock utility of MOE was used to generate a database of the top 30 scoring poses from each compound, followed by overall ranking of poses for the entire database. The top scoring compounds were then selected for further analysis. Poses with more negative scores than the best scoring of our control (i.e., an actual positive in the ROC curve context) were then subjected to the SAEM docking procedure detailed earlier. Following SAEM, all of the poses were ranked and evaluated. This resulted in nine compounds from the AnalytiCon database that had poses with scores better than those of the internal control. The best scoring structure was ordered from AnalytiCon Discovery GmbH (Postsdam, Germany) for in vitro inhibition studies. This compound, **NP-004255**, is a macrocycle ester, consisting of three trihydroxylated phenolic moieties; it is called *corilagin*, which belongs to a family of secondary plant metabolites called *ellagitannins*. Importantly, the hit was validated by both NMR WaterLOGSY and the FRET-based method described earlier, as fully described in Hengel et al. (2016), as shown in Fig. 9.

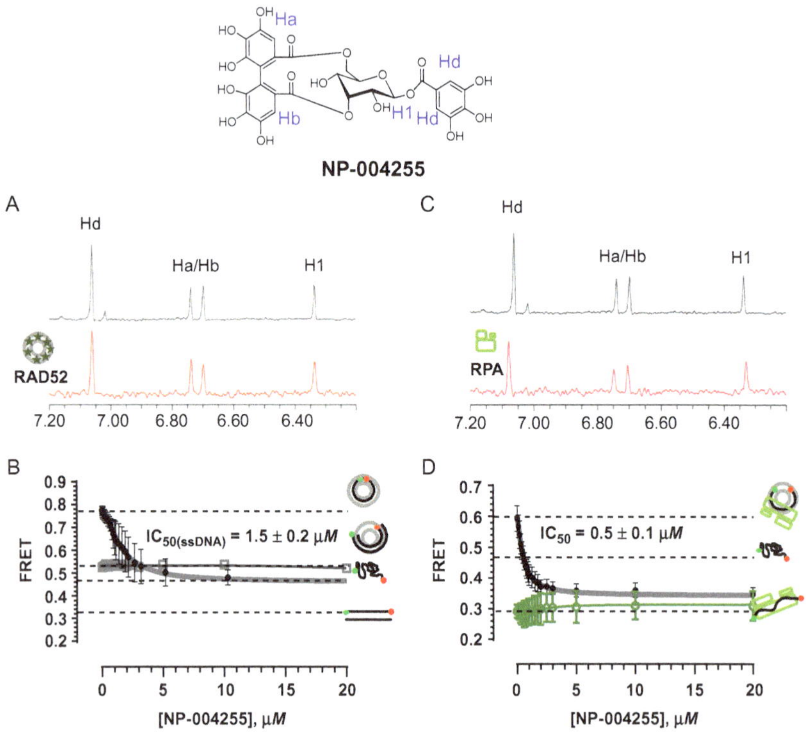

Fig. 9 Biochemical validation of **NP-004255**. (A) Aromatic region of the 1D ^{1}H NMR spectrum of compound **NP-004255** alone (*black*) and the WaterLOGSY spectrum of 40 μ*M* compound **NP-004255** in the presence of 3.3 μ*M* RAD52 (*red*). The non-exchangeable proton peaks (*blue*) using atom names are shown on the structure of compound **NP-004255**. (B) IC_{50} values for inhibition of ssDNA binding and wrapping were determined using FRET-based assays that follow the change in geometry of a Cy3-dT30-Cy5 substrate (*black circles*). The computed IC_{50} value is shown above the curve. Titration of the RAD52–dsDNA with **NP-004255** (*gray boxes*) shows that this inhibitor does not perturb the RAD52–dsDNA interaction. (C) Aromatic region of the 1D ^{1}H NMR spectrum of compound **NP-004255** alone (*black*) and the WaterLOGSY spectrum of 40 μ*M* compound **NP-004255** in the presence of 3.3 μ*M* RPA (*red*). (D) Titration of the RAD52-RPA-Cy3-dT_{30}-Cy5 complex with **NP-004255** (*black circles*). The computed IC_{50} value is shown below the curve. *Green squares* show titration of the RPA-Cy3-dT_{30}-Cy5 complex with **NP-004255** indicating **NP-004255** does not perturb the RPA–ssDNA. *Originally published as fig. 9 (https://doi.org/10.7554/eLife.14740) in Hengel, S. R., Malacaria, E., Folly da Silva Constantino, L., Bain, F. E., Diaz, A., Koch, B. G., et al. (2016). Small-molecule inhibitors identify the RAD52-ssDNA interaction as critical for recovery from replication stress and for survival of BRCA2 deficient cells.* eLife, 5; *used under a CC-BY license.*

NP-004255 was shown to bind to RAD52 and compete for ssDNA binding. Two assays were performed to obtain IC_{50} values, competition with ssDNA only and competition in the presence of the ssDNA-binding protein RPA, yielding values of $1.5 \pm 0.2\,\mu M$ and $0.5 \pm 0.1\,\mu M$, respectively. The importance of multiple forms of orthogonal validation of any hit, whether from an in vitro or in silico HTS campaign, cannot be overemphasized. It is especially important to have at least one assay that establishes direct binding to the target (i.e., not a functional assay, but an assay that shows direct interaction between the proposed hit and the target). Methods for direct binding that are frequently employed include surface plamon resonance, NMR WaterLOGSY, and crystallography. WaterLOGSY was an excellent approach in this case, since the method derives from the magnetization transfer from solvent to a bound ligand, which often proceeds through interstitial water contacts. Fig. 10 illustrates the placement of **NP-004255** within the RAD52–ssDNA-binding groove, as a comparison to the HTS hits shown (Fig. 5).

NP-004255 binds to RAD52 in an analogous way as "**1**" and "**6**," employing a buried interstitial water network, and adopts a helical-like conformation that fits nicely into the ssDNA-binding groove (Fig. 10). Its complexation produces hydrogen-bonding patterns ideally located to prevent ssDNA access to residues that are important for complexation (Hengel et al., 2016; Lloyd et al., 2005).

4. IMPLICATIONS FOR EMPLOYING ROC AND MD-INFORMED WORKFLOWS IN THE DISCOVERY OF NOVEL PNI INHIBITORS

The workflow described in this chapter was highly successful in the challenging domain of prospectively discovering the pharmacophore of the PNI of RAD52. The data show that choice of workflow, employing experimental, docking, and MD simulations can be designed to almost perfectly separate hits from negatives. The use of ROC curves allowed us to optimize the threshold to meet our very low risk tolerance (for false positives), such that we could focus on screening natural product libraries. We believe that this approach should work well for generally disrupting challenging macromolecular–protein targets, as long as the system under investigation is amenable to some of the biophysical validation methods described earlier. However, our choice of a very strict cutoff threshold raises some potential caveats. Namely, we may have generated several false

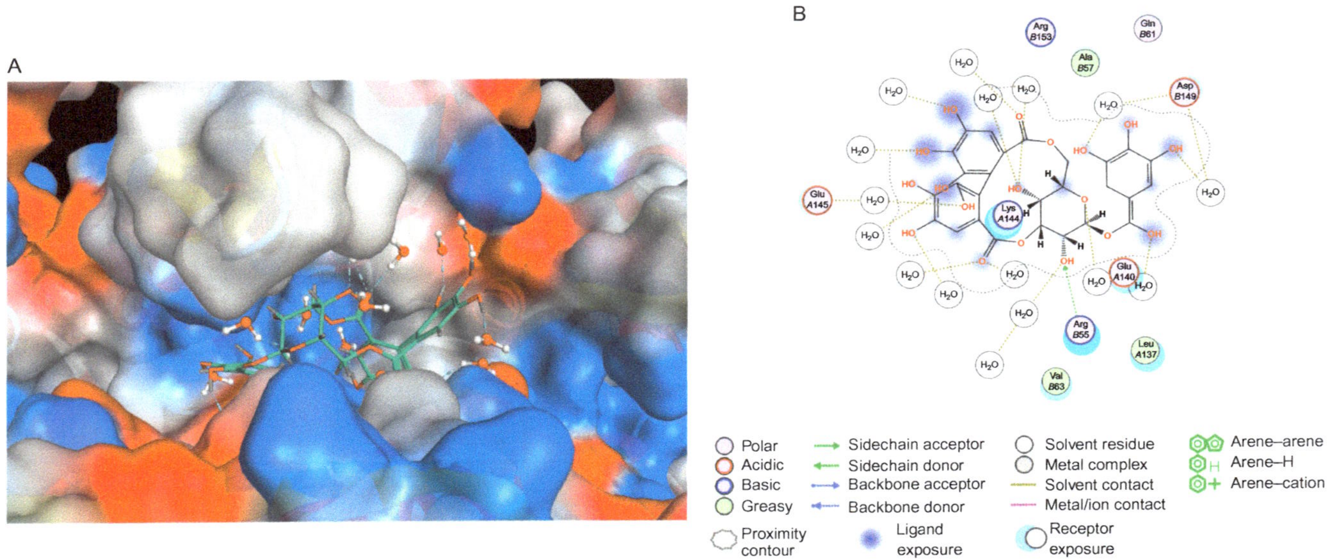

Fig. 10 Placement of RAD52 **NP-004255** within the ssDNA-binding groove. (A) Electrostatic potential surface of three monomers of the RAD52-NTD (PDB 1KN0) positioning **NP-004255** within the ssDNA-binding groove. (B) MOE ligand map of **NP-004255** when binding RAD52. *Originally published as fig. 8D and E (https://doi.org/10.7554/eLife.14740) in Hengel, S. R., Malacaria, E., Folly da Silva Constantino, L., Bain, F. E., Diaz, A., Koch, B. G., et al. (2016). Small-molecule inhibitors identify the RAD52-ssDNA interaction as critical for recovery from replication stress and for survival of BRCA2 deficient cells.* eLife, 5; *used under a CC-BY license.*

negatives in the natural product screen, which might have been excellent inhibitors. This could be alleviated by employing cheaper, standard synthetic diversity lead libraries, and using a moderate cutoff threshold. Nevertheless, it is our belief that natural product macrocycles, such as **NP-004255** provide a unique opportunity to target PNIs and PPIs. Their stereochemical complexity and preorganized ring structures provide a functionality that is difficult to replicate in classical synthetic leads (Driggers, Hale, Lee, & Terrett, 2008). As evidenced by Hengel et al. (2016), and the data provided in this chapter, the extended nature of the PNI-binding site is complementary to the preorganized macrocycle ligand. It is almost certainly the case that a significant contributing factor to its efficacy is reduction in the entropic penalty in protein binding, due to this preorganization. While these features often provide excellent complementarity to protein targets, they are also associated with the, surprisingly, sufficient bioavailability of macrocycles. Interestingly, natural products, especially larger ones, are known to have complex conformational ensembles in solution, which may contribute to their unique properties. The nature of the workflow described in this chapter, and in Hengel et al., is not optimized to capture these natural product properties. This is largely because, although it employs MD simulations of the protein–ligand complex, it still relies in part on docking; for classic drug-like compounds this should not be a problem, but for large macrocycles it would fail to capture true breadth their solution conformational ensembles. One possible solution to this is to use a free energy calculation that is amenable to medium-throughput analysis, such as extended linear response, which has been highly successful in explicitly capturing the contributions of interstitial waters to ligand binding (Whalen et al., 2013).

ACKNOWLEDGMENT

This work was supported by National Institutes of Health (R01-GM097373), the University of Iowa College of Medicine pilot HTS grant to MAS and MS, the NIH S10 (1S10RR029274-01) grant supporting the UIHTS facility. LFdaSC is supported by Coordenação de Aperfeiçoamento de Pessoal de Nível Superior (CAPES) Brazil. The funders had no role in study design, data collection, analysis, decision to publish, or preparation of the chapter.

REFERENCES

Alexandre Varnek, A. T. (2008). *Chemoinformatics approaches to virtual screening*. Royal Society of Chemistry.

Chandramouly, G., Mcdevitt, S., Sullivan, K., Kent, T., Luz, A., Glickman, J. F., et al. (2015). Small-molecule disruption of RAD52 rings as a mechanism for precision medicine in BRCA-deficient cancers. *Chemistry & Biology*, *22*, 1491–1504.

Chemical Computing Group Inc. (2013). *Molecular Operating Environment* (8th ed.). 1010 Sherbooke St. West, Suite #910, Montreal, QC, Canada, H3A 2R7.
De Vega, M. J., Martin-Martinez, M., & Gonzalez-Muniz, R. (2007). Modulation of protein-protein interactions by stabilizing/mimicking protein secondary structure elements. *Current Topics in Medicinal Chemistry*, *7*, 33–62.
Dean, S. F., Whalen, K. L., & Spies, M. A. (2015). Biosynthesis of a novel glutamate racemase containing a site-specific 7-hydroxycoumarin amino acid: Enzyme-ligand promiscuity revealed at the atomistic level. *ACS Central Science*, *1*, 364–373.
DeLong, E. R., DeLong, D. M., & Clarke-Pearson, D. L. (1988). Comparing the areas under two or more correlated receiver operating characteristic curves: A nonparametric approach. *Biometrics*, *44*, 837–845.
Driggers, E. M., Hale, S. P., Lee, J., & Terrett, N. K. (2008). The exploration of macrocycles for drug discovery—An underexploited structural class. *Nature Reviews. Drug Discovery*, *7*, 608–624.
Eyrisch, S., & Helms, V. (2007). Transient pockets on protein surfaces involved in protein-protein interaction. *Journal of Medicinal Chemistry*, *50*, 3457–3464.
Grimme, J. M., Honda, M., Wright, R., Okuno, Y., Rothenberg, E., Mazin, A. V., et al. (2010). Human Rad52 binds and wraps single-stranded DNA and mediates annealing via two hRad52-ssDNA complexes. *Nucleic Acids Research*, *38*, 2917–2930.
Grimme, J. M., & Spies, M. (2011). FRET-based assays to monitor DNA binding and annealing by Rad52 recombination mediator protein. *Methods in Molecular Biology*, *745*, 463–483.
Head, M. (2010). *Docking: A domesday report*. Cambridge University Press.
Hengel, S. R., Malacaria, E., Folly da Silva Constantino, L., Bain, F. E., Diaz, A., Koch, B. G., et al. (2016). Small-molecule inhibitors identify the RAD52-ssDNA interaction as critical for recovery from replication stress and for survival of BRCA2 deficient cells. *eLife*, *5*, e14740. https://doi.org/10.7554/eLife.14740.
Huang, N., Shoichet, B. K., & Irwin, J. J. (2006). Benchmarking sets for molecular docking. *Journal of Medicinal Chemistry*, *49*, 6789–6801.
Huggins, D. J., Marsh, M., & Payne, M. C. (2011). Thermodynamic properties of water molecules at a protein-protein interaction surface. *Journal of Chemical Theory and Computation*, *7*, 3514–3522.
Kagawa, W., Kurumizaka, H., Ishitani, R., Fukai, S., Nureki, O., Shibata, T., et al. (2002). Crystal structure of the homologous-pairing domain from the human Rad52 recombinase in the undecameric form. *Molecular Cell*, *10*, 359–371.
Kitchen, D. B., Decornez, H., Furr, J. R., & Bajorath, J. (2004). Docking and scoring in virtual screening for drug discovery: Methods and applications. *Nature Reviews. Drug Discovery*, *3*, 935–949.
Krieger, E., Darden, T., Nabuurs, S. B., Finkelstein, A., & Vriend, G. (2004). Making optimal use of empirical energy functions: Force-field parameterization in crystal space. *Proteins*, *57*, 678–683.
Krieger, E., Koraimann, G., & Vriend, G. (2002). Increasing the precision of comparative models with YASARA NOVA—A self-parameterizing force field. *Proteins*, *47*, 393–402.
Lloyd, J. A., Forget, A. L., & Knight, K. L. (2002). Correlation of biochemical properties with the oligomeric state of human rad52 protein. *The Journal of Biological Chemistry*, *277*, 46172–46178.
Lloyd, J. A., McGrew, D. A., & Knight, K. L. (2005). Identification of residues important for DNA binding in the full-length human Rad52 protein. *Journal of Molecular Biology*, *345*, 239–249.
Makley, L. N., & Gestwicki, J. E. (2013). Expanding the number of "Druggable" targets: Non-enzymes and protein–protein interactions. *Chemical Biology & Drug Design*, *81*, 22–32.

Malo, N., Hanley, J. A., Cerquozzi, S., Pelletier, J., & Nadon, R. (2006). Statistical practice in high-throughput screening data analysis. *Nature Biotechnology*, *24*, 167–175.

Metz, C. E. (1978). Basic principles of ROC analysis. *Seminars in Nuclear Medicine*, *8*, 283–298.

Mortensen, U. H., Erdeniz, N., Feng, Q., & Rothstein, R. (2002). A molecular genetic dissection of the evolutionarily conserved N terminus of yeast Rad52. *Genetics*, *161*, 549–562.

Naim, M., Bhat, S., Rankin, K. N., Dennis, S., Chowdhury, S. F., Siddiqi, I., et al. (2007). Solvated interaction energy (SIE) for scoring protein-ligand binding affinities. 1. Exploring the parameter space. *Journal of Chemical Information and Modeling*, *47*, 122–133.

Scott, D. E., Bayly, A. R., Abell, C., & Skidmore, J. (2016). Small molecules, big targets: Drug discovery faces the protein-protein interaction challenge. *Nature Reviews. Drug Discovery*, *15*, 533–550.

Singleton, M. R., Wentzell, L. M., Liu, Y., West, S. C., & Wigley, D. B. (2002). Structure of the single-strand annealing domain of human RAD52 protein. *Proceedings of the National Academy of Sciences of the United States of America*, *99*, 13492–13497.

Sperandio, O., Miteva, M. A., Segers, K., Nicolaes, G. A., & Villoutreix, B. O. (2008). Screening outside the catalytic site: Inhibition of macromolecular inter-actions through structure-based virtual ligand screening experiments. *Open Biochemistry Journal*, *2*, 29–37.

Steinbrecher, T., & Labahn, A. (2010). Towards accurate free energy calculations in ligand protein-binding studies. *Current Medicinal Chemistry*, *17*, 767–785.

Subramanyam, S., Jones, W. T., Spies, M., & Spies, M. A. (2013). Contributions of the RAD51 N-terminal domain to BRCA2-RAD51 interaction. *Nucleic Acids Research*, *41*, 9020–9032.

Trott, O., & Olson, A. J. (2010). AutoDock Vina: Improving the speed and accuracy of docking with a new scoring function, efficient optimization, and multithreading. *Journal of Computational Chemistry*, *31*, 455–461.

Venter, J. C., Adams, M. D., Myers, E. W., et al. (2001). The sequence of the human genome. *Science*, *291*, 1304–1351. https://doi.org/10.1126/science.1058040.

Wang, W., & Kollman, P. A. (2000). Free energy calculations on dimer stability of the HIV protease using molecular dynamics and a continuum solvent model. *Journal of Molecular Biology*, *303*, 567–582.

Whalen, K. L., Chau, A. C., & Spies, M. A. (2013). In silico optimization of a fragment-based hit yields biologically active, high-efficiency inhibitors for glutamate racemase. *ChemMedChem*, *8*, 1681–1689.

Whalen, K. L., Tussey, K. B., Blanke, S. R., & Spies, M. A. (2011). Nature of allosteric inhibition in glutamate racemase: Discovery and characterization of a cryptic inhibitory pocket using atomistic MD simulations and pKa calculations. *The Journal of Physical Chemistry. B*, *115*, 3416–3424.

Yap, J. L., Chauhan, J., Jung, K. Y., Chen, L. J., Prochownik, E. V., & Fletcher, S. (2012). Small-molecule inhibitors of dimeric transcription factors: Antagonism of protein-protein and protein-DNA interactions. *Medicinal Chemistry Communications*, *3*, 541–551.

Zhang, J. H., Chung, T. D. Y., & Oldenburg, K. R. (1999). A simple statistical parameter for use in evaluation and validation of high throughput screening assays. *Journal of Biomolecular Screening*, *4*, 67–73.

CHAPTER TWELVE

Monitoring Recombination During Meiosis in Budding Yeast

Shannon Owens*[,†,‡], Shangming Tang*[,†,‡], Neil Hunter*[,†,1]
*Howard Hughes Medical Institute, University of California, Davis, Davis, CA, United States
[†]University of California, Davis, Davis, CA, United States
[‡]Biochemistry, Molecular, Cellular and Developmental Biology Graduate Group, University of California, Davis, Davis, CA, United States
[1]Corresponding author: e-mail address: nhunter@ucdavis.edu

Contents

Abstract

Homologous recombination is fundamental to sexual reproduction, facilitating accurate segregation of homologous chromosomes at the first division of meiosis, and creating novel allele combinations that fuel evolution. Following initiation of meiotic

Methods in Enzymology, Volume 601
ISSN 0076-6879
https://doi.org/10.1016/bs.mie.2017.12.005

recombination by programmed DNA double-strand breaks (DSBs), homologous pairing and DNA strand exchange form joint molecule (JM) intermediates that are ultimately resolved into crossover and noncrossover repair products. Physical monitoring of the DNA steps of meiotic recombination in *Saccharomyces cerevisiae* (budding yeast) cultures undergoing synchronous meiosis has provided seminal insights into the molecular basis of meiotic recombination and affords a powerful tool for dissecting the molecular roles of recombination factors. This chapter describes a suit of electrophoretic and Southern hybridization techniques used to detect and quantify the DNA intermediates of meiotic recombination at recombination hotspots in budding yeast. DSBs and recombination products (crossovers and noncrossovers) are resolved using one-dimensional electrophoresis and distinguished by restriction site polymorphisms between the parental chromosomes. Psoralen cross-linking is used to stabilize branched JMs, which are resolved from linear species by native/native two-dimensional electrophoresis. Native/denaturing two-dimensional electrophoresis is employed to determine the component DNA strands of JMs and to measure the processing of DSBs. These techniques are generally applicable to any locus where the frequency of recombination is high enough to detect intermediates by Southern hybridization.

1. INTRODUCTION

Meiosis is the specialized cell division that serves as the conduit for heredity, producing haploid gametes from diploid germline cells through two successive cell divisions (Hunter, 2015). Homologous chromosomes (homologs) segregate during the first division, and sister chromatids separate at the second division. Accurate segregation during meiosis-I requires that homologs become connected by chiasmata—the conjunction of interhomolog crossing over and sister chromatid cohesion (Fig. 1). These de novo chromosome connections enable stable bipolar attachment of homolog pairs to the spindle, which is the essential prelude to accurate disjunction. Crossover recombination also produces new combinations of gene alleles upon which natural selection can act. Thus, meiotic crossing-over underlies the fundamental laws of heredity.

Chiasma formation is the culmination of an elaborate series of interdependent chromosomal events that include programmed homologous recombination, and the intimate pairing and synapsis of homologs (Fig. 1). Meiotic recombination is initiated by programmed DNA double-strand breaks (DSBs), numbering ~200–300 per nucleus in budding yeast, mouse, and human (Lam & Keeney, 2014). Ensuing recombinational interactions promote the pairing of homologous chromosomes, and their intimate synapsis by zipper-like structures called synaptonemal complexes (SCs)

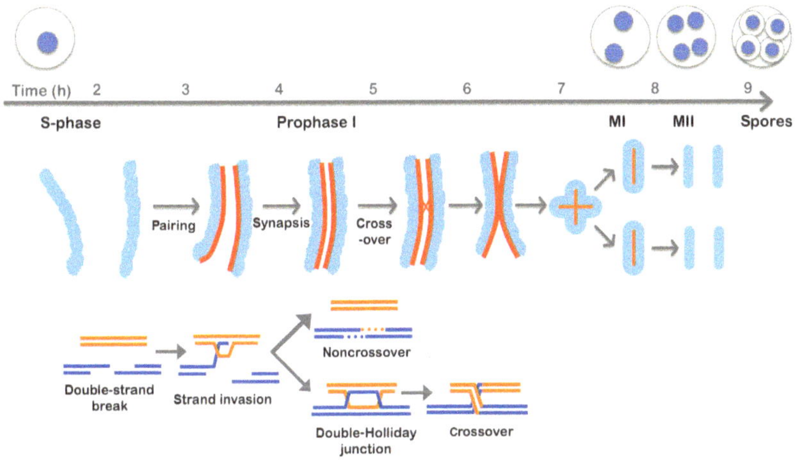

Fig. 1 Meiosis timeline in budding yeast illustrating the major nuclear, chromosomal, and DNA events of meiotic prophase.

(Fraune, Schramm, Alsheimer, & Benavente, 2012; von Wettstein, Rasmussen, & Holm, 1984; Zickler & Kleckner, 1999). Within the context of SCs, a few recombinational interactions mature into crossovers. After SCs are removed, homologs remain connected at their crossover points allowing them to condense into bivalent structures that can stably biorient on the spindle, and faithfully disjoin at the first meiotic division (Petronczki, Siomos, & Nasmyth, 2003; Watanabe, 2012).

The primary role of crossing-over in homolog disjunction dictates that each pair of homologs becomes connected by at least one exchange. Regulatory processes help ensure that each homolog pair obtains this obligatory crossover (crossover assurance), while total crossover levels remain low (on the order of one crossover per chromosome arm) (Jones, 1984). Crossover numbers appear to be minimized by a process called interference, which prevents two crossovers from forming in close proximity between the same chromosomes. Interference acts over huge physical distances. Consequently, on small chromosomes interference can preclude the formation of a second crossover, while on longer chromosomes interference causes adjacent crossovers to be widely and evenly spaced (Zickler & Kleckner, 2015).

Historically, fungal genetics has dominated studies of meiotic recombination because the four haploid products of individual meioses can be recovered in asci and subject to genetic analysis. Fungal genetics deduced the key features that define contemporary models of meiotic recombination including initiation by DSBs, strand exchange to form heteroduplex DNA and joint molecule (JM) intermediates called double-Holliday junctions (dHJs) (Szostak, Orr-Weaver, Rothstein, & Stahl, 1983). These DNA intermediates were subsequently confirmed in budding yeast, *Saccharomyces cerevisiae*, using the gel electrophoresis and Southern hybridization methods described in this chapter (Allers & Lichten, 2001a, 2001b; Cao, Alani, & Kleckner, 1990; Hunter & Kleckner, 2001; Schwacha & Kleckner, 1994, 1995; Sun, Treco, Schultes, & Szostak, 1989).

A working model of meiotic recombination is shown in Fig. 2. DSB formation is catalyzed by the Spo11 transesterase (Lam & Keeney, 2014). DSB-ends undergo nucleolytic resection to remove ~800 nt of the 5′ strand yielding long single-stranded 3′-terminal tails (Mimitou, Yamada, & Keeney, 2017; Zakharyevich et al., 2010), which then assemble into nucleoprotein complexes comprising an oligomeric filament of a RecA-family protein (Dmc1 and/or Rad51) (Brown & Bishop, 2015). Dmc1 filaments catalyze pairing and strand exchange between the DSB-end and a homologous duplex to form a nascent JM called a displacement loop (D-loop). The D-loop structure provides a primer-template for the local DNA synthesis that is essential to repair the DSB and restore sequences removed during resection. Meiotic recombination ensues via one of two major pathways, distinguished by the fate of the nascent D-loop (Fig. 2). In the synthesis-dependent strand-annealing (SDSA) pathway, the extended strand is dissociated and anneals to complimentary sequences on the second DSB-end (McMahill, Sham, & Bishop, 2007; Nassif, Penney, Pal, Engels, & Gloor, 1994; Paques & Haber, 1999). SDSA always results in a noncrossover outcome. In the dHJ pathway, a metastable one-ended strand exchange intermediate called a Single-End Invasion (SEI) is formed, likely through extension of heteroduplex (Hunter & Kleckner, 2001). The second DSB-end is then engaged to form a dHJ intermediate in which recombining duplexes are connected by two exchange or "Holliday" junctions (Schwacha & Kleckner, 1995; Szostak et al., 1983). In theory, dHJ resolution can yield either a crossover or noncrossover, but in meiosis resolution is biased to yield crossovers (Allers & Lichten, 2001a, 2001b; Clyne et al., 2003; Zakharyevich, Tang, Ma, & Hunter, 2012).

Central to the physical monitoring of recombination intermediates are synchronous meiotic time course experiments that allow detection of transient events (Fig. 1). Genomic DNA is extracted from cells sampled at various time throughout meiosis and subject to gel electrophoresis and Southern analysis to detect and quantify DNA events at defined recombination/DSB hotspots, such as the *HIS4::LEU2* and *ERG1* loci (Hunter &

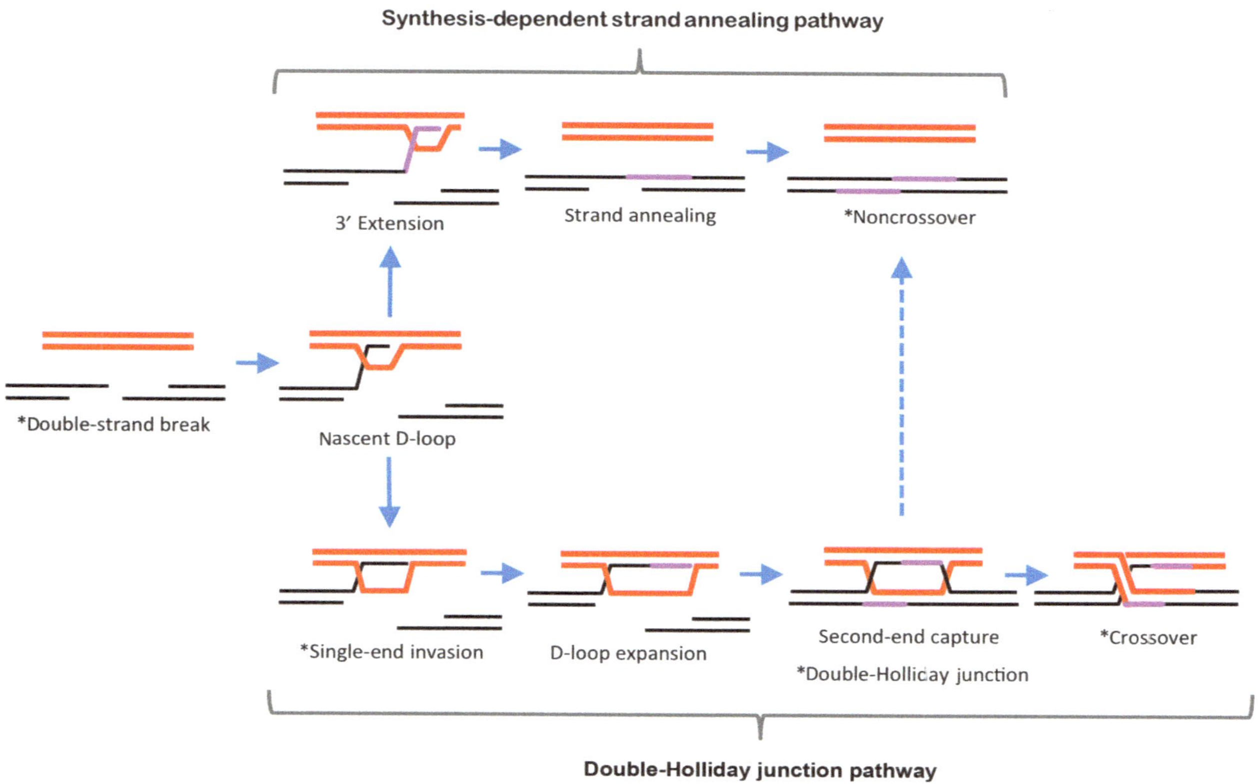

Fig. 2 Model of meiotic recombination (see text for details). *Asterisks* indicate DNA species detectable using the assays described here. Recombination-associated DNA synthesis is highlighted in *magenta*.

Kleckner, 2001; Lao, Cloud, et al., 2013; Thacker, Mohibullah, Zhu, & Keeney, 2014). Restriction enzyme site polymorphisms produce DNA fragments that are diagnostic of parental chromosomes, JMs formed between homologs or sister chromatids, crossovers, and noncrossovers (Fig. 3). JM intermediates may be lost through spontaneous branch migration, which is exacerbated by DNA purification treatments that employ heat and EDTA to inhibit nucleases (Panyutin & Hsieh, 1994). To preserve JMs, cells sampled from meiotic time courses are treated with psoralen and UV light to induce interstrand DNA cross-links (Frobel, Reiffers, Torres

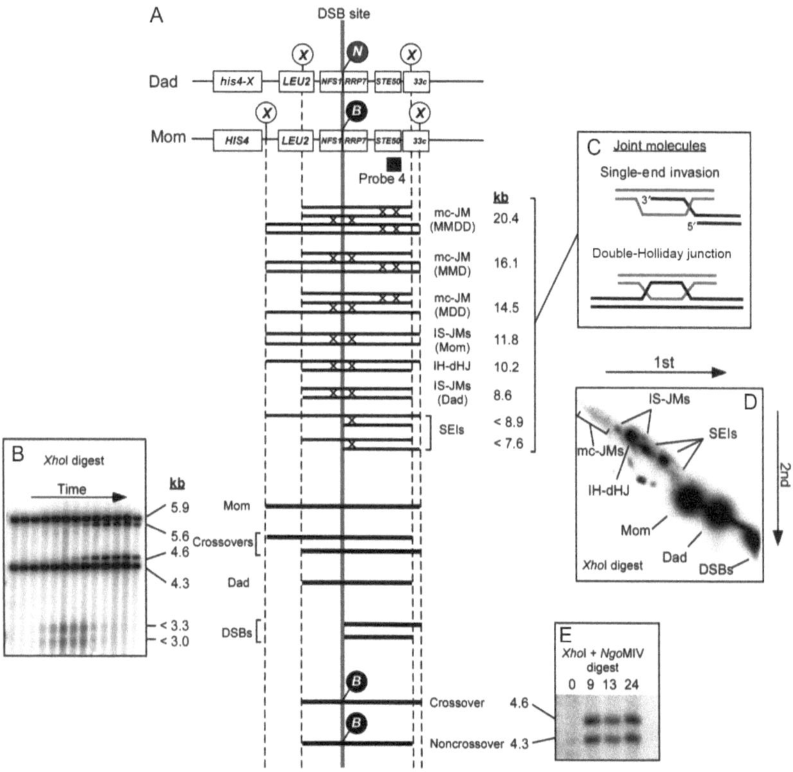

Fig. 3 Physical assay system for monitoring meiotic recombination. (A) Map of the *HIS4::LEU4* locus showing diagnostic restriction sites and position of the probe. DNA species detected are shown. *B*, *Bam*HI; *DSB*, double-strand break; *IH-JM*, interhomolog joint molecule; *IS-JM*, intersister joint molecule; *mc-JM*, multichromatid joint molecule; *N*, *Ngo*MIV; *SEI*, single-end Invasion; *X*, *Xho*I. (B) Southern image of one-dimensional gel analysis of crossovers and DSBs. (C) Inferred structures of SEI and dHJ intermediates. (D) Southern image of native/native 2D gel analysis of JMs. Species detailed in (A) are highlighted. (E) Noncrossover analysis. Southern image of *Xho*I + *Ngo*MIV double-digested genomic DNA showing diagnostic bands (see Fig. 4). Note that the noncrossover assay detects a representative subset of total recombinants.

Ziegenbein, & Gilch, 2015). One- and two-dimensional (1D and 2D) electrophoresis is applied to resolve DNA species of interest. Conditions optimized for 2D analysis resolve DNA molecules according to molecular weight in the first dimension but exaggerate any shape component in the second dimension such that branched DNA molecules, such as JMs, migrate more slowly than linear molecules of the same molecular weight (Bell & Byers, 1983) (Fig. 3).

The suit of assays described here allow the following steps to be monitored: DSB formation and resection, strand exchange to form SEIs, second-end capture to form dHJs, and resolution to form crossovers and noncrossover products (Cao et al., 1990; Hunter & Kleckner, 2001; Schwacha & Kleckner, 1995; Zakharyevich et al., 2010). In addition, interhomolog bias (template choice), at both SEI (Kim et al., 2010) and dHJ (Lao, Cloud, et al., 2013; Schwacha & Kleckner, 1994) stages, can be determined by measuring the ratio of interhomolog to intersister JMs. Nonallelic (ectopic) crossing over can also be monitored between the *HIS4::LEU2* locus and homologous sequences located ~25 kb away (Grushcow et al., 1999).

Native/denaturing 2D gel electrophoresis allows the strand composition of DSBs and JMs to be determined (Lao, Tang, & Hunter, 2013; Oh et al., 2009; Schwacha & Kleckner, 1995). In this case, DNA molecules are denatured prior to running the second-dimension gel under denaturing conditions. This method has been utilized to measure the extent of DSB resection (Zakharyevich et al., 2010) and to infer the structures of JMs including SEIs, dHJs, recombinant JMs (containing crossover and parental length strands), and aberrant JMs containing three and four interacting duplexes (multichromatid JMs) (Allers & Lichten, 2001a, 2001b; Hunter & Kleckner, 2001; Oh et al., 2007; Oh, Lao, Taylor, Smith, & Hunter, 2008; Schwacha & Kleckner, 1995). For JM analysis by native/denaturing 2D gels, psoralen cross-links must be removed in situ, within gel slices, before running samples in the second dimension under denaturing conditions.

2. SYNCHRONOUS MEIOTIC CULTURES, PSORALEN CROSS-LINKING, AND PREPARATION OF GENOMIC DNA

2.1 Meiotic Time Course—Pregrowth, Sporulation, and Psoralen Cross-Linking

S. cerevisiae strains that sporulate efficiently, such as SK1 (Kane & Roth, 1974), can be induced to undergo meiosis with reasonable synchrony enabling temporal analysis of meiotic events. For a full temporal analysis of the DNA events of meiosis, cultures are sampled at 0, 2, 2.5, 3, 3.5, 4, 4.5, 5, 6, 7, 8, 9, and 10 h after transfer to sporulation medium. These times may need to be adjusted for mutants that delay the progression of meiosis. Standard culture growth and sporulation methods are detailed later. Sharper synchrony may be achieved using a recently developed method that triggers meiosis by inducing expression of *IME1*, the master transcriptional regulator of meiosis (Chia & van Werven, 2016).

2.1.1 Equipment

Suppliers are listed, but equivalent products can be substituted.

1. 2.8-L triple-baffled Fernbach flasks (Bellco, Part # 2551-02800).
2. Orbital platform shaker in 30°C warm room (New Brunswick Innova or similar)
3. 500-mL centrifuge bottles (Corning, Manufacturer # 431123).
4. 50-mL disposable conical plastic tubes (Falcon, Manufacturer # 352070).
5. 60 × 15-mm Petri dishes (Corning, Manufacturer # 430166).
6. Long-wave UV transilluminator (UVP High Performance UV Transilluminator, Variable Intensity Model, 365 nm, or similar model).
7. Benchtop centrifuge (Sorvall Legend RT or similar)

2.1.2 Culture Media, Buffers, and Other Reagents

Suppliers are given where deemed important; in the absence of an indication, any high-quality source will do.

1. YPG agar: 1% (w/v) Bacto™ yeast extract (BD, Franklin Lakes, NJ), 2% (w/v) AMRESCO peptone (VWR), 1.5% agar (molecular biology grade, Teknova), 2% glycerol.
2. YPD broth: 1% (w/v) Bacto™ yeast extract, 2% (w/v) AMRESCO peptone, 2% D-glucose, 0.005% adenine hemisulfate (Sigma-Aldrich, A9126). Prepare broth without glucose or adenine as a 1.05 × solution and autoclave, mix with sterile 40% (w/v) D-glucose in a 19:1 ratio, and sterile 0.5% adenine hemisulfate in a 99:1 ratio before use. YPD agar is identical but contains 2% agar.
3. Presporulation broth (SPS): 1% (w/v) AMRESCO peptone (VWR), 0.5% (w/v) Bacto™ yeast extract, 1% (w/v) potassium acetate, 0.17% (w/v) Difco™ yeast nitrogen base without amino acids, 0.5% (w/v) ammonium sulfate, 1.5% (w/v) potassium hydrogen phthalate. Adjust pH to 5.5 with 10 N KOH, as sodium inhibits sporulation.

Make fresh and autoclave. Add two drops of Antifoam 204 (Sigma-Aldrich, A6426) per liter of media with a P200 pipette.

4. Sporulation medium (KAc): 1% (w/v) potassium acetate, 0.02% (w/v) D-(+)-raffinose pentahydrate (low glucose; US Biological). Add two drops of Antifoam 204 per liter of media from a P200 pipette. Additional nutrients (amino acids, purines, pyrimidines) required to supplement auxotrophic requirements of the strain being sporulated are added at 1/5th the levels used in vegetative growth medium.
5. DAPI stock: 1 μg/mL 4′,6-diamidino-2-phenylindole.
6. Fixation solution: 70% ethanol.
7. 10% (w/v) sodium azide.
8. 5 × psoralen stock solution: 0.5 mg/mL Trioxsalen dissolved in ethanol (200 proof). Store at 4°C wrapped in aluminum foil. To completely dissolve prior to use, shake at room temperature overnight.
9. 1 × psoralen working stock: diluted in 50 m*M* Tris–HCl pH 8.0, 50 m*M* EDTA pH 8.0. Keep on ice, wrapped in aluminum foil. Make fresh on the day of the experiment.
10. 50 m*M* EDTA, 50 m*M* Tris–HCl pH 8.0.

2.1.3 Procedure

Day 1 (5 days prior to the time course experiment)

1. Patch desired diploid strain from −80°C glycerol stock onto YPG plate.
2. Incubate at 30°C overnight for <17 h to select against petite colonies.

Day 2

3. Streak for single colonies on YPD plates.

Day 4

4. Inoculate single diploid colonies into 5 mL YPD broth. Grow to saturation at 30°C overnight.
5. Prepare and autoclave SPS (500 mL for each strain to be sporulated, plus a few mL extra to use as a blank in the spectrophotometer), and 1% SPM (1 L per strain). Autoclave three 2.8-L baffled flasks, one 1-L graduated cylinder, and two 500-mL graduated cylinders per strain to be sporulated.

Day 5

6. Measure 250 mL of SPS into each of two 2.8-L baffled flask.
7. 17 h before the start of the time course experiment, dilute cells 1/800 and 1/1000 from YPD overnight cultures into 250 mL SPS.
8. Incubate with vigorous shaking (350 RPM) at 30°C for 17 h. Make sure that the flasks are tightly clamped.

9. Measure 500 mL SPM into a 2.8-L baffled flask (one for each strain to be sporulated). Prewarm flasks at 30°C and keep remaining 1% SPM at 30°C. Note that volumes up to 1 L can be sporulated in 2.8-L baffled flasks, but may not aerate properly causing inefficient/asynchronous meiosis. 500 mL maintains ideal aeration.

Day 6

10. After 17 h of growth, measure the OD_{600} of each culture. Select cultures with an OD_{600} ~1.0 (but not higher than 1.1). Transfer 0.5 mL of selected cultures into a 1.5-mL microfuge tube, harvest the cells by centrifugation, and resuspend in 0.5 mL ice-cold 70% ethanol. Stain cells with DAPI by mixing 5 μL cells with 5 μL DAPI stock. Place 4–5 μL of stained cells onto a slide, cover with a coverslip, and view cells by fluorescence microscopy. Choose cultures with ≥80% of G1/G0 cells (defined as cells with no bud and with a single DAPI-staining body).
11. Transfer selected cultures into 500-mL centrifuge bottles and harvest cells at 3200 × *g*, at 28°C, for 3 min. Wash cells by resuspending in an equal volume of 1% SPM. Harvest once more and resuspend in the 500 mL prewarmed SPM in the 2.8-L baffled flask. Take the zero time-point sample (see next step) and incubate the culture with vigorous shaking (350 RPM) at 30°C.
12. Sampling volumes and times will depend on the goal of the experiment and the specific strain being analyzed. For each sample, collect at least 25 mL of culture in a 50-mL conical tube containing a 1:100 dilution of 10% sodium azide. 50 mL of culture should be collected for the zero time point, as DNA replication has not yet occurred and DNA content is half that of subsequent samples. Also transfer 0.5 mL of culture into a 1.5-mL microfuge tube, harvest the cells by centrifugation, and resuspend in 0.5 mL ice-cold 70% ethanol. Freeze at −20°C or process immediately for DAPI staining to monitor the progression of meiotic divisions.
13. Centrifuge the larger samples in the 50-mL tubes, at 3200 × *g* for 3 min. Pour off the supernatant and resuspend cell pellets in 2.5 mL ice-cold 1 × psoralen solution.
14. Transfer the cell/psoralen mixture to a 60-mm Petri dish and place on the long-wave UV box on the "high" setting. Allow cells to expose for 10 min, swirling the cells twice during this period.
15. Transfer the cells back into the original 50-mL tube. Recover any remaining cells in the Petri dish by washing with 2 mL ice-cold 50 m*M* EDTA/50 m*M* Tris–HCl solution and add to the original tube.

16. Harvest the cells at 3200 × *g* for 3 min and drain well. Spin again briefly and pull off last traces of supernatant with a P200 and freeze at −20°C.

2.2 DNA Extraction and Purification—Guanidine/Sarkosyl Preparation of Psoralen Cross-Linked DNA

2.2.1 Equipment

1. 1.5-mL microfuge tubes.
2. 37°C water bath; 65°C water bath.
3. Wide-orifice 1-mL micropipette tips (Rainin, RT-L1000W).

2.2.2 Buffers and Reagents

1. Spheroplasting buffer: 1 *M* sorbitol, 50 m*M* potassium phosphate pH 7.0, 10 m*M* EDTA pH 7.5. Filter sterilize and store at 4°C.
2. β-Mercaptoethanol.
3. Zymolyase 100T (US Biological).
4. Guanidine lysis solution pH 8.0: 4.5 *M* guanidine-HCl, 0.1 *M* EDTA, 0.15 *M* NaCl, 0.05% sodium lauroyl sarcosinate (Sarkosyl). Adjust to pH 8.0 with 50% NaOH.
5. RNase stock solution: 10 × TE pH 8.0 (100 m*M* Tris–HCl, 10 m*M* EDTA pH 8.0), 50 μg/mL RNase (DNase-free). Store at −20°C.
6. Proteinase-K solution: 20 mg/mL proteinase-K, 20 m*M* $CaCl_2$, 10 m*M* Tris–HCl pH 7.5, 50% glycerol. Store at −20°C.
7. Phenol pH 8.0/chloroform/isoamyl alcohol mixed in a 25:24:1 ratio.
8. 200 Proof ethanol.
9. 3 *M* sodium acetate pH 5.2.
10. 1 × TE pH 8.0 (10 m*M* Tris–HCl, 1 m*M* EDTA pH 8.0).

2.2.3 Procedure

1. Prepare sufficient spheroplasting buffer to allow 0.5 mL per cell pellet sample. Add 0.25 mg zymolyase 100T and 1% β-mercaptoethanol per sample. To process 12 samples (25 mL of SPM culture) from a typical time course requires 6 mL spheroplasting buffer with 3 mg zymolyase and 60 μL β-mercaptoethanol. For samples from larger volumes of culture, increase the spheroplasting buffer volume (for example, 0.45 mL of culture will require 0.9 mL spheroplasting buffer).
2. Thoroughly resuspend the cell pellets from Section 2.1.3 (step 15) in 0.5 mL spheroplasting buffer. Incubate at 37°C for 15 min. Gently flick

the tubes with fingers to mix the sample (finger-vortexing) twice during the incubation. See Note 1.

3. Harvest spheroplasted cells by centrifugation for 4 min, at 1500 × *g*. Using a P1000 pipette, carefully remove as much of the supernatant as possible without disturbing the loose cell pellet.
4. Add 1.5 mL guanidine lysis solution and resuspend the stringy spheroplast pellet by gentle finger-vortexing. Place at 65°C for 20 min. Finger vortex several times during this incubation to completely disperse the pellet and lyse the cells (incubation time can be extended 10–20 min to ensure the pellet is completely lysed).
5. Cool tubes on ice. Add 1.5 mL ethanol. Mix tubes well by inversion and store at −20°C overnight (alternatively, incubate at −20°C for at least 20 min before proceeding).
6. Pellet the precipitated material by centrifugation for 15 min at 3200 × *g* in a benchtop centrifuge. Drain well, centrifuge briefly, and carefully remove the last traces of supernatant with a pipette.
7. Add 0.5 mL RNase solution. Dislodge the pellet from the bottom of the tube by finger vortexing and allow the pellet to sit in the RNase solution for 5 min. Once the pellet has loosened, gently break up the pellet by finger vortexing (do not use a pipette at this stage). Incubate in a 37°C water bath for 1 h. Finger vortex several times during this incubation to completely disperse the pellet.
8. Add 15 μL proteinase-K solution and incubate at 65°C for 1 h. Finger vortex several times during this incubation. Spin tubes briefly to collect all solution to the bottom, and then transfer to a microfuge tube using wide-orifice pipette tips. At this point, cell lysates may be frozen at −20°C overnight before proceeding to the next step.
9. Add 0.5 mL phenol/chloroform/isoamyl alcohol (25:24:1). Shake and invert for ~30 s to mix thoroughly, let stand for 3 min, and shake again (do not vortex). Centrifuge at full speed (15,000 RPM) for 10 min in a microcentrifuge. *Carefully* remove the upper aqueous layer using a wide-orifice pipette tip and dispense into a new microcentrifuge tube, avoiding the white interface.
10. Extract a second time with 0.5 mL phenol/chloroform/isoamyl alcohol. Centrifuge at full speed in a microcentrifuge for 10 min. Carefully remove the upper, aqueous layer to a fresh microcentrifuge tube. See Note 2.
11. Ethanol precipitate the DNA: add 1/20th volume of 3 *M* sodium acetate pH 5.2 and two volumes of 100% ethanol. Mix well by

inversion, and let the samples stand for 20 min. DNA should form a visible precipitate. Centrifuge briefly and decant the supernatant. If no spool is visible or the spool is very small, centrifuge for 5 min at full speed in a microcentrifuge.

12. Rinse the DNA pellets by adding 1 mL 70% ethanol, centrifuge briefly, decant the supernatant, and drain the tubes. Pulse-spin in the centrifuge and remove the last traces of ethanol from the DNA pellets with a pipette.
13. Air dry the DNA pellets for at least 10 min (or until pellet looks dry). See Note 3. Add 50 μL 1 × TE and allow the DNA to hydrate in the refrigerator overnight. Mix well by flicking the tube. Do not vortex.
14. Determine DNA concentration of sample and adjust as required (see Section 2.3).
15. Store samples at −20°C.

2.3 Quantification of DNA Concentration

2.3.1 Equipment

1. Promega GLOMAX multidetection system (or equivalent 96-well plate reader).
2. 96-Well assay plate (Flat bottom black polystyrene; COSTAR®).

2.3.2 Buffers and Reagents

1. 10× TNE pH 8: 0.5 *M* Tris–HCl, 1 *M* NaCl, 10 m*M* EDTA.
2. Hoechst 33342 stock solution: 1 mg/mL in water. Store in the dark at 4°C.

2.3.3 Procedure

1. Prepare Hoechst working solution. For 50 mL: 5 mL 10× TNE buffer, 5 μL Hoechst 33342 stock solution and 45 mL Milli-Q water. Mix well and wrap in aluminum foil at room temperature. Prepare fresh each day.
2. Add 199 μL Hoechst working solution to all wells that will be used. To the first 8–10 wells, add 1 μL each from a dilution series of DNA standards ranging from 50 to 600 ng/μL. We dilute DNA standards from a 10-mg/mL stock of salmon sperm DNA (Sigma-Aldrich, D1626) and measure their concentrations on a NanoDrop spectrophotometer. Fluorescence readings from these samples are used to prepare a standard curve in Microsoft Excel by plotting concentration vs absorbance, and adding a trendline (best fit). The equation from this trendline is used

to calculate the concentrations of all DNA samples from their fluorescence readings.

3. Starting in the next row, add 1 μL of each meiotic DNA sample to the Hoechst working solution using a P2 pipette and mix all samples by pipetting up and down with a P200 pipette.
4. Read the fluorescence intensity of each sample using the Hoechst settings in the Promega GLOMAX multidetection system. Average the fluorescence readings for all blanks wells (wells that only contain Hoechst working solution) to obtain an average background signal. Subtract this signal from the fluorescence readings of all DNA samples.
5. Use Microsoft Excel to calculate the DNA concentration according to the standard curve. Adjust DNA concentrations to 100–500 ng/μL with TE buffer (we typically adjust to 200 ng/μL).

3. ONE-DIMENSIONAL NATIVE GEL ELECTROPHORESIS

3.1 DNA Digestion and One-Dimensional Gel Electrophoresis

3.1.1 Equipment

1. Buffer Puffer™ gel box, Model A5 (Owl Separation Systems) with a 36-well comb (Fig. 5).
2. Long-wave UV transilluminator (UVP High Performance UV Transilluminator, Variable Intensity Model, 365 nm, or similar model).

3.1.2 Buffers and Reagents

1. *Xho*I (20,000 U/mL, #R0146L) and *Ngo*MIV (10,000 U/mL, #R0564L) restriction enzymes and CutSmart buffer (New England Biolabs)
2. SeaKem LE agarose (Lonza).
3. TBE: 90 m*M* Tris base, 90 m*M* boric acid, 2 m*M* EDTA. Prepare as 10 × stock (900 m*M* Tris base, 900 m*M* boric acid, 20 m*M* EDTA) and dilute to 1 × as needed.
4. Loading buffer: 100 μL 6 × loading dye + 60 μL 10 × NEB3 restriction enzyme buffer (New England Biolabs; the additional salt prevents samples from drifting out of the wells).
5. 6 × loading dye: 0.25% (w/v) bromophenol blue, 0.25% (w/v) xylene cyanol FF, 15% (w/v) Ficoll in water.

6. λ*Bst*EII and λ*Hin*dIII DNA molecular weight markers (New England Biolabs).
7. 10 mg/mL ethidium bromide in dH_20. Store at room temperature wrapped in aluminum foil.

3.1.3 Procedure

Day 1

1. For *DSB and crossover* analysis, digest 2 μg of DNA in 80 μL total volume with 5 μL *Xho*I, and 8 μL CutSmart buffer. Incubate at 37°C for 4 h. This releases parental, crossover, and double-strand break fragments (Fig. 3).
2. For noncrossover analysis, digest 2 μg of DNA in 120 μL total volume with 5 μL *Xho*I, 5 μL *Ngo*MIV, and 12 μL CutSmart buffer. Incubate at 37°C for 4 h. This combination of enzymes produces a fragment that is diagnostic for a subset of noncrossovers formed via conversion of a *Bam*HI /*Ngo*MIV polymorphism located directly at the DSB site (Fig. 4).
3. Precipitate digested DNA by adding 190 μL ethanol and 5 μL 3 *M* sodium acetate pH 5.2; for the noncrossover digest with 120 μL volume, add 285 μL ethanol and 7.5 μL 3 *M* sodium acetate pH 5.2. Let sit for 20 min at room temperature.
4. Centrifuge for 10 min at full speed in a microcentrifuge.
5. Decant the supernatant and wash the pellet by adding 100 μL 70% ethanol.
6. Centrifuge for 5 min at full speed in a microcentrifuge.
7. Decant, pulse spin tubes and pipette off residual ethanol.
8. Air dry pellets for ~10 min at room temperature.
9. Resuspend pellets in 15 μL 1 × TE. Allow DNA to hydrate in the refrigerator overnight. Mix well by flicking the tube.

Day 2

1. Prepare a 0.6% agarose gel: add 2.1 g SeaKem LE agarose to 350 mL 1 × TBE (*without* ethidium bromide). Dissolve completely by heating in the microwave and boiling steadily for 30 s (approximately 4 min total). Cool to 55°C and pour into a clean, level gel tray in the cold room. Allow gel to harden for at least 30 min.
2. Add 5 μL loading buffer to each digested DNA sample, mix gently.
3. Carefully load gel and run in ~2 L 1 × TBE (*without* ethidium bromide) at 70 V (~2 V/cm) for 26 h at room temperature.

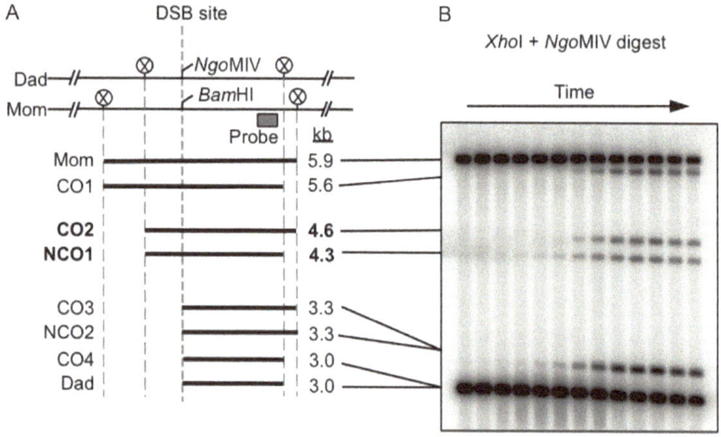

Fig. 4 Noncrossover assay. Map of the *HIS4::LEU2* hotspot highlighting the DSB site, diagnostic restriction sites, and position of the probe used in Southern analysis. Sizes of diagnostic fragments are shown. *Circled Xs* indicate *Xho*I sites. Note that NCO1 (in *bold*) is the only band that is unambiguously diagnostic of a noncrossover product. NCO1 is quantified as percentage of total hybridization signal. NCO1 and CO2 are compared to assess changes in the CO/NCO ratio.

Fig. 5 First dimension of 2D gel electrophoresis. (A) Digested DNA samples loaded onto a 0.4% gel in the Buffer Puffer™ gel apparatus. (B) After electrophoresis, the gel is stained with ethidium bromide and placed on a long-wave UV transilluminator to visualize DNA, check for even loading and efficient digestion, and excise gel slices for second-dimension gels.

Day 3

1. Stain gel in 1 L dH_2O with 0.5 μg/mL (50 μL of 10 mg/mL stock) ethidium bromide in a clean Pyrex dish with gentle shaking for 30 min.
2. View gel on UV box to check that DNA digestion is complete and that loading is even (Fig. 5).
3. Blot the gel by alkaline transfer overnight (see Section 6).

4. TWO-DIMENSIONAL GEL ELECTROPHORESIS (NATIVE/NATIVE)

4.1 Equipment

1. Buffer Puffer™ gel box, Model A5 (Owl Separation Systems) with a 36-well comb.
2. Large pyrex dishes (39 × 45 cm) for staining gels.
3. Platform shaker.
4. Long-wave UV transilluminator (UVP High Performance UV Transilluminator, Variable Intensity Model, 365 nm, or similar model).
5. Clean razor blades.
6. Fluorescent ruler.

4.2 Buffers and Reagents

1. SeaKem Gold Agarose (Lonza); SeaKem LE Agarose (Lonza).
2. 1 × TBE: 90 m*M* Tris base, 90 m*M* boric acid, 2 m*M* EDTA. Prepare as 10 × stock (900 m*M* Tris base, 900 m*M* boric acid, 20 m*M* EDTA) and dilute to 1 × as needed.
3. λ*Bst*EII and λ*Hin*dIII DNA molecular weight markers (New England Biolabs).
4. 10 mg/mL ethidium bromide in dH_20. Store at room temperature wrapped in aluminum foil.

4.3 Procedure

Day 1

1. Prepare a 0.4% agarose gel: add 1.4 g SeaKem Gold agarose to 350 mL 1 × TBE (*without* ethidium bromide). Dissolve completely by heating in the microwave and boiling steadily for 30 s (approximately 4 min). Cool to 55°C and pour into a clean, leveled gel tray in the cold room. Allow gel to harden for at least 30 min.
2. Add 5 μL loading buffer to each digested DNA sample, mix gently (for sample digestion, see Section 3.1.3).
3. Carefully load the gel, leaving at least one lane space between samples (Fig. 5). Also load λ*Bst*EII and λ*Hin*dIII DNA ladder on either side of the gel. Run the first dimension gel for 21 h at 35 V (1 V/cm) at room temperature.
4. Prepare two 5-L batches of 1 × TBE containing 0.5 μg/mL (250 μL of 10 mg/mL stock solution) ethidium bromide and put in the cold room to chill overnight.

Day 2

5. Stain gel in 1 L prechilled 1 × TBE + 0.5 μg/mL ethidium bromide for 30 min at room temperature with gentle shaking. To analyze 12 samples (a standard time course) prepare two 400 mL 0.8% agarose gels: add 3.2 g SeaKem LE agarose to 400 mL 1 × TBE. Dissolve completely by heating in the microwave and boil steadily for 30 s. Add ethidium bromide to 0.5 μg/mL, mix and place to cool in a 55°C water bath.
6. Lay a piece of Saran Wrap on top of a *long wave* UV box and gently slide the gel on top. Visualize the DNA and carefully cut ~9.5-cm slices from the lanes to cover the size range of interest, typically ~0.5 cm from the wells down to the 2.2-kb band of the λ*Bst*EII ladder (Fig. 6). Start by cutting off the top and bottom of the gel and then excise each lane as cleanly and as quickly as possible.
7. Using a flexible plastic ruler, place the excised lanes horizontally in three rows of two in a 20 × 25-cm gel tray (rows spaced ~8 cm apart) (Fig. 6). In the cold room, carefully pour the 0.8% agarose around the slices to *just* cover them. Allow gel to harden for 30 min. Do the same for the second gel (12 samples for a full time course analysis).

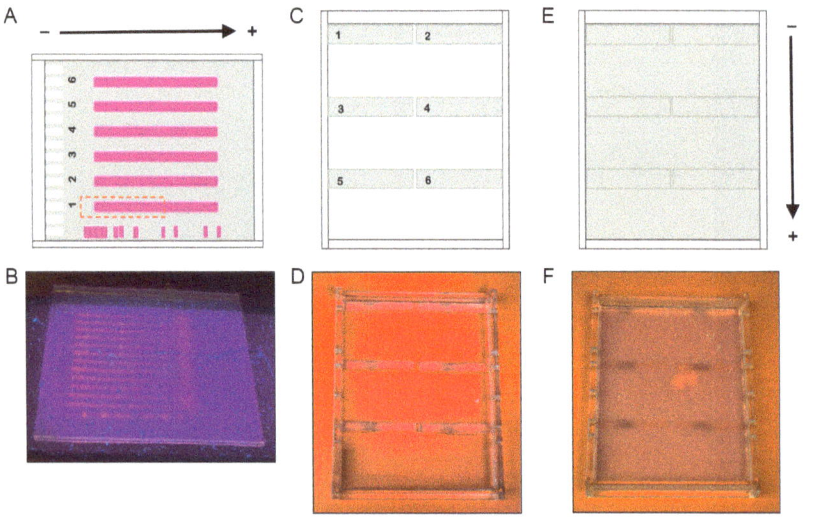

Fig. 6 2D Gel Setup. (A and B) Digested DNA electrophoresed in the first dimension is visualized using long-wave UV light. (C and D) Each lane slice is cut to the appropriate size and arranged perpendicularly in a clean gel tray, with the higher molecular weight DNA oriented to the *left*. (E and F) Molten agarose gel is poured around the lane slices to just cover them.

8. Run the gels in the 4°C cold room using the prechilled 1× TBE with 0.5 μg/mL ethidium bromide at 170 V (6 V/cm) for 6 h.
9. Blot gels by alkaline transfer overnight (see Section 6).

5. TWO-DIMENSIONAL GEL ELECTROPHORESIS (NATIVE/DENATURING)

This procedure is used to analyze the strand composition of recombination intermediates, including JMs and DSBs (Oh et al., 2008; Zakharyevich et al., 2010) (Fig. 7). For JMs, psoralen cross-links must be reversed prior to running the second-dimension gel under denaturing conditions (Oh et al., 2009). Steps unique to this procedure are highlighted in *italics*. Strand analysis of DSBs (e.g., to measure 5′ resection) is typically performed on DNA samples that were not cross-linked with psoralen (Lao, Tang, et al., 2013).

5.1 Equipment

1. Buffer Puffer™ gel box, Model A5 (Owl Separation Systems) with a 36-well comb.
2. Large pyrex dishes (39 × 45 cm) for staining and equilibrating gels.
3. Platform shaker.
4. Long-wave (350 nm) UV transilluminator box.
5. Clean razorblades.
6. Flexible plastic ruler.
7. A two-toothed comb to create wells for the molecular weight standards, see Note 4.

5.2 Buffers and Reagents

1. SeaKem Gold Agarose; SeaKem LE Agarose (Lonza).
2. TBE: 90 m*M* Tris base, 90 m*M* boric acid, 2 m*M* EDTA. Prepare as 10× stock.
3. TE: 10 m*M* Tris–HCl pH 8.0, 1 m*M* EDTA. Prepare as 10× stock.
4. DNA molecular weight standards covering the size range of interest.
5. 10 mg/mL ethidium bromide in dH_2O. Store at room temperature wrapped in aluminum foil.
6. 6× Gel Loading Dye: 0.25% (w/v) bromophenol blue, 0.25% (w/v) xylene cyanol FF, 15% (w/v) Ficoll (Type 400, Pharmacia) in dH_2O. Store at room temperature.

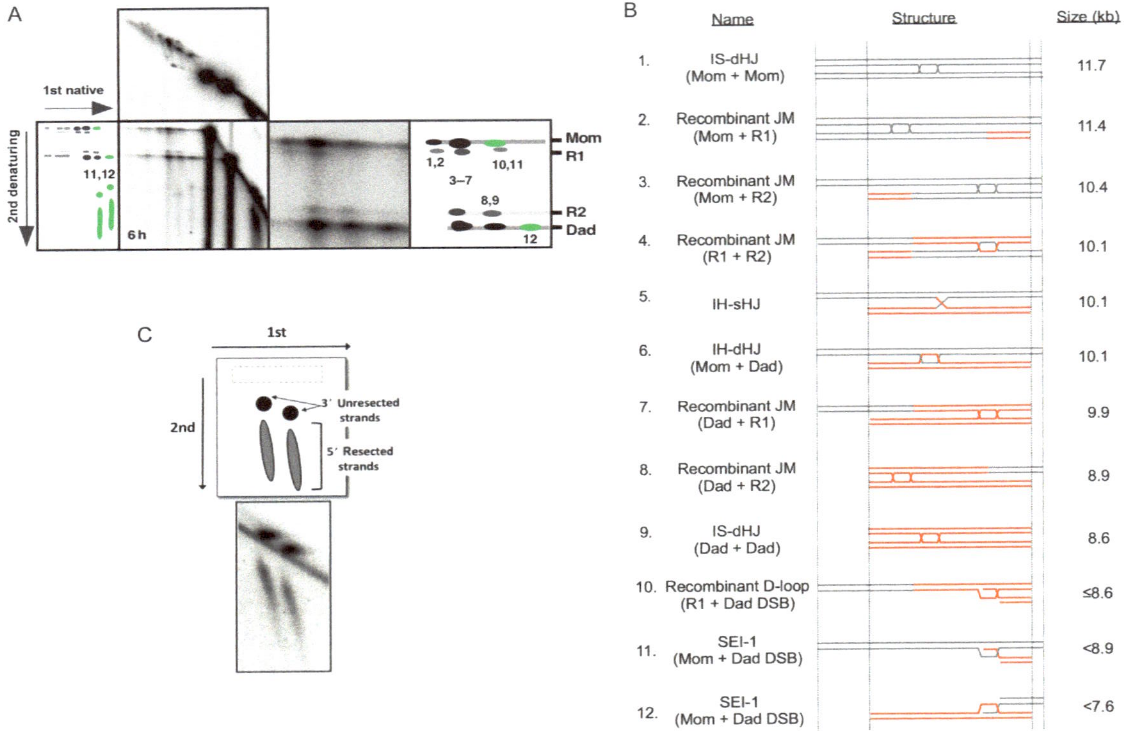

Fig. 7 Native/denaturing two-dimensional gel electrophoresis. (A) A full image of the native/denaturing 2D blot is shown together with a blowup of the region of interest, to the *right*. The corresponding native/native 2D panel is shown to orient native JM species with their component strands. Interpretative diagrams highlight signals of interest, and their corresponding JMs are shown in (B) (Oh et al., 2008). Signals colored *green* correspond to the component strands of SEIs (Hunter & Kleckner, 2001). (C) Native/denaturing 2D gel analysis of DSBs revealing discreet 3′-terminal strand and heterogeneously resected 5′-terminal strands (Zakharyevich et al., 2010).

7. Native Gel Loading Buffer: 100 μL 6 × loading dye + 60 μL 10 × NEB3 restriction enzyme buffer (New England Biolabs)
8. Equilibration Buffer: 5 m*M* Na_3PO_4. Warm solution to 65°C and adjust to pH 12 using 50% NaOH.
9. Cross-link Reversal Buffer: 1 m*M* Na_3PO_4, 3 *M* urea. Warm solution to 65°C and adjust to pH 12 using 50% NaOH.
10. 5 × Alkaline Running Buffer: 250 m*M* NaOH, 5 m*M* EDTA in dH_2O. Make fresh for each use.
11. 1 × Alkaline Running Buffer: 50 m*M* NaOH, 1 m*M* EDTA in dH_2O. Make fresh for each use.
12. 6 × Alkaline Gel Loading Buffer: 300 m*M* NaOH, 6 m*M* EDTA, 18% (w/v) Ficoll (Type 400, Pharmacia), 0.15% (w/v) bromocresol green, 0.25% (w/v) xylene cyanol in dH_2O. Store in plastic tubes at 4°C.

5.3 Procedure

Day 1

1. To release DNA fragments of interest, digest an appropriate amount of sample genomic DNA to completion. For analysis of DSB processing in budding yeast, we digest 4–10 μg of genomic DNA at a concentration of 1 μg per 20 μL, with a fourfold excess of restriction enzyme. *For analysis of JM species, we digest ≥10 μg of DNA.*
2. Ethanol precipitate, air dry, and resuspend DNA in 15 μL 1 × TE. Add 5 μL native loading buffer and load the samples into a 0.6% SeaKem *Gold* agarose gel (*0.4% for JM analysis*) prepared as described in Section 4.3. Leave one lane space between different samples and also between the molecular weight standards and the samples. Each second-dimension gel can fit six samples total, with two columns of samples, and three rows.
3. Run the first dimension at 70 V (1 V/cm) for 24 h.
4. Chill at least 7 L dH_2O in the cold room (for preparation of the alkaline running buffer). Also chill 1 L dH_2O + 0.5 μg/mL ethidium bromide.

Day 2

5. Prewarm a shaking water bath to 65°C. *Prepare 500 mL Equilibration Buffer and 500 mL Cross-link Reversal Buffer for JM analysis.*
6. Stain gel in 1 L prechilled dH_2O + 0.5 μg/mL ethidium bromide for 30 min at room temperature with gentle shaking.
7. Prepare a 1.2% alkaline agarose gel (*1% gel for JM analysis*): Dissolve 5.4 g (*4.5 g*) SeaKem LE agarose in 427.5 mL 1 m*M* EDTA and cool

in a 50°C bath (keep a stir bar in the flask). See Note 5 regarding running conditions. Separately, prepare 50 mL of 1 *M* NaOH (fresh, using solid NaOH).

8. Lay a piece of Saran Wrap on top of the long wave UV box and carefully slide the gel on top. Visualize the DNA and carefully cut slices from the lanes to cover the size range of interest. Cut off one corner of the gel slice to keep track of the orientation. Weigh the gel slices to estimate their volume.
9. In a small plastic container (one per gel slice), rinse the gel slices in ≥10 volumes cold dH_2O with gentle shaking for 15 min (≥10 volumes are used in all subsequent rinses and incubations, steps 6–14). Very carefully decant the wash and repeat.
10. *Incubate as above with prewarmed Equilibration Buffer in the 65°C water bath. Repeat this wash. During steps 5–7, the gel slices become fragile and slippery, and great care must be taken when decanting buffers.*
11. *Incubate in prewarmed Cross-link Reversal Buffer for 2–4 h at 65°C. Repeat this incubation.*
12. *Briefly rinse gel slice in 500 mL dH_2O at room temperature.*
13. *Wash twice with ≥ 10 volumes ice-cold dH_2O for 15 min to remove remaining urea.*
14. Wash twice, for 15 min, with ice-cold 5 × alkaline running buffer to denature the DNA.
15. Wash twice, for 15 min, with ice-cold 1 × alkaline running buffer to equilibrate gel slices.
16. Place gel slices in a gel tray perpendicular to first dimension. Add 25 mL of the 1 *M* NaOH solution to the cooled gel (for a final concentration of 50 m*M* NaOH) while stirring.
17. Promptly pour the agarose to just cover the gel slice. Allow the gel to harden in the cold room for 30 min.
18. Add 1 × alkaline running buffer to just cover the gel. Load molecular weight standard in denaturing loading buffer, see Note 6. Lay saran wrap on top of the buffer to exclude air. Run at 55 V (~1.7 V/cm) for 30 h with a change of buffer after 15 h, see Note 7. *For JM analysis, run at 55 V (~1.7 V/cm) for 50 h with a change of buffer after ~24 h.* Run times may need to be optimized to resolve molecules of interest.
19. Rinse the gel twice in 1 L of dH_2O with gentle shaking.
20. Blot gel by alkaline transfer overnight (see Section 6).

6. SOUTHERN TRANSFER AND HYBRIDIZATION

6.1 Southern Transfer

6.1.1 Equipment

1. Large pyrex dishes (39 × 45 cm) for staining and equilibrating gels, and to use as a reservoir for the Southern transfer.
2. Platform shaker.
3. Glass plate (22.5 × 43 cm) to span the large pyrex dish and serve as a platform for the transfer.
4. Whatman 3MM paper (46 × 57-cm sheets).
5. GeneScreen Plus (from PerkinElmer) or ZetaProbe nylon hybridization transfer membrane—20 × 25 cm for each gel.
6. Single-fold paper towels.
7. Plastic/plexiglass plate (22 × 28 cm) to place on top of the transfer stack.

6.1.2 Buffers and Reagents

1. 0.25 *M* HCl (1 L per gel). Make immediately before use.
2. 0.4 *M* NaOH (3 L per gel). Make immediately before use.
3. 1 *M* sodium phosphate buffer pH 7.2.

6.1.3 Procedure

Day 1

1. Soak the gel in 1 L dH_2O for 10 min. All subsequent steps are in 1 L volumes with gentle shaking. Prepare 0.25 *M* HCl and 0.4 *M* NaOH solutions. You will need sufficient NaOH to soak the gel and do the transfer, i.e., 3 L per gel.
2. Flip the gel so that the bottom, flat side is facing up. To do this, gently sandwich the gel between two flat, gel-sized plastic plates, quickly flip, and slide the gel gently back into the dish containing dH_2O.
3. Soak gel in 0.25 *M* HCl for 20 min with gentle shaking.
4. Pour the HCl into a large beaker and rinse the gel briefly with 1 L dH_2O. Soak the gel in 0.4 *M* NaOH for 30 min with gentle shaking.
5. During steps 3–4, prepare the following: four pieces of Whatman blotting paper the same size as the gel (25 × 20 cm), one piece of Whatman paper suitable for a wick (see Note 8), and one piece of nylon membrane the same size as the gel (see Fig. 8). Use gloves and a clean dry surface when measuring and cutting the membrane (do not remove the protective paper; keep the membrane free of dust; avoid any

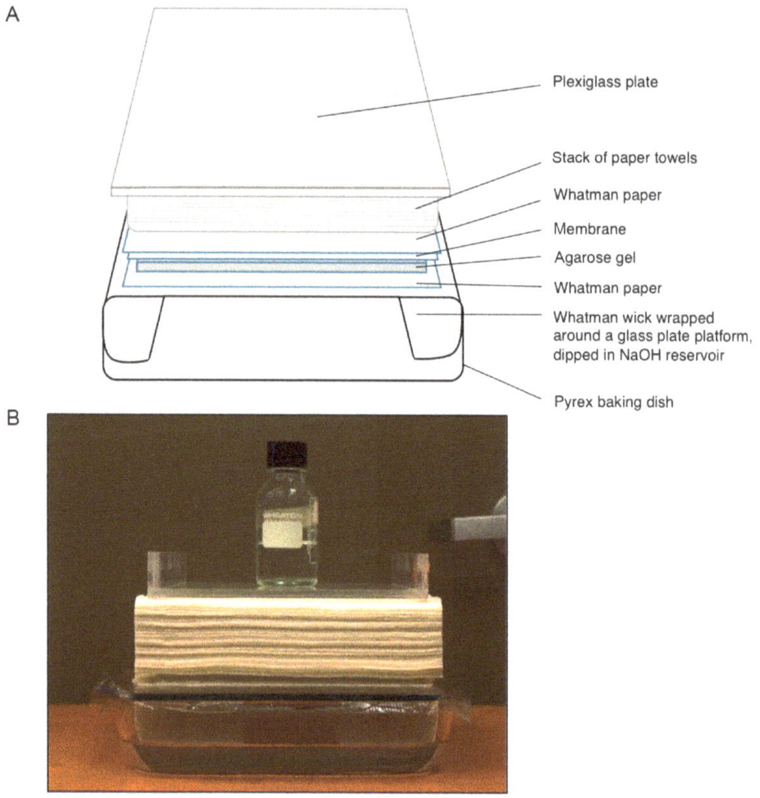

Fig. 8 Southern transfer. (A) Schematic of Southern transfer setup. (B) Photo of transfer setup. See text for details.

crumpling or other mechanical stress on the membrane; all will increase background in subsequent hybridization). Use a fine-tipped sharpie or pencil to write the date and identifier on the top right hand side of the membrane.

6. Set up the blot. Fold the wick evenly across a clean glass plate (see Note 9). Place the glass plate and wick on top of a clean pyrex dish with both ends of the wick in the tray. Wet the surface of the wick with 0.4 *M* NaOH and flatten by rolling with a 25-mL glass pipette. Add more 0.4 *M* NaOH to the dish to create a reservoir. Place two gel-size pieces of Whatman in the center of the wick, wet well with NaOH and roll flat (see Fig. 8). You will use ~1 L total NaOH solution in this step.
7. Carefully slide the gel from the tray to the wick. Gently push any trapped air bubbles from under the gel with gloved fingers. Empty the NaOH used for gel soaking into the beaker containing the used HCl (see Note 10). Rinse the dish and fill with dH_2O.

8. Wet the nylon membrane in the dH_2O for 5 min (see Note 11). Minimize handling and hold only at the edges. Carefully align the top edges of the membrane and the gel and lower the membrane onto the gel. Do not move the membrane around once it contacts with gel.
9. Soak the two additional gel-sized pieces of Whatman in the dH_2O and carefully place on top of the membrane. Take a 25-mL pipette and carefully, but firmly, roll out any bubbles working out from the center.
10. Place a large piece of plastic wrap across the whole construction and cut around the gel with a sharp razor blade. The plastic wrap should seal around the edges of the gel so that capillary action occurs only via the gel and membrane.
11. Take a pack of single-fold paper towels; unfold several towels and lay them flat on top of the Whatman paper. Evenly split the rest of the stack into two and arrange on top of the gel stack horizontally in a side-by-side arrangement. Place a plexiglass plate (not glass) and a *small* weight (small bottle containing less than 50 mL of liquid) on top of the blot (see Fig. 8).
12. Blot for at least 6 h, generally overnight.

Day 2

13. Prepare 1 L, 50 m*M* sodium phosphate pH 7.2 and pour into a clean pyrex dish that can accommodate the nylon membrane. Deconstruct the blot: remove the weight, plate, paper towels, and top two pieces of Whatman paper. Immediately peel off the membrane and place into the sodium phosphate solution to neutralize. Do not allow the membrane to dry out prior to neutralization as this will result in high background. Gently shake membrane for 10 min with one change of buffer.
14. Move directly to the prehybridization step (Section 6.2.3, step 1), or briefly drain the membrane on a piece of Whatman paper and store the damp membrane in plastic wrap at 4°C (short term) or −20°C (longer than 2 days).

6.2 Hybridization—See Note 12

6.2.1 Equipment

1. Radioactive workstation with appropriate protective screening, Geiger counter, microcentrifuge, 37°C and 95°C heating blocks and waste containers for dry and liquid radioactive waste.
2. Large pyrex dish (39 × 45 cm) to accommodate the Southern blot membrane.
3. Hybridization oven and bottles (Autoblot mini hybridization oven, BELLCO glass, Inc. or equivalent).

4. ProbeQuant G-50 Micro Columns (GE Healthcare).
5. Large shaking water bath at 65°C.
6. Large plastic container (Tupperware or similar) with a sealing lid to accommodate the Southern blot membrane and 1 L of wash buffer.
7. Long round-nosed forceps to remove membrane from hybridization bottle.
8. Whatman paper.
9. Plastic wrap.

6.2.2 Buffers and Reagents

1. 1 *M* sodium phosphate pH 7.2.
2. Hybridization Buffer: 0.25 *M* sodium phosphate pH 7.2, 0.25 *M* NaCl, 1 m*M* EDTA, 7% SDS, 5% Dextran Sulfate. For 500 mL: 125 mL 1 *M* Na-phosphate pH 7.2, 25 mL 5 *M* NaCl, 1 mL 0.5 *M* EDTA, 35 g SDS, 25 g Dextran Sulfate. Add liquids to a 1-L beaker. Make up to ~300 mL with sterile deionized water. Slowly add SDS and Dextran Sulfate while stirring. Heat to 65°C and stir until fully dissolved. Aliquot into 50-mL plastic tubes. Store at room temperature. SDS will come out of solution so heat to 65°C and mix well before use.
3. Sheared Salmon sperm DNA (10 mg/mL, Sigma) in 0.2 mL aliquots. Store at −20°C.
4. Prime-It® RmT Random Primer Labeling Kit (Agilent Technologies).
5. Radiolabeled nucleotide: ^{32}P αdCTP (PerkinElmer BLU013H250UC or equivalent). Buy only as needed and store at −20°C.
6. 20 × Saline Sodium Citrate (SSC) stock solution (3 *M* NaCl, 300 m*M* sodium citrate, adjust to pH 7.0).
7. 20% SDS solution.
8. Low Stringency Wash: 2 × SSC, 0.1% SDS. For 1 L, place 100 mL 20 × SSC in a 1-L measuring cylinder and make up to 995 mL. Add 5 mL 20% SDS.
9. High Stringency Wash: 0.1 × SSC, 0.1% SDS. For 1 L, place 990 mL water in 1-L measuring cylinder. Add 5 mL 20% SDS and 5 mL 20 × SSC.

6.2.3 Procedure

1. Prewarm the Hybridization Solution to 65°C and mix well by inverting the bottle several times. Turn on the hybridization oven and set to 65°C. Rinse the hybridization bottles, lids, and O-rings (if used) with dH_2O. Prewarm bottles in the oven.

2. Add 10 mL Hybridization Solution per small bottle or 20 mL per large bottle. Place bottles back into the oven.
3. Denature a 0.2-mL aliquot of sheared Salmon sperm DNA by heating in a 95°C dry block for 5 min. Immediately quench on ice/water and then store on ice.
4. Submerge the membrane to be hybridized, with the DNA-side facing up, in 50 m*M* sodium phosphate pH 7.2 in a clean pyrex dish. Carefully roll up the membrane with DNA face on the inside (avoid touching the DNA-side during the rolling process) and drain out excess buffer by touching one edge of the rolled-membrane on paper towel. Place membrane into prewarmed hybridization bottle and slowly unroll by turning the bottle, keeping the opening face wet with hybridization solution. Avoid bubbles. Add the Salmon sperm DNA, promptly return to the oven and prehybridize for >30 min, preferably several hours.
5. Remove radiolabeled nucleotide from the freezer and allow it to thaw (this can be done during the prehybridization step). Sign out the amount of radiolabel to be used in the logbook. Sign into the radioactive workstation and monitor the area using a Geiger counter.
6. Label 25 ng of probe DNA using a Prime-It® RmT Random Primer Labeling Kit (or equivalent) and ^{32}P αdCTP according to the manufacturers protocol: Add appropriate amounts of probe and water to total 42 μL of probe mix with 25 ng of probe. Add 5 μL ^{32}P αdCTP, 3 μL polymerase, and pipette up and down briefly. Incubate at 37°C for at least 1 h (see Note 13).
7. Separate the probe from unincorporated nucleotides using a ProbeQuant G-50 Micro Column. Remove storage liquid from column by centrifuging at 3000 RPM for 1 min. Add probe-labeling reaction mix to column, spin at 3000 RPM for 2 min, and collect flow-through in a new tube.
8. Denature the probe in a 95°C heating block or boiling water bath for 2 min. Quench on ice for 2 min (see Note 14).
9. Add the denatured probe to the hybridization bottle and promptly return to the oven. Monitor the workstation with a Geiger counter, clean up any contamination immediately, and sign out.
10. Hybridize for >6 h, typically overnight.
11. Turn on a shaking water bath and prewarm to 65°C. Sign into the radioactive workstation.
12. Make up 1 L Low Stringency Wash and add to a 1-L flask. Heat to 65°C (for 1 L this takes ~3.5 min in a microwave on full power).

Check the temperature carefully—do NOT heat above 65°C or you will strip the probe.

13. Carefully pour off the hybridization mix into the radioactive liquid waste (record all volumes of liquid waste added). Fill the bottle ~1/2 full with preheated wash and return promptly to the oven. Incubate for 10 min.
14. Pour the wash into the radioactive liquid waste and record the volume. Repeat the low-stringency wash. Prepare 2 L High-Stringency wash and heat to 65°C.
15. Pour the second low stringency wash into liquid waste and record the volume. Fill the bottle ~1/2 full with preheated High Stringency wash and return promptly to the oven. Incubate for 30 min.
16. Pour the wash into the radioactive liquid waste and record the volume. Carefully and quickly remove the membrane from the tube with long round-nosed forceps and place into a clean plastic container with 1 L prewarmed High Stringency wash. Cover the container and incubate in a 65°C shaking water bath for 30 min with gentle shaking. To prevent the container from floating, you will need to add a weight (we use a piece of lead flashing).
17. Repeat the high-stringency wash.
18. Monitor the blot—the edges should be essentially free of counts (~2 × background with the Geiger counter on its most sensitive setting). If there are significant background counts, wash again. Distinct signals corresponding to specific hybridizing species should be detectable.
19. Drain excess wash from the blot onto a wad of paper towels. Carefully wrap the damp blot in plastic wrap. Using small pieces of tape, mount onto a piece of Whatman paper with the DNA side facing out.

6.3 Imaging and Analysis

6.3.1 Equipment

1. Phosphorimager (Typhoon™, Storm™, or PhosphorImager™; GE Healthcare)
2. Phosphorimager screens and cassettes (GE Healthcare)
3. ImageQuant™ (GE Healthcare) and Excel™ (Microsoft) software packages. Open-source software such as ImageJ (http://rsb.info.nih.gov/ij/) can also be used for image analysis.

6.3.2 Procedure

1. Expose the hybridized Southern blots to storage phosphor screens for >30 min to overnight and scan the screens on a Phosphorimager.

To obtain the optimum exposure for quantification, we aim to have the strongest signals in the upper third of the dynamic range of the phosphorimager.

2. Open the Southern image in ImageQuant. To quantify the relative levels of hybridizing species, integrated signal intensities are measured in by drawing shapes around individual bands/spots and subtracting background signal from shapes of equivalent size. The level of a given species is expressed as a fraction of the total hybridization signal (the sum of all analyzed bands/spots minus background) or "% DNA."
3. For the analysis of DSB processing during recombination, we calculate a "resection profile," which describes the distribution of resection lengths for the 5-strands of DNA double-strand breaks. To do this, first draw a line tracing through the lane containing the DNA molecular weight standards.
4. Create a graph using the analysis tool. The graph will show distance in mm vs radioactive signal counts. Distinct peaks correspond to the individual fragments of the DNA standards. The position of the wells is the reference point for migration distance and is indicated by a tiny peak located above the largest signal peak for the DNA standards.
5. Import the ImageQuant graph data into Excel. Assign a size (bp) to the migration distance (mm) of each band of the DNA standards.
6. Obtain a linear regression equation by graphing migration distances (x) vs *log* (size in bp) (y).
7. For each signal of interest, calculate the migration distance (x). Enter this value into the regression equation and solve for y. To determine the fragment size in bp, take the inverse *log* of y.
8. To create the "resection profile," the signal intensities for 5′ strands within successive length bins of 100 nt are calculated and plotted in a bar graph as percentages of the total hybridizing signal (Zakharyevich et al., 2010). To correct for background signal, the same process is repeated for an adjacent region of the Southern blot without DSB signals.

7. NOTES

1. Time this step carefully. Over-spheroplasted cells will not form a tight pellet and may require repeated centrifugation. Under-spheroplasted cells will form a dense pellet (resembling intact cells) and should be gently resuspended and incubated for an additional 2–5 min.

2. If a distinct white interface is visible in any of the extracted solutions, perform another extraction.
3. DNA pellets should not be over dried. DNA can be rehydrated for several hours to overnight at 4°C.
4. A makeshift comb for DNA standards can be made by turning a regular gel comb over in its holder (comb side facing up) and attaching well-sized plastic sticks using adhesive tape. Alternatively, teeth can be broken from an existing comb to leave wells at desired positions (typically at the two ends). When running two rows of samples, marker lanes should be staggered.
5. Note that agarose percentage may need to be optimized to resolve the size range of interest. Stated concentrations are optimal for analysis of component DNA strands at the *HIS4::LEU2* locus
6. To accurately estimate the size of DNA strands, take special care to align the gel slices with the marker wells. The migration distance of the marker will be used to calculate the sizes of component strands being analyzed.
7. NaOH buffer is rapidly depleted, which will slow electrophoresis and lead to overheating. Buffer should be replaced at least once.
8. For a wick, cut a large sheet of Whatman paper in half.
9. The wick will drape into the reservoir filled with NaOH, ensuring the entire blot setup will remain alkaline for transfer.
10. Check the pH of the final acid/base mixture and neutralize for disposal.
11. Float the membrane on top of the water and then tilt the tray up and down to completely submerge.
12. Southern hybridization with ^{32}P–radiolabeled DNA probes is performed using routine methods (Green & Sambrook, 2012; Ruven, Seelen, Lohman, Mullenders, & van Zeeland, 1994). Probes should hybridize to the ends of the DNA fragments being analyzed to allow unambiguous estimation of DNA strand lengths. To detect the molecular weight standards, we add 0.05 ng of the DNA standard to probe-labeling reactions. If strand-specific probes are required, we amplify and radiolabel probes using linear PCR, as described (Ruven et al., 1994).
13. To maximize specificity, DNA probes should be highly purified. Cloned probes should be gel purified from the vector backbone. PCR amplified probes should be prepared by nested PCR and gel purified.
14. Lid locks may be used to boil the labeled probe to prevent aerosol radioactive contamination from tubes popping open.

ACKNOWLEDGMENTS

Work in the Hunter lab is supported by NIH NIGMS award GM074223. S.T. and S.O. were supported, respectively, by the NIH NIEHS T32 Advanced Training in Environmental Health Sciences Training Grant 5T32ES007059 and the NIH NIGMS T32 Training Program in Molecular and Cellular Biology 5T32GM007377. S.O. is currently supported by an F31 Ruth L. Kirschstein National Research Service Award 1F31GM125106. N.H. is an Investigator of the Howard Hughes Medical Institute.

REFERENCES

Allers, T., & Lichten, M. (2001a). Differential timing and control of noncrossover and crossover recombination during meiosis. *Cell*, *106*(1), 47–57.

Allers, T., & Lichten, M. (2001b). Intermediates of yeast meiotic recombination contain heteroduplex DNA. *Molecular Cell*, *8*(1), 225–231.

Bell, L., & Byers, B. (1983). Separation of branched from linear DNA by two-dimensional gel electrophoresis. *Analytical Biochemistry*, *130*, 527–535.

Brown, M. S., & Bishop, D. K. (2015). DNA strand exchange and RecA homologs in meiosis. *Cold Spring Harbor Perspectives in Biology*, 7(1), a016659.

Cao, L., Alani, E., & Kleckner, N. (1990). A pathway for generation and processing of double-strand breaks during meiotic recombination in S. cerevisiae. *Cell*, *61*(6), 1089–1101.

Chia, M., & van Werven, F. J. (2016). Temporal expression of a master regulator drives synchronous sporulation in budding yeast. *G3 (Bethesda)*, *6*(11), 3553–3560.

Clyne, R. K., Katis, V. L., Jessop, L., Benjamin, K. R., Herskowitz, I., Lichten, M., et al. (2003). Polo-like kinase Cdc5 promotes chiasmata formation and cosegregation of sister centromeres at meiosis I. *Nature Cell Biology*, *5*(5), 480–485.

Fraune, J., Schramm, S., Alsheimer, M., & Benavente, R. (2012). The mammalian synaptonemal complex: Protein components, assembly and role in meiotic recombination. *Experimental Cell Research*, *318*(12), 1340–1346.

Frobel, S., Reiffers, A., Torres Ziegenbein, C., & Gilch, P. (2015). DNA intercalated psoralen undergoes efficient photoinduced electron transfer. *Journal of Physical Chemistry Letters*, *6*(7), 1260–1264.

Green, M. R., & Sambrook, J. (2012). *Molecular cloning: A laboratory manual*. Cold Spring Harbor Laboratory Press.

Grushcow, J. M., Holzen, T. M., Park, K. J., Weinert, T., Lichten, M., & Bishop, D. K. (1999). Saccharomyces cerevisiae checkpoint genes MEC1, RAD17 and RAD24 are required for normal meiotic recombination partner choice. *Genetics*, *153*(2), 607–620.

Hunter, N. (2015). Meiotic recombination: The essence of heredity. *Cold Spring Harbor Perspectives in Biology*, 7(12), a016618.

Hunter, N., & Kleckner, N. (2001). The single-end invasion: An asymmetric intermediate at the double-strand break to double-holliday junction transition of meiotic recombination. *Cell*, *106*(1), 59–70.

Jones, G. H. (1984). The control of chiasma distribution. *Symposia of the Society for Experimental Biology*, *38*, 293–320.

Kane, S. M., & Roth, R. (1974). Carbohydrate metabolism during ascospore development in yeast. *Journal of Bacteriology*, *118*(1), 8–14.

Kim, K. P., Weiner, B. M., Zhang, L., Jordan, A., Dekker, J., & Kleckner, N. (2010). Sister cohesion and structural axis components mediate homolog bias of meiotic recombination. *Cell*, *143*(6), 924–937.

Lam, I., & Keeney, S. (2014). Mechanism and regulation of meiotic recombination initiation. *Cold Spring Harbor Perspectives in Biology*, 7(1), a016634.

Lao, J. P., Cloud, V., Huang, C. C., Grubb, J., Thacker, D., Lee, C. Y., et al. (2013). Meiotic crossover control by concerted action of Rad51-Dmc1 in homolog template bias and robust homeostatic regulation. *PLoS Genetics*, *9*(12), e1003978.

Lao, J. P., Tang, S., & Hunter, N. (2013). Native/denaturing two-dimensional DNA electrophoresis and its application to the analysis of recombination intermediates. *Methods in Molecular Biology*, *1054*, 105–120.

McMahill, M. S., Sham, C. W., & Bishop, D. K. (2007). Synthesis-dependent strand annealing in meiosis. *PLoS Biology*, *5*(11), e299.

Mimitou, E. P., Yamada, S., & Keeney, S. (2017). A global view of meiotic double-strand break end resection. *Science*, *355*(6320), 40–45.

Nassif, N., Penney, J., Pal, S., Engels, W. R., & Gloor, G. B. (1994). Efficient copying of nonhomologous sequences from ectopic sites via P-element-induced gap repair. *Molecular and Cellular Biology*, *14*(3), 1613–1625.

Oh, S. D., Jessop, L., Lao, J. P., Allers, T., Lichten, M., & Hunter, N. (2009). Stabilization and electrophoretic analysis of meiotic recombination intermediates in Saccharomyces cerevisiae. *Methods in Molecular Biology*, *557*, 209–234.

Oh, S. D., Lao, J. P., Hwang, P. Y., Taylor, A. F., Smith, G. R., & Hunter, N. (2007). BLM ortholog, Sgs1, prevents aberrant crossing-over by suppressing formation of multichromatid joint molecules. *Cell*, *130*(2), 259–272.

Oh, S. D., Lao, J. P., Taylor, A. F., Smith, G. R., & Hunter, N. (2008). RecQ helicase, Sgs1, and XPF family endonuclease, Mus81-Mms4, resolve aberrant joint molecules during meiotic recombination. *Molecular Cell*, *31*(3), 324–336.

Panyutin, I. G., & Hsieh, P. (1994). The kinetics of spontaneous DNA branch migration. *Proceedings of the National Academy of Sciences of the United States of America*, *91*(6), 2021–2025.

Paques, F., & Haber, J. E. (1999). Multiple pathways of recombination induced by double-strand breaks in Saccharomyces cerevisiae. *Microbiology and Molecular Biology Reviews*, *63*(2), 349–404.

Petronczki, M., Siomos, M. F., & Nasmyth, K. (2003). Un menage a quatre: The molecular biology of chromosome segregation in meiosis. *Cell*, *112*(4), 423–440.

Ruven, H. J., Seelen, C. M., Lohman, P. H., Mullenders, L. H., & van Zeeland, A. A. (1994). Efficient synthesis of 32P-labeled single-stranded DNA probes using linear PCR; application of the method for analysis of strand-specific DNA repair. *Mutation Research*, *315*(2), 189–195.

Schwacha, A., & Kleckner, N. (1994). Identification of joint molecules that form frequently between homologs but rarely between sister chromatids during yeast meiosis. *Cell*, *76*(1), 51–63.

Schwacha, A., & Kleckner, N. (1995). Identification of double Holliday junctions as intermediates in meiotic recombination. *Cell*, *83*(5), 783–791.

Sun, H., Treco, D., Schultes, N. P., & Szostak, J. W. (1989). Double-strand breaks at an initiation site for meiotic gene conversion. *Nature*, *338*(6210), 87–90.

Szostak, J. W., Orr-Weaver, T. L., Rothstein, R. J., & Stahl, F. W. (1983). The double-strand-break repair model for recombination. *Cell*, *33*(1), 25–35.

Thacker, D., Mohibullah, N., Zhu, X., & Keeney, S. (2014). Homologue engagement controls meiotic DNA break number and distribution. *Nature*, *510*(7504), 241–246.

von Wettstein, D., Rasmussen, S. W., & Holm, P. B. (1984). The synaptonemal complex in genetic segregation. *Annual Review of Genetics*, *18*, 331–413.

Watanabe, Y. (2012). Geometry and force behind kinetochore orientation: Lessons from meiosis. *Nature Reviews. Molecular Cell Biology*, *13*(6), 370–382.

Zakharyevich, K., Ma, Y., Tang, S., Hwang, P. Y., Boiteux, S., & Hunter, N. (2010). Temporally and biochemically distinct activities of Exo1 during meiosis: Double-strand break resection and resolution of double Holliday junctions. *Molecular Cell*, *40*(6), 1001–1015.

Zakharyevich, K., Tang, S., Ma, Y., & Hunter, N. (2012). Delineation of joint molecule resolution pathways in meiosis identifies a crossover-specific resolvase. *Cell*, *149*(2), 334–347.

Zickler, D., & Kleckner, N. (1999). Meiotic chromosomes: Integrating structure and function. *Annual Review of Genetics*, *33*, 603–754.

Zickler, D., & Kleckner, N. (2015). Recombination, pairing, and synapsis of homologs during meiosis. *Cold Spring Harbor Perspectives in Biology*, 7(6), a016626.

CHAPTER THIRTEEN

S1-seq Assay for Mapping Processed DNA Ends

Eleni P. Mimitou*[,2], **Scott Keeney***[,†,1]
*Molecular Biology Program, Memorial Sloan Kettering Cancer Center, New York, NY, United States
†Howard Hughes Medical Institute, Memorial Sloan Kettering Cancer Center, New York, NY, United States
[1]Corresponding author: e-mail address: s-keeney@ski.mskcc.org

Contents

Abstract

During meiosis, the specialized cell division giving rise to gametes, numerous DNA double-strand breaks (DSBs) are introduced at multiple places throughout the genome by the topoisomerase-like protein Spo11. Homologous recombination, a highly conserved DSB repair pathway, is employed for their repair and ensures the formation of chiasmata and the proper segregation of homologous chromosomes. In the initial steps

[2] *Current address*: New York Genome Center, New York, NY, United States.

Methods in Enzymology, Volume 601
ISSN 0076-6879
https://doi.org/10.1016/bs.mie.2017.11.031

of recombination, end resection takes place, wherein Spo11 is endonucleolytically released and the 5′-terminal strands of each DSB are exonucleolytically processed, exposing the ssDNA necessary to identify a homologous repair template. DNA removed by DSB processing is reconstituted by DNA synthesis, which copies genetic information from the intact homologous template. We developed a next-generation sequencing assay, termed S1-seq, to study DNA end resection genome-wide at high-spatial resolution during yeast meiotic recombination. The assay relies on the fact that removal of the ssDNA tails of resected DSBs marks the position where resection stopped. Molecular features of resection are revealed by sequencing of these ssDNA-to-dsDNA junctions and comparison to high-resolution Spo11 DSB maps. We describe the experimental and computational methods for S1-seq as applied to meiosis in the SK1 strain of budding yeast *Saccharomyces cerevisiae* and discuss how it can also be applied to map DSBs and recombination intermediates.

1. INTRODUCTION

Meiotic homologous recombination plays a pivotal role in sexual reproduction and genetic diversity by promoting proper segregation of homologous chromosomes and by disrupting linkage groups (Hunter, 2015; Thacker & Keeney, 2016). The event that initiates exchange of genetic information between parental chromosomes is the formation of DNA double-strand breaks (DSBs) by the transesterase Spo11 (Lam & Keeney, 2014) (Fig. 1). These DSBs tend to cluster in narrow regions called hotspots (de Massy, 2013; Petes, 2001). Following break formation, the DNA ends undergo resection, the $5' \rightarrow 3'$ degradation that gives rise to $3'$ single-stranded DNA (ssDNA) overhangs (Mimitou & Symington, 2009). This step is important to reveal the sequence proximal to the break, which when coated with the strand exchange proteins Rad51 and Dmc1 can invade the repair template in the homologous chromosome and initiate repair DNA synthesis, thus restoring the information lost during resection (Neale & Keeney, 2006). Shortly after the DSB is formed, the highly conserved Mre11–Rad50–Xrs2 (MRX) complex provides the Mre11 nuclease, which cooperates with Sae2 to nick the Spo11-bound DNA strands near the break (Cannavo & Cejka, 2014; Keeney, Giroux, & Kleckner, 1997; Moreau, Ferguson, & Symington, 1999; Neale, Pan, & Keeney, 2005). Mre11 then degrades the nicked strand toward the break via its $3' \rightarrow 5'$ exonuclease activity, releasing Spo11 bound to short oligos (Garcia, Phelps, Gray, & Neale, 2011; Zakharyevich et al., 2010). The Mre11 nicks also constitute entry points for the $5' \rightarrow 3'$ exonuclease, Exo1, which extends resection to an average of 822nt (Mimitou, Yamada, & Keeney, 2017; Zakharyevich et al., 2010).

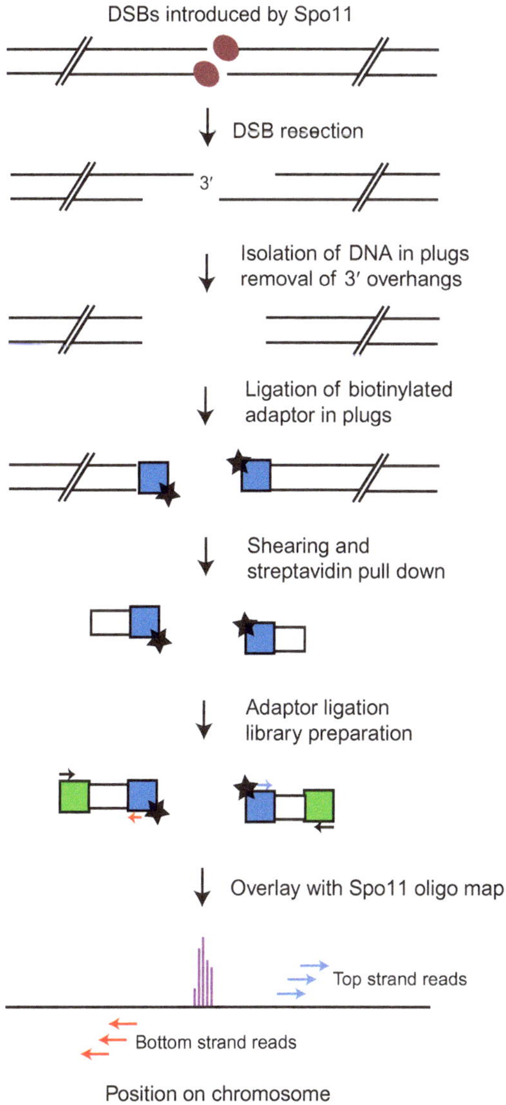

Fig. 1 Schematic of the key steps employed in S1-seq: DNA from meiotic cells bearing processed Spo11 DSBs is embedded in agarose plugs to protect from shearing. Treatment with S1 nuclease and T4 DNA polymerase removes the 3′ overhang and prepares the ends for ligation with the 5′ biotinylated adaptor. Following extraction from the plugs and sonication, the biotinylated fragments are bound on streptavidin beads, and the second adaptor is ligated. Low-cycle amplification by PCR creates the library that is submitted for next-generation sequencing and eventually mapped to the reference genome.

During recombination in vegetative cells, a parallel pathway resects DSBs redundantly with Exo1 and depends on the concerted action of the Sgs1–Top3–Rmi1 complex and the Dna2 endonuclease (Mimitou & Symington, 2008; Zhu, Chung, Shim, Lee, & Ira, 2008).

2. OVERVIEW OF S1-seq

The sensitivity of the 3′ ssDNA of resected DSBs to ssDNA-specific endonucleases provides a tool to map resection endpoints by sequencing the region where ssDNA transitions to dsDNA (Fig. 1) (Mimitou et al., 2017). We isolate genomic DNA of meiotic cells in low-melting point (LMP) agarose plugs to protect the DNA from random shearing. The plugs are treated with S1 nuclease and T4 DNA polymerase, then the blunt-ended chromosome fragments are ligated to a 5′-biotinylated dsDNA adaptor. Following extraction of DNA from the agarose plugs and sonication, we affinity purify the biotinylated fragments and ligate different adaptors to the opposite ends. The DNA is then amplified by low-cycle PCR and deep sequenced, then the reads are mapped to a reference genome. Overlay of the sequencing reads with mapped Spo11 DSB sites (Pan et al., 2011) reveals features of resection genome wide (Fig. 1).

At least two biological replicates for each background and time point should be sequenced. In our hands, the assay displays high reproducibility (typical Pearson's $r=0.98$) (Fig. 2A and B). Several features of the data

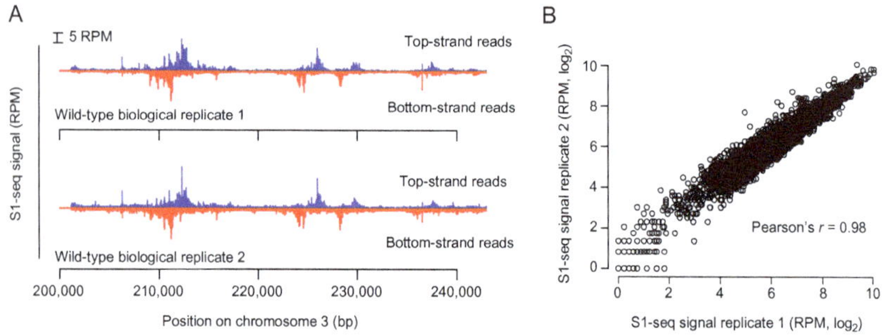

Fig. 2 Reproducibility of S1-seq: (A) Spatial reproducibility of S1-seq maps. Uniquely mapped S1-seq reads were normalized to reads per million (RPM). A region of chromosome III is shown as an example of the reproducibility between two wild-type datasets. (B) Quantitative reproducibility of S1-seq maps. Normalized, uniquely mapped reads were summed in 1-kb nonoverlapping windows for two wild-type biological replicates. *Reproduced from Mimitou, E. P., Yamada, S., & Keeney, S. (2017). A global view of meiotic double-strand break end resection.* Science, *355(6320), 40–45. https://doi.org/10.1126/science.aak9704.*

confirm S1-seq as a readout of bona fide resection endpoints. First, each DSB hotspot gives rise to surrounding reads that have the expected polarity: resection to the right results in endpoints that map to the top strand, whereas resection to the left gives rise to endpoints that map to the bottom strand (Fig. 3A and B). Second, premeiotic samples or samples from the catalytically inactive *spo11-YF* mutant lack enrichment of signal around hotspots, confirming that the S1-seq signal is DSB dependent (Fig. 3B). Third, S1-seq read counts correlate well with Spo11-oligo counts (Fig. 3C). Finally, the S1-seq signal displays the proper genetic control: it clusters more closely to hotspot midpoints in an *exo1* nuclease-defective mutant and spreads further away in the *dmc1Δ* mutant, in which DSBs experience hyperresection (Bishop, Park, Xu, & Kleckner, 1992) (Fig. 3D). S1-seq therefore is a selective, sensitive, and quantitative measure of DSB resection tract distributions. It should also be feasible to apply it to create high-resolution resection tract maps in other settings or systems.

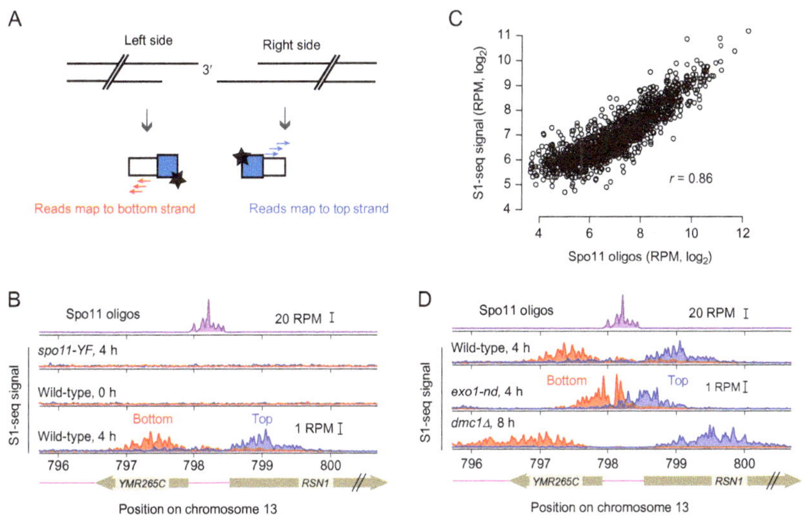

Fig. 3 Validation of S1-seq: (A) Schematic of the S1-seq signal polarity. Resection to the *right* or *left* should yield reads mapping to the top or bottom strands, respectively. (B) An example of Spo11-oligo and strand-specific S1-seq reads (51-bp smoothing) surrounding the *RSN1* hotspot. (C) S1-seq correlates well with Spo11 oligos: S1-seq signal (wild type, 4h) was summed from +200 to +1600 (top strand) and −200 to −1600 (bottom strand) relative to hotspot midpoints and plotted against the sum of the Spo11-oligo hitcount for each corresponding hotspot. For this analysis, hotspots with no other hotspot within 2kb were used. (D) Spo11-oligo and strand-specific S1-seq reads as in (B). *Panels (B–D) reproduced from Mimitou, E. P., Yamada, S., & Keeney, S. (2017). A global view of meiotic double-strand break end resection.* Science, 355*(6320), 40–45. https://doi.org/10.1126/science.aak9704.*

3. ADAPTATIONS OF S1-seq

3.1 Mapping Unresected DSBs

We have shown that the sequential treatment with S1 nuclease and T4 DNA polymerase can degrade the 2-nt 5′ overhang of unresected DSBs to which Spo11 is still covalently bound (Mimitou et al., 2017). We therefore reasoned that S1-seq can provide a facile method for nucleotide-resolution mapping of the unresected Spo11 DSBs that accumulate in *sae2Δ* or *rad50S* mutants (Alani, Padmore, & Kleckner, 1990; McKee & Kleckner, 1997; Prinz, Amon, & Klein, 1997). As a proof of principle, we performed S1-seq in *sae2Δ* mutants and observed DSB patterns similar to those revealed by traditional Spo11-oligo mapping (Fig. 4A). For DSB mapping, S1-seq can be applied without any modification to the protocol below, as long as it is performed using cells of *sae2Δ* or *rad50S* background.

3.2 Mapping Recombination Intermediate Branchpoints

Along with resection signal, there is a weaker S1-seq signal with the "wrong" polarity, e.g., top-strand reads mapping to the left of hotspots and bottom-strand reads mapping to the right of hotspots (Fig. 4B). As shown previously (Mimitou et al., 2017) this signal is Dmc1 dependent and derives from S1-sensitive recombination intermediates (RIs), probably displacement (D) loops from strand exchange. It is therefore possible to implement S1-seq toward the study of the spatial, temporal, and quantitative features of RI branchpoints by filtering S1-seq maps for reads of polarity opposite to resection-derived reads. A schematic summarizing the species detected by S1-seq is presented in Fig. 4C.

4. GENOMIC DNA ISOLATION

4.1 Yeast Culture

Cells of desired genotype are induced into the meiotic program by nitrogen starvation, and samples are withdrawn to follow DSB formation and processing.

4.1.1 Material, Solutions, and Reagents

1. YPD plates: 1% Bacto yeast extract, 2% Bacto peptone, 2% dextrose, 2% Bacto agar. Autoclave.
2. YPD: 1% Bacto yeast extract, 2% Bacto peptone, 2% dextrose. Autoclave.

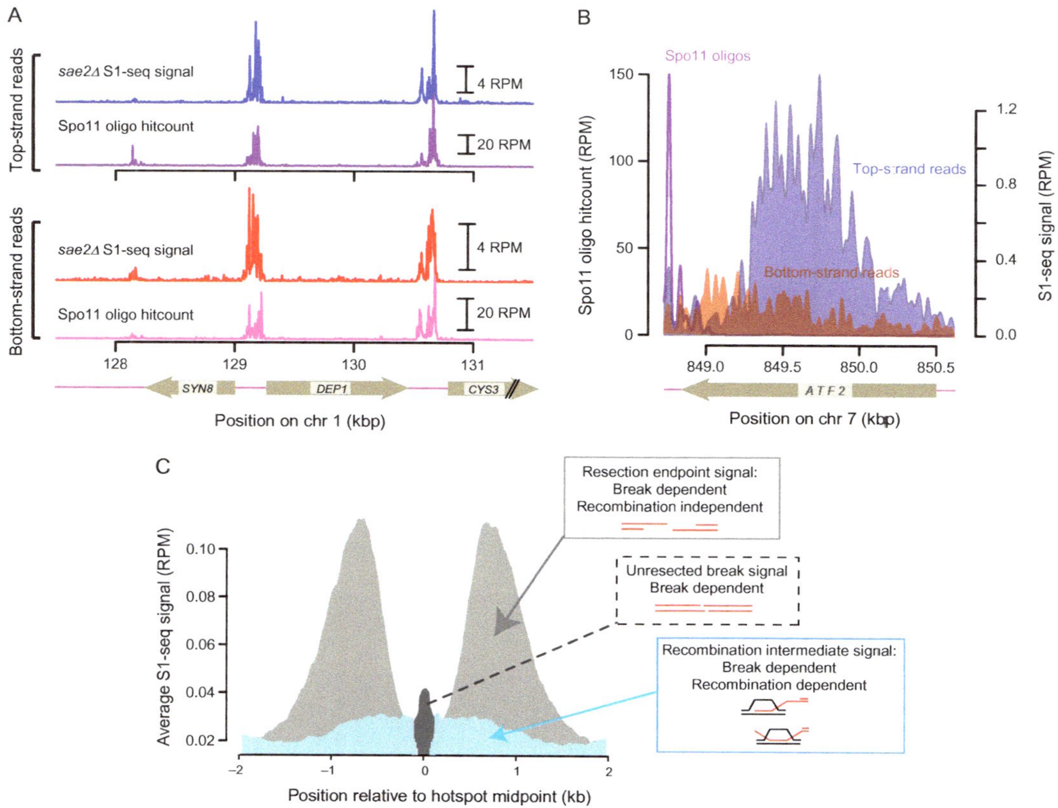

Fig. 4 Adaptations of S1-seq: (A) S1-seq in a *sae2Δ* background can reveal unresected Spo11 DSBs that correlate well with DSBs mapped by traditional Spo11-oligo sequencing. (B) A snapshot of top- and bottom-strand S1-seq reads on the *right side* of a Spo11 hotspot. The most abundant top-strand reads represent resection endpoints, whereas the bottom-strand signal originates from RIs. (C) A graphical representation of the different species detected by S1-seq around hotspots: resection endpoints (*light gray*), RIs (*blue*), and unresected DSBs (*dark gray*). *Adapted from Mimitou, E. P., Yamada, S., & Keeney, S. (2017). A global view of meiotic double-strand break end resection.* Science, 355*(6320), 40–45. https://doi.org/10.1126/science.aak9704.*

3. YPA: 1% Bacto yeast extract, 2% Bacto peptone, 2% potassium acetate, 0.001% antifoam. Autoclave.
4. SPM: 2% potassium acetate, 0.001% antifoam. Autoclave.
5. 2.8-L wide-mouth baffled Fernbach flasks.
6. 50 m*M* EDTA pH 8.0.
7. 15-mL conical tubes.

4.1.2 Procedure

1. Streak diploid cells from a frozen stock onto YPD plates and allow to grow at 30°C until single colonies are visible (approximately 36–48 h).
2. Inoculate a single colony into 20 mL liquid YPD medium and grow for >24 h at 30°C.
3. Using a spectrophotometer, measure cell density (OD_{600}) of a 1:10 dilution of the overnight culture. Subculture into the appropriate volume of YPA medium to OD_{600} of 0.2 and grow in 2.8-L baffled flasks at 250 rpm at 30°C for 14 h.
4. Harvest cells by centrifuging at 2000 × *g* for 3 min at 4°C and wash once with sporulation medium.
5. Resuspend cells in SPM in a 2.8-L baffled flask at an OD_{600} of 6.0 and incubate at 30°C, 250 rpm. Harvest cells at indicated time points: 75 mL for 0 and 2 h, 66 mL for 4 h, 60 mL for 6 h, and 54 mL for 8 h. See Note 1: Remove small volume samples to monitor progression of cell division by DAPI staining.
6. Wash cells with one half volume of 50 m*M* EDTA pH 8.0 and harvest by centrifugation.
7. Wash cells with 10 mL 50 m*M* EDTA and move into 15-mL falcon tubes.
8. Harvest cells by centrifugation, remove the liquid, freeze on dry ice, and store at −80°C until plug preparation.

4.2 Isolation of DNA in Plugs

Isolation of genomic DNA from the yeast cell samples is performed, while the cells are embedded in LMP agarose to yield intact high-molecular weight DNA.

4.2.1 Material, Solutions, and Reagents

1. 500 m*M* EDTA pH 8
2. 10 m*M* Tris–HCl pH 7.5
3. 2-Mercaptoethanol

4. SCE solution: 1 *M* sorbitol, 0.1 *M* sodium citrate, 0.06 *M* EDTA, pH 7
5. Zymolyase 100T (US Biological)
6. LMP agarose (Lonza SeaPlaque)
7. Agarose plug molds (Biorad)
8. RNAse A 10 mg/mL (Thermo Scientific)
9. 20% SDS
10. Proteinase K, recombinant, PCR grade (Roche)
11. Glycerol
12. Solution 1: 5% 2-Mercaptoethanol, 1 mg/mL Zymolyase 100T in SCE solution
13. Solution 2: 0.45 *M* EDTA pH 8, 0.01 *M* Tris–HCl pH 7.5, 7.5% 2-Mercaptoethanol, 10 μg/mL RNase A
14. Solution 3: 0.25 *M* EDTA pH 8, 0.01 *M* Tris–HCl pH 7.5, 1% SDS, 1 mg/mL Proteinase K
15. Plug storage solution: 0.05 *M* EDTA pH 8, 50% glycerol
16. Low-retention filter tips
17. 15-mL conical tubes

4.2.2 Procedure

1. Prepare fresh 2% LMP in 125 m*M* EDTA pH 8 and keep at 65°C.
2. Prepare fresh Solution 1 and keep at 40°C.
3. Resuspend each pellet in 150 μL 50 m*M* EDTA pH 8 and allow to thaw at 40°C.
4. Add 102 μL Solution 1 to the cells and mix well, so no clumps are present. See Note 2.
5. Add 498 μL 2% LMP agarose, mix well, and quickly pipette into plug molds (~90 μL each).
6. Allow to solidify at 4°C for ≥30 min.
7. Express agarose plugs (three at a time) into 3 mL of Solution 2 in 15-mL conical tubes and incubate at 37°C for 1 h.
8. Carefully aspirate Solution 2 and replace with 3 mL of Solution 3.
9. Incubate at 50°C overnight.
10. Wash plugs with 3 mL 50 m*M* EDTA pH 8, three times for 20 min each, with gentle shaking.
11. Aspirate last wash and add 3 mL plug storage solution.
12. Store at −20°C until all plugs are ready to be processed for library preparation.

5. LIBRARY PREPARATION

5.1 Plug Processing for Overhang Removal and Adaptor Ligation

All DNA treatments in this step are performed in the plugs to avoid pipetting and to minimize nonspecific shearing of genomic DNA. For efficient in-plug overhang removal and adaptor ligation, a sequential treatment with S1 nuclease and T4 DNA polymerase is required before ligation. Even though S1 nuclease treatment is sufficient to remove the ssDNA tails, the subsequent ligation step is not efficient unless a clean-up reaction with T4 DNA polymerase is performed (E.P. Mimitou, unpublished observations).

Optional step: Some processed plugs can be set aside to serve as a control reaction, wherein a nonbiotinylated P5 adaptor is ligated to the blunt ends after S1/T4 treatment. This reaction is then processed throughout the protocol in parallel with the biotinylated samples and serves to monitor background from nonspecific binding of DNA to the beads.

5.1.1 Material, Solutions, and Reagents

1. 1 × TE: 10 m*M* Tris, 1 m*M* EDTA
2. 500 m*M* EDTA pH 8
3. 10 × S1 buffer: 500 m*M* sodium acetate (pH 4.5), 2.8 *M* NaCl, 45 m*M* $ZnSO_4$
4. S1 Nuclease (Promega)
5. 10 mg/mL (100 ×) BSA (NEB, currently discontinued, can be replaced with 20 mg/mL BSA)
6. 100 m*M* dNTPs set (Sigma-Aldrich)
7. T4 DNA polymerase reaction buffer: 1 × T4 DNA ligase buffer, 1 × BSA, 100 μ*M* dNTPs (each)
8. T4 DNA polymerase (NEB)
9. 10 × T4 ligase reaction buffer (NEB)
10. T4 DNA ligase, 2000 U/μL (NEB)
11. P5-top oligo 100 μ*M*, dissolved in 10 m*M* Tris–HCl pH 7.5, 50 m*M* NaCl, 1 m*M* EDTA pH 8 (see Table 1)
12. P5-bottom oligo 100 μ*M*, dissolved in 10 m*M* Tris–HCl pH 7.5, 50 m*M* NaCl, 1 m*M* EDTA pH 8 (see Table 1)
13. Low-retention filter tips
14. 2-mL conical tubes

Table 1 Oligonucleotide Sequences Used to Prepare the Adaptors and Primers Used to Amplify the Fragment Library

Oligos/Primers	Sequence
P5-bio-top	5BiosG/ACACTCTTTCCCTACACGACGCTCTTCCGATCT
P5-top	ACACTCTTTCCCTACACGACGCTCTTCCGATCT
P5-bottom	5Phos/AGATCGGAAGAGCGTCGTGTAGGGAAAGAGTGT/3InvdT
P7-top	5Phos/GATCGGAAGAGCACACGTCTGAACTCCAGTCAC/3InvdT
P7-bottom	GTGACTGGAGTTCAGACGTGTGCTCTTCCGATC
P5 primer	AATGATACGGCGACCACCGAGATCTACACTCTTTCCCTACACGACGCTCTTCCGATCT
P7 primer	CAAGCAGAAGACGGCATACGAGAT*NNNNNN*GTGACTGGAGTTCAGACGTGTG

NNNNNN denotes TruSeq barcode sequence.

5.1.2 Procedure

1. For each sample or time point to be analyzed, prepare 10 2-mL conical tubes with 500 μL 1 × TE.
2. Place one plug in each tube to rinse away the plug storage buffer.
3. Aspirate 1 × TE and replace with 500 μL 1 × S1 buffer. Equilibrate for 30 min. See Note 3.
4. Repeat step 3 three times, so total equilibration time in 1 × S1 buffer is 2 h.
5. Aspirate and replace with 500 μL 1 × S1 buffer containing 9 U of S1 nuclease.
6. Place on ice for 15 min to allow the enzyme diffuse into the plugs.
7. Incubate at 37°C for 20 min.
8. Inactivate S1 nuclease by addition of EDTA pH 8.0 to a final concentration of 10 m*M* and incubate on ice for 15 min.
9. Rinse plugs with 500 μL 1 × TE.
10. Equilibrate in 500 μL T4 polymerase reaction buffer, four times for 30 min each.
11. At this point prepare the P5 adaptor (biotinylated and optionally nonbiotinylated), by mixing oligos P5-top (biotinylated and nonbiotinylated, 100 μ*M*) and P5-bottom (100 μ*M*) at equimolar concentration, boiling for 5 min, and cooling at room temperature for at least 1 h.
12. Aspirate and add 500 μL T4 polymerase reaction buffer containing 30 U of T4 DNA polymerase.
13. Incubate on ice for 15 min, followed by 30 min incubation at 12°C.
14. Inactivate T4 polymerase by addition of EDTA pH 8.0 to a final concentration of 10 m*M* and incubate on ice for 15 min.
15. Aspirate buffer and rinse plugs with 1 × TE.
16. Perform a quick spin and remove all buffer.
17. Incubate at 75°C for 20 min to fully inactivate T4 polymerase.
18. Allow plugs to gradually cool and solidify at room temperature.
19. Equilibrate in 250 μL 1 × T4 ligase buffer, four times for 15 min each.
20. Aspirate 200 μL of the buffer leaving each plug immersed in ~50 μL 1 × T4 DNA ligase buffer.
21. Add 1 μL of 50 μ*M* adaptor and 1 μL of 2000 U/μL T4 DNA ligase to each plug.
22. Incubate at 16°C for ≥18 h.

5.2 Gel Extraction and Size Selection

This step aims at releasing total DNA from the agarose plugs and further separating large genomic DNA fragments from residual unligated P5 adaptor. We found that including the size chromatography step reduced the background at later steps in this protocol (PCR amplification) so we deem it important (see Note 4).

5.2.1 Material, Solutions, and Reagents

1. GELase Agarose Gel-Digesting Preparation (Epicentre, currently discontinued, can be replaced with NEB β-agarase)
2. Phenol:Chloroform pH 8 (Amresco)
3. 3 *M* sodium acetate pH 5.5
4. 100% Ethanol
5. 70% Ethanol
6. 6 × Loading dye (NEB)
7. 10 m*M* Tris–HCl pH 7.5
8. 50 × Tris–Acetate–EDTA (Fisher)
9. LMP agarose (Lonza SeaPlaque)
10. Scalpels or razor blades
11. Equipment for UV visualization and photography of DNA gel
12. 5 mg/mL ethidium bromide
13. Chroma Spin +TE-1000 columns (Clontech)
14. Low-retention filter tips
15. Low-retention tubes: Siliconized G-tube microcentrifuge tubes, 1.5 mL (VWR)

5.2.2 Procedure

1. Add 3 μL of 50 × GELase buffer to each plug and incubate at 70°C until plug has thoroughly melted, approximately 5–10 min.
2. Pool plugs of the same sample together and equilibrate at 45°C for ≥5 min.
3. Add 1 unit GELase per plug (i.e., 10 U total) and incubate at 45°C for 30–45 min.
4. Add 1:1 ratio of phenol, mix by inversion, and spin at max speed for 10 min.
5. Carefully transfer the aqueous phase to new tubes.
6. Add 0.1 volume of sodium acetate pH 5.5 and 2.5 volumes 100% ethanol, mix by inversion 5–6 times, and place at −20°C for ≥1 h.
7. Centrifuge at full speed for 15 min and wash pellet once with 70% ethanol.
8. Dry pellet at room temperature for 15 min.
9. Solubilize in 120 μL 10 m*M* Tris–HCl pH 7.5.

10. Add 24 μL 6 × loading dye and mix well by gently flicking the tube (see Note 5).
11. Meanwhile prepare a 1% LMP 1 × TAE agarose gel, containing 0.5 μg/mL ethidium bromide.
12. Carefully load the genomic DNA into the wells (see Note 5) and subject to electrophoresis at 80 V/cm for 2 h.
13. Visualize the gel; a typical run is shown in Fig. 5A. Excise the high-molecular weight band migrating beneath the wells.
14. Weigh the gel slice, add GELase buffer to 1 × final concentration, and melt the gel slice thoroughly at 70°C.
15. After equilibration at 45°C for 5 min add 1 unit GELase for every 200 mg gel slice and incubate at 45°C for 30 min.
16. Add phenol in 1:1 ratio, mix by inversion, and spin at max speed for 10 min. Remove aqueous phase and place in fresh tubes.
17. Add 0.1 volumes of sodium acetate pH 5.5 and 2.5 volumes 100% ethanol, mix by inversion 5–6 times, and place at −20°C for ≥1 h.
18. Centrifuge at full speed for 15 min and wash pellet once with 70% ethanol.
19. Dry pellet at room temperature for 15 min.
20. Solubilize pellet in 100 μL 1 × TE.
21. Pass each sample through Chroma Spin-1000 columns.
22. Repeat step 21 two more times.
23. Subject the final eluates to shearing according to the Covaris instrument protocol to DNA fragment sizes ranging between 200 and 500 bp. To verify the size range of the library, a dilution of the library is analyzed on a Bioanalyzer Instrument (Agilent Technologies) according to manufacturer's instruction. An example is shown in Fig. 5B.

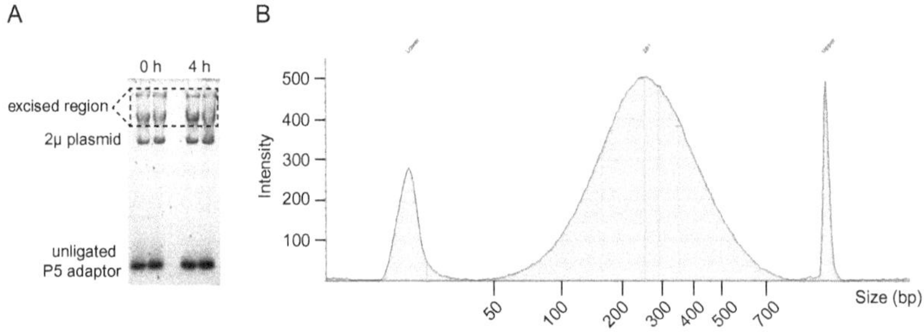

Fig. 5 Gel electrophoresis to separate high-molecular weight DNA from the unligated adaptor: (A) Example of the agarose gel after electrophoresis. The *boxed region* is excised and treated with agarase to retrieve the DNA. (B) Bioanalyzer (Agilent) lane profile showing the distribution of the genomic DNA after Covaris shearing (~100–700 bp).

5.3 Streptavidin Purification, End Repair, and Second-End Ligation

This step enriches for fragments containing the biotinylated P5 adaptor followed by ligation of the P7 adaptor. Upon capture, all subsequent steps up to PCR amplification can be performed on beads.

5.3.1 Material, Solutions, and Reagents

1. Dynabeads M-280 Streptavidin, (ThermoFisher)
2. Magnet for capturing Dynabeads
3. 1× TE: 10 m*M* Tris, 1 m*M* EDTA
4. 2× B&W buffer: 10 m*M* Tris–HCl pH 7.5, 1 m*M* EDTA, 2 *M* NaCl
5. End-It DNA End-Repair Kit (Epicentre)
6. 1 *M* Tris–HCl pH 7.5
7. 10× T4 ligase reaction buffer (NEB)
8. T4 DNA ligase, 2000 U/μL (NEB)
9. P7-top oligo 100 μM, dissolved in 10 m*M* Tris–HCl pH 7.5, 50 m*M* NaCl, 1 m*M* EDTA pH 8 (see Table 1)
10. P7-bottom oligo 100 μ*M*, dissolved in 10 m*M* Tris–HCl pH 7.5, 50 m*M* NaCl, 1 m*M* EDTA pH 8 (see Table 1)
11. Low-retention filter tips
12. Low-retention tubes: Siliconized G-tube microcentrifuge tubes, 1.5 mL (VWR)

5.3.2 Procedure

1. Vortex the beads to fully resuspend and pipette the appropriate volume in a fresh eppendorf tube (you will need 50 μL beads per sample).
2. Capture beads with magnet and wash twice with 1× TE and twice with 1× B&W (wash with at least double the volume of beads).
3. Capture beads and resuspend in 2× B&W, using double the volume withdrawn at step 1.
4. Aliquot 100 μL beads into fresh tubes and add equal volume (adjust with 1× TE if less than 100 μL) of sheared DNA sample.
5. Allow binding by incubating at 20°C for 30 min.
6. During the incubation prepare P7 adaptor: anneal P7-top and P7-bottom in 1:1 ratio, boil for 5 min, and allow to cool down at room temperature for at least 1 h.
7. Wash twice with 500 μL 1× B&W.
8. Wash twice with 500 μL 10 m*M* Tris–HCl pH 7.5.

9. During the last wash, prepare the end-repair reaction mix, using the End-it DNA Repair kit (Epicentre). Each reaction will be 100 μL, scale accordingly: 68 μL ddH_2O, 10 μL 10× buffer, 10 μL dNTPs, 10 μL ATP, and 2 μL enzyme mix.
10. Resuspend beads in 100 μL reaction mix and incubate at 20°C for 45 min.
11. Capture beads and wash twice with 500 μL 1× B&W.
12. Wash twice with 500 μL 10 m*M* Tris–HCl pH 7.5.
13. Wash once with 100 μL 1× T4 ligase buffer.
14. Meanwhile prepare the ligation reaction mix; per reaction: 87.8 μL ddH_2O, 10 μL 10× ligase buffer, 0.2 μL 2000 U/μL T4 ligase.
15. Resuspend beads in 98 μL of ligation reaction mix and add 2 μL 50 μ*M* P7 adaptor.
16. Incubate at 20°C for 3–4 h.
17. Collect beads, wash three times with 500 μL 1× TE and three times with 500 μL 10 m*M* Tris–HCl.
18. Resuspend beads in 20 μL 10 m*M* Tris–HCl pH 7.5 and store at 4°C.

5.4 PCR Amplification and Size Selection

5.4.1 Material, Solutions, and Reagents

1. PCR tubes (e.g., PCR 8-strips (BrandTech Scientific))
2. Phusion High-Fidelity DNA Polymerase (Thermo Scientific)
3. 100 m*M* dNTPs set (Sigma-Aldrich)
4. Barcoded sequencing primers for Illumina library (TruSeq Indexed Primer P7, TruSeq Universal Primer P5). See Table 1
5. Thermocycler
6. Equipment for UV visualization and photography of DNA gel
7. 7.5 *M* ammonium acetate
8. 5% Mini-Protean TBE gel (Biorad)
9. 1× TBE diluted from 10× TBE (890 m*M* Tris, 890 m*M* boric acid, 20 m*M* EDTA)
10. 0.6-mL tubes (Fisher)
11. BD PrecisionGlide Single-use Needles 19 Gauge (Fisher)
12. Spin-X centrifuge Tube filter, 0.45 μm cellulose acetate (Corning costar)
13. Molecular weight markers (e.g., Low-Molecular Weight Marker (NEB)).
14. Ethidium bromide
15. Platform shaker and tray for gel staining with ethidium bromide
16. Dedicated bottle of ethanol for ethanol precipitation
17. 10 m*M* Tris–HCl pH 8.0

18. Spin-X column: Corning Incorporated Costar Centrifugal Devices, cellulose acetate membrane (Fisher)
19. Scalpels or razor blades
20. Saran wrap
21. Thermomixer
22. Low-retention filter tips
23. Low-retention tubes: Siliconized G-tube microcentrifuge tubes, 1.5 mL (VWR)

5.4.2 Procedure

1. Split each 20 μL bead sample into two PCR tubes, using 10 μL as template:

Component	Per Reaction (μL)	Final Concentration
Beads in 10 m*M* Tris–HCl	10	
5 × Phusion HF buffer	10	1 ×
10 m*M* dNTPs	1	0.2 m*M*
Phusion high-fidelity 2 U/μL	0.5	1 U
10 μ*M* Primer P5	2	0.4 μ*M*
10 μ*M* Primer P7 (indexed)	2	0.4 μ*M*
ddH_2O	24.5	
Total volume	50	

2. Run the PCR with the following conditions: 98°C for 30 s; 16 cycles of 98°C for 10 s, 65°C for 20 s, and 72°C for 20 s and a final extension step at 72°C for 5 min.
3. Handle PCR products on a different bench from the one used to set up the reactions and use designated pipettes and tips, as well as reagent aliquots reserved for PCR products. UV-irradiate both sides of the pipettes using 1200 J/m^2 before use.
4. Combine the two reactions coming from the same sample into one 1.5-mL tube.
5. Capture beads and move the supernatant (100 μL) into fresh 1.5-mL tubes.
6. Precipitate the DNA by adding 50 μL 7.5 *M* ammonium acetate and 250 μL 100% ethanol.

7. Incubate at −20°C overnight.
8. Centrifuge at maximum speed for ≥30 min at 4°C.
9. Wash the pellet with cold 70% ethanol. A pellet should be clearly visible at this step. Centrifuge at maximum speed for 10 min at 4°C. Remove all supernatant, and perform a quick spin to gather residual ethanol and remove with a pipette. Air dry for 10–15 min.
10. Dissolve the pellets by adding 20 μL of 10 m*M* Tris–HCl pH 7.5.
11. Set up the 5% Mini-Protean TBE gel in the Biorad tank and add 1 × TBE as running buffer.
12. Add 4 μL 6 × DNA loading dye to each PCR product and load everything into one lane.
13. Run the gel at 200 V until just before the dye front migrates out of the gel (~30 min).
14. Stain the gel with 0.5 μg/mL ethidium bromide for 15 min and image the gel. Make sure to wipe all surfaces and place the gel on Saran wrap (not directly on any surface) to avoid contamination. A smear spanning from 200 to 800 bp should be visible, with primer dimers migrating at ~100 bp. An example is shown in Fig. 6A.
15. Set up a centrifuge device by punching a hole at the bottom of a 0.6-mL tube using a 19G1/2 needle and placing it into an open 1.5-mL tube.
16. Using a UV transilluminator and wearing appropriate eye protection, cut out the 200–600 bp smeared region and transfer to the centrifuge device.

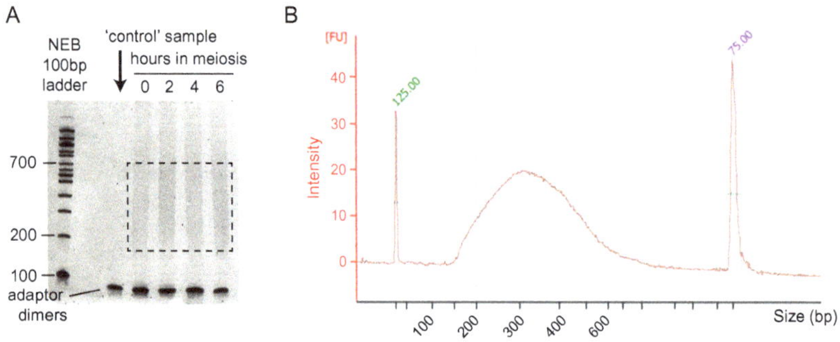

Fig. 6 PCR (sequencing libraries) of a wild-type time course. (A) Samples from different time points (0, 2, 4, and 6 h) were resolved on a 5% nondenaturing polyacrylamide gel. The first lane is a control where the nonbiotinylated P5 adaptor was used. The smear at 200–700 bp (*boxed region*) indicates PCR-amplified adaptor-ligated fragments and is excised from the gel. The DNA is extracted and ethanol precipitated prior to next-generation sequencing. (B) Bioanalyzer (Agilent) lane profile showing the S1-seq library size distribution (~150–600 bp).

17. Centrifuge at maximum speed for 2 min. Discard the 0.6-mL tube.
18. Add 300 μL 10 m*M* Tris–HCl pH 7.5.
19. Elute overnight in a 37°C thermomixer with shaking at 800 rpm.
20. Transfer the gel eluate mixture from one tube to the Spin-X column. Centrifuge at maximum speed for 2 min.
21. Add 150 μL 7.5 *M* ammonium acetate and 750 μL 100% ethanol to the ~300 μL eluate. Mix by inverting the tube. Precipitate at −20°C overnight.
22. Centrifuge at maximum speed for ≥30 min at 4°C.
23. Wash with 300 μL cold 70% ethanol. Centrifuge at maximum speed for 10 min at 4°C. Remove supernatant, then quickly spin and pipette out all remaining ethanol. Air dry for 10–15 min.
24. Dissolve the pellet in 20 μL 10 m*M* Tris–HCl pH 7.5.
25. To verify the size range of the library, a dilution of the library is analyzed on a Bioanalyzer Instrument (Agilent Technologies) according to manufacturer's instruction. An example is shown in Fig. 6B.
26. Proceed with next-generation sequencing (50 bp single-end reads) or store at −20°C until ready (see Note 6).

6. BIOINFORMATICS ANALYSIS

6.1 Mapping of Reads

Mapping of the reads onto the S288c reference genome (SacCer2) is performed using the SHRiMP mapper (Rumble et al., 2009) (gmapper-ls) with arguments:

-E -U -n 1 -Q –sam-unaligned –strata -o 10001 -N 20

Before mapping, adaptor sequences are removed using fastx_clipper (http://hannonlab.cshl.edu/fastx_toolkit/) and a custom script. Code used for read processing and mapping is available online at https://github.com/soccin/S1-seq. After mapping, the reads are separated into unique and multiple mapping sets, but only uniquely mapping reads are analyzed.

6.2 Curation of Data

Before any analysis, maps are curated by masking 500 bp at the ends of each chromosome, because telomeres represent naturally occurring resected DNA ends and therefore give a high frequency of reads. We further mask regions that give a high frequency of meiotic DSB-independent reads (Mimitou et al., 2017). A table of mask coordinates is provided at https://github.com/soccin/S1-seq. Moreover, reads mapping to mitochondrial DNA or the 2 μ plasmid are excluded. Each map is normalized to reads per million remaining mapped reads and then biological replicates are averaged.

6.3 Data Analysis

Analyses can be performed using the R package. Our analyses use a hotspot list compiled from a combination of multiple independent wild-type Spo11-oligo maps (Mohibullah & Keeney, 2017). For quantitative and modeling analyses we use a subset of hotspots for which no other hotspot is located within 3 kb, and for which hotspot width is less than 400 bp ($n=405$). S1-seq reads of polarity opposite to expectation for resection endpoints are subtracted to correct for RI signal and any negative values that arise are set to NA.

7. NOTES

1. To monitor the progression of cell divisions, collect a small volume ($\geq$25 μL) of the sporulation culture at least every 2 h (e.g., $t=0$, 2, 4, 6, 8 h, and one at a later time point, e.g., $t=24$ h for *Saccharomyces cerevisiae* SK1). Harvest cells and fix in 100 μL 40% ethanol in 0.1 *M* sorbitol. Add 1 μL 10 mg/mL DAPI (4′,6-diamidino-2-phenylindole) and score mono-, bi-, and tetranucleate cells by fluorescence microscopy. Fixed cells can be stored at 4°C until ready to count.
2. The cell suspension is very dense at this point, so avoid moving the cells from tube to tube. Instead, add solution 1 and LMP to the cells, either as a premix or sequentially and quickly mix and cast into the plugs.
3. Perform all buffer exchanges carefully—so as not to disrupt the plugs—by guiding the pipette tip to the side of the tube. Incubations can be performed with gentle shaking, on a horizontal platform mixer, for example, but we do not find this necessary.
4. We noticed substantial DNA loss after the Chroma Spin column step, but omitting it altogether resulted in the presence of background bands after PCR amplification. More recent S1-seq libraries in our laboratory replaced this step with sonicator shearing (200–500 bp), followed by SPRI size selection (e.g., Ampure XP beads). In this case the Covaris shearing step is also omitted.
5. The DNA solution at this point is extremely viscous because it contains high-molecular weight fragments. Be patient: pipette gently and slowly

and mix by flicking the tube. Moreover, caution should be exerted while loading the samples onto the gel wells. Deposit sample inside the well slowly, avoiding air bubbles, and make sure the tip is no longer connected to the viscous solution when you remove it from the well.

6. The latest S1-seq libraries in our laboratory were submitted for paired-end sequencing and the improved mapping helped remove some of the nonspecific signal, reducing the number of positions in the genome that need to masked.

ACKNOWLEDGMENTS

This work was supported by National Institutes of Health (NIH) Grants R01 GM058673 and R35 GM118092 (to S.K.) and a fellowship from the Helen Hay Whitney Foundation (to E.P.M.). MSKCC core facilities were supported by NIH Grant P30 CA008748. S.K. is a Howard Hughes Medical Institute Investigator. Nicholas D. Socci (MSKCC Bioinformatics Core Facility) developed the sequence mapping pipeline, and the MSKCC Integrated Genomics Operation core facility performed Covaris shearing and NGS. We are grateful to Shintaro Yamada for assistance in the bioinformatics analysis and Pei-Ching Huang (Keeney lab, MSKCC) for comments on this manuscript.

REFERENCES

Alani, E., Padmore, R., & Kleckner, N. (1990). Analysis of wild-type and rad50 mutants of yeast suggests an intimate relationship between meiotic chromosome synapsis and recombination. *Cell*, *61*(3), 419–436.

Bishop, D. K., Park, D., Xu, L., & Kleckner, N. (1992). *DMC1*: A meiosis-specific yeast homolog of *E. coli recA* required for recombination, synaptonemal complex formation, and cell cycle progression. *Cell*, *69*(3), 439–456. https://doi.org/10.1016/0092-8674(92)90446-J.

Cannavo, E., & Cejka, P. (2014). Sae2 promotes dsDNA endonuclease activity within Mre11-Rad50-Xrs2 to resect DNA breaks. *Nature*, *514*(7520), 122–125. https://doi.org/10.1038/nature13771.

de Massy, B. (2013). Initiation of meiotic recombination: How and where? Conservation and specificities among eukaryotes. *Annual Review of Genetics*, *47*, 563–599. https://doi.org/10.1146/annurev-genet-110711-155423.

Garcia, V., Phelps, S. E., Gray, S., & Neale, M. J. (2011). Bidirectional resection of DNA double-strand breaks by Mre11 and Exo1. *Nature*, *479*(7372), 241–244. https://doi.org/10.1038/nature10515.

Hunter, N. (2015). Meiotic recombination: The essence of heredity. *Cold Spring Harbor Perspectives in Biology*, 7(12), 361–395. https://doi.org/10.1101/cshperspect.a016618.

Keeney, S., Giroux, C. N., & Kleckner, N. (1997). Meiosis-specific DNA double-strand breaks are catalyzed by Spo11, a member of a widely conserved protein family. *Cell*, *88*(3), 375–384. https://doi.org/10.1016/S0092-8674(00)81876-0.

Lam, I., & Keeney, S. (2014). Mechanism and regulation of meiotic recombination initiation. *Cold Spring Harbor Perspectives in Biology*, 7(1), a016634. https://doi.org/10.1101/cshperspect.a016634.

McKee, A. H., & Kleckner, N. (1997). A general method for identifying recessive diploid-specific mutations in Saccharomyces cerevisiae, its application to the isolation of mutants

blocked at intermediate stages of meiotic prophase and characterization of a new gene SAE2. *Genetics*, *146*(3), 797–816.
Mimitou, E. P., & Symington, L. S. (2008). Sae2, Exo1 and Sgs1 collaborate in DNA double-strand break processing. *Nature*, *455*(7214), 770–774. https://doi.org/10.1038/nature07312.
Mimitou, E. P., & Symington, L. S. (2009). DNA end resection: Many nucleases make light work. *DNA Repair*, *8*(9), 983–995. https://doi.org/10.1016/j.dnarep.2009.04.017.
Mimitou, E. P., Yamada, S., & Keeney, S. (2017). A global view of meiotic double-strand break end resection. *Science*, *355*(6320), 40–45. https://doi.org/10.1126/science.aak9704.
Mohibullah, N., & Keeney, S. (2017). Numerical and spatial patterning of yeast meiotic DNA breaks by Tel1. *Genome Research*, *27*(2), 278–288. https://doi.org/10.1101/gr.213587.116.
Moreau, S., Ferguson, J. R., & Symington, L. S. (1999). The nuclease activity of Mre11 is required for meiosis but not for mating type switching, end joining, or telomere maintenance. *Molecular and Cellular Biology*, *19*(1), 556–566.
Neale, M. J., & Keeney, S. (2006). Clarifying the mechanics of DNA strand exchange in meiotic recombination. *Nature*, *442*(7099), 153–158. https://doi.org/10.1038/nature04885.
Neale, M. J., Pan, J., & Keeney, S. (2005). Endonucleolytic processing of covalent protein-linked DNA double-strand breaks. *Nature*, *436*(7053), 1053–1057. https://doi.org/10.1038/nature03872.
Pan, J., Sasaki, M., Kniewel, R., Murakami, H., Blitzblau, H. G., Tischfield, S. E., et al. (2011). A hierarchical combination of factors shapes the genome-wide topography of yeast meiotic recombination initiation. *Cell*, *144*(5), 719–731. https://doi.org/10.1016/j.cell.2011.02.009.
Petes, T. D. (2001). Meiotic recombination hot spots and cold spots. *Nature Reviews. Genetics*, *2*(5), 360–369. https://doi.org/10.1038/35072078.
Prinz, S., Amon, A., & Klein, F. (1997). Isolation of COM1, a new gene required to complete meiotic double-strand break-induced recombination in Saccharomyces cerevisiae. *Genetics*, *146*(3), 781–795.
Rumble, S. M., Lacroute, P., Dalca, A. V., Fiume, M., Sidow, A., & Brudno, M. (2009). SHRiMP: Accurate mapping of short color-space reads. *PLoS Computational Biology*, *5*(5), e1000386. https://doi.org/10.1371/journal.pcbi.1000386.
Thacker, D. K., & Keeney, S. (2016). DNA replication, recombination, and repair. In F. Hanaoka & K. Sugasawa (Eds.), *DNA replication, recombination, and repair* (pp. 131–151). Springer.
Zakharyevich, K., Ma, Y., Tang, S., Hwang, P. Y., Boiteux, S., & Hunter, N. (2010). Temporally and biochemically distinct activities of Exo1 during meiosis: Double-strand break resection and resolution of double Holliday junctions. *Molecular Cell*, *40*(6), 1001–1015. https://doi.org/10.1016/j.molcel.2010.11.032.
Zhu, Z., Chung, W. H., Shim, E. Y., Lee, S. E., & Ira, G. (2008). Sgs1 helicase and two nucleases Dna2 and Exo1 resect DNA double-strand break ends. *Cell*, *134*(6), 981–994. https://doi.org/10.1016/j.cell.2008.08.037.

CHAPTER FOURTEEN

Methods for Controlled Protein Depletion to Study Protein Function during Meiosis

Hardeep Kaur, Jasvinder S. Ahuja, Michael Lichten[1]
Laboratory of Biochemistry and Molecular Biology, Center for Cancer Research, National Cancer Institute, Bethesda, MD, United States
[1]Corresponding author: e-mail address: michael.lichten@nih.gov

Contents

Abstract

Proteins with potential roles in meiotic recombination often have essential or important functions during the mitotic cell cycle. In addition, proteins may have different functions at different times during meiosis. In such cases, it can be challenging to precisely determine protein function during meiosis using null or hypomorphic mutants. One example is the Sgs1–Top3–Rmi1 helicase–decatenase complex, which is required for normal vegetative growth and genome stability. In such cases, conditional loss-of-function mutants can be useful. In this chapter, we describe the construction of two types of conditional mutants, meiotic depletion alleles and auxin-induced degradation alleles, that allow protein depletion specifically during budding yeast meiosis, and illustrate their use with Sgs1. We also describe a modified method for the isolation of meiotic recombination intermediates that combines previous psoralen cross-linking and cetyltrimethylammonium bromide isolation methods.

Methods in Enzymology, Volume 601
ISSN 0076-6879
https://doi.org/10.1016/bs.mie.2017.11.032
2018 Published by Elsevier Inc.

ABBREVIATIONS

3XHA triple influenza hemagglutinin epitopes
CoHex hexamminecobalt(III) chloride
CTAB cetyltrimethylammonium bromide
DAPI 4′,6-diamidino-2-phenylindole dihydrochloride
DMSO dimethyl sulfoxide
HJ Holliday junction
HR homologous recombination
KAc potassium acetate sporulation medium
PBS phosphate-buffered saline
SPS presporulation medium
STR Sgs1–Top3–Rmi1
TCA trichloroacetic acid
TE 10 m*M* Tris, 1 m*M* EDTA, pH 7.5
TEMED tetramethylethylenediamine
YEPAD yeast extract, peptone, adenine, dextrose growth medium

1. INTRODUCTION

In the budding yeast *Saccharomyces cerevisiae*, meiotic recombination intermediate metabolism and the crossover/noncrossover pathway choice are regulated by an evolutionarily conserved complex that contains the BLM helicase homolog, Sgs1, and the Top3–Rmi1 topoisomerase/decatenase (STR complex; De Muyt et al., 2012; Jessop & Lichten, 2008; Jessop, Rockmill, Roeder, & Lichten, 2006; Kaur, De Muyt, & Lichten, 2015; Oh et al., 2007; Oh, Lao, Taylor, Smith, & Hunter, 2008; Rockmill, Fung, Branda, & Roeder, 2003; Tang, Wu, Zhang, & Hunter, 2015). This complex (and its homologs in other species) plays important roles in homologous recombination (HR) and genome stability during the mitotic cell cycle (Larsen & Hickson, 2013). Null mutants lacking any member of the STR complex display growth and genome stability defects that preclude their use in analyses of meiotic recombination (Gangloff, McDonald, Bendixen, Arthur, & Rothstein, 1994; Jessop et al., 2006; Mullen, Kaliraman, Ibrahim, & Brill, 2001). To address this problem, conditional expression and depletion approaches have been developed and applied to the study of STR meiotic function. These allow functional STR to be present during premeiotic growth, while allowing subsequent events of meiosis to occur in its absence. In this chapter, we describe the application of two different approaches to conditional STR protein expression: promoter replacement, so that a gene of interest is expressed only during mitotic growth;

and auxin-induced degradation, which allows the time-controlled conditional degradation of a target protein during meiosis. The use of these approaches for conditional depletion of Sgs1 is described as an example, but these approaches should be applicable to the analysis of meiotic function of many proteins that are normally present in both mitotic and meiotic cells.

2. MEIOSIS-SPECIFIC PROTEIN DEPLETION

Many cellular proteins perform important functions during both meiosis and the mitotic cell cycle. Because of this, the null mutants and heat-sensitive conditional mutants usually used to study mitotic functions are often impractical for the study of their meiotic functions. For example, the poor growth and genome instability shown by null mutants lacking Sgs1, Top3, or Rmi1, particularly in diploids in the widely used SK1 background, are so severe that the mating-type heterozygosity necessary for meiosis cannot be maintained during the growth of bulk cultures (Jessop et al., 2006; Kaur et al., 2015). Moreover, because many yeast strains, including SK1, sporulate poorly above 34°C, and higher temperatures affect meiotic recombination (Börner, Kleckner, & Hunter, 2004; Chan, Borts, & Hoffmann, 2009; Honigberg & Esposito, 1994), the use of heat-sensitive mutants is often impractical. For these reasons, several strategies have been adopted to allow conditional depletion of proteins during meiosis.

2.1 Strategies for Meiotic Protein Depletion

2.1.1 Meiotic Depletion Alleles

The term "meiotic depletion allele" is generally used to refer to constructs, where the promoter of a gene is replaced by one that is expressed only during the mitotic cell cycle. In practice, promoters from *MCD1* (also called *SCC1*; Clyne et al., 2003) and from *CLB2* (Lee & Amon, 2003) have been used. Both genes are poorly expressed during meiosis (Chu et al., 1998; Primig et al., 2000), but are expressed at different times during the mitotic cell cycle. *MCD1* is expressed during S-phase, while *CLB2* is first expressed in late S-phase and continues to be expressed during G2 (Cho et al., 1998; Grandin & Reed, 1993). Both promoters are available in plasmids containing the relevant 5′ upstream region (−840 to −1 for *MCD1*; −1000 to −1 for *CLB2*) and an upstream drug-resistance cassette (Goldstein & McCusker, 1999; Wach, 1996) to allow selection of transformants (see Sections 2.2 and 2.3, for construct and transformation details). We and

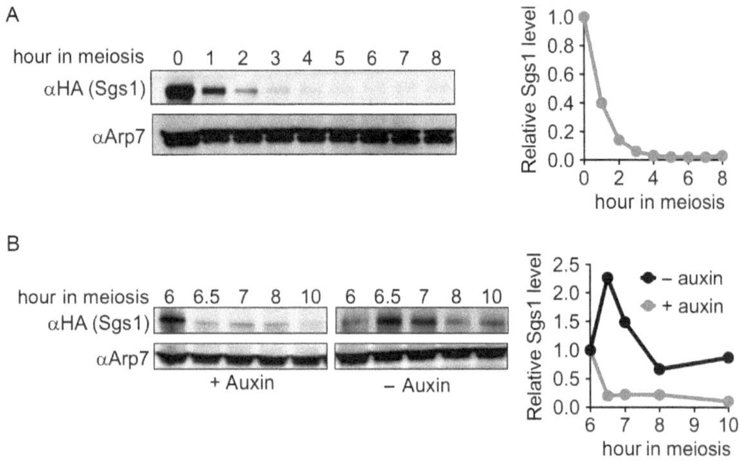

Fig. 1 Western blots illustrating meiotic depletion of Sgs1. (A) Meiotic depletion, using a strain where an HA epitope-tagged Sgs1 protein is expressed from the *CLB2* promoter (*kanMX-pCLB2-3HA-SGS1*, see Fig. 2A). Arp7 is used as a loading control. Arp7-normalized levels of Sgs1, relative to the premeiotic culture (0h), are plotted. (B) Auxin-induced depletion, using a strain where Sgs1 protein contains C-terminal HA epitope tags and the IAA17 auxin recognition domain (*SGS1-3HA-IAA17-hygMX*, see Fig. 2C), and the osTIR1 F-box protein is expressed from the *TEF1* promoter (*pTEF1-osTIR1*). At 6h in meiosis, the culture was split, auxin (3-indoleacetic acid) was added to one half, and vehicle (DMSO) was added to the other. Arp7 is used as a loading control. Sgs1 levels, relative to the time of auxin addition (6h), are plotted.

others have used such alleles successfully to deplete Sgs1, Top3, and Rmi1 (Jessop et al., 2006; Kaur et al., 2015; Oh et al., 2008; Tang et al., 2015; see Fig. 1, for an example), the repair protein Mms4 (Jessop & Lichten, 2008; Oh et al., 2008), and polo kinase Cdc5 (Clyne et al., 2003; Lee & Amon, 2003). Others have used meiotic depletion alleles to deplete many other proteins; a partial list includes the replicative helicase loader Cdc6 (Blitzblau & Hochwagen, 2013), the kinetochore proteins Iml3, Dsn1, and Mtw1 (Vincenten et al., 2015), SUMO-specific proteases Ulp1 and Ulp2 (Leung et al., 2015), the spindle pole body protein Mps3 (Li, Shao, Jin, & Yu, 2015), topoisomerase II (Zhang et al., 2014), the Smc5/6 complex proteins Smc5 and Nse4 (Copsey et al., 2013), the ATR kinase Mec1 (Gray, Allison, Garcia, Goldman, & Neale, 2013), the PP2a subunit Cdc55 (Bizzari & Marston, 2011), the cohesin loading and maintenance factors Scc2, Scc3, and Pds5 (Jin, Guacci, & Yu, 2009; Lin, Jin, Liu, Hampton, & Yu, 2011; Lin, Wang, Jin, & Yu, 2011), the condensin component Brn1 (Brito, Yu, & Amon, 2010), the APC/C adaptor Cdc20

(Lee & Amon, 2003), the Aurora B kinase Ipl1 (Yu & Koshland, 2007), and the centromeric cohesin protector Sgo1 (Katis et al., 2010).

Several considerations are important to the success of this promoter-replacement strategy. First, it relies on endogenous processes to deplete the protein of interest before it can function during meiosis, and one cannot assume that the protein will be absent during meiosis simply because the replacement promoter is not active. Thus, the protein of interest must either have a short half-life or be specifically degraded during the mitotic division prior to initiation of meiosis; long-lived proteins that remain at high levels during meiosis in the absence of further transcription are not amenable to this approach. It is important to monitor levels of the protein of interest during meiosis to confirm effective depletion, either using antibodies against the endogenous protein or with antibodies against an antigenic tag that is included in the promoter-replacement construct. In the case of the *pCLB2-SGS1* allele described here, the target protein is still present at significant levels during the first 2h of meiosis, when DNA replication occurs, but is substantially depleted at later times, when meiotic recombination occurs (Fig. 1A).

Second, it is important to ensure that the promoter replacement does not have collateral consequences on gene function during the mitotic cell cycle that might carry over into meiosis. This is best determined empirically, by examining promoter-replacement strains to see if they display mutant phenotypes, during vegetative growth, expected for full or partial loss-of-function alleles. In practice, this has not been a concern for *pCLB2-SGS1* promoter-replacement constructs (as well as for *pCLB2-TOP3* and *pCLB2-RMI1* constructs), even though the time of peak expression from the *CLB2* promoter (late S- and G2-phases) comes after the time (S-phase) when Sgs1–Top3–Rmi1 complex activity is important. However, it is conceivable that use of a *CLB2* promoter might create a mutant phenotype, especially for a gene whose protein product is degraded during mitosis but is required earlier in the cell cycle. In such a case, the *MCD1* promoter may be more suitable.

Finally, it is important to recognize that deletion of a gene's promoter and/or its replacement with the *CLB2* or *MCD1* promoter may impact expression of the gene immediately upstream of the target gene, particularly when the two genes are divergently transcribed. While this has proven not to be an issue in the constructs we have used, it remains a formal concern. One possible way to address this concern would be to insert the mitosis-specific promoter immediately upstream of the gene of interest, without deleting any upstream sequences, and to rely on transcription terminators present in the drug-resistance cassette, used to select transformants, to block read-through transcription.

2.1.2 Conditional Depletion Using Auxin-Inducible Degrons

While meiotic depletion alleles generally confer loss of function for a target protein throughout meiosis, some analyses (for example, execution point analysis) require wild-type function during early stages of meiosis, followed by target protein depletion at later stages. Two different conditional depletion strategies have been used in studies of yeast meiosis. The first approach, called anchor-away (Haruki, Nishikawa, & Laemmli, 2008), removes a protein from the nucleus by tethering it to a cytoplasmic protein via a rapamycin-dependent dimerization module. This approach has been used to conditionally relocate several nuclear proteins (Zip1, Rad54, Psy2) during yeast meiosis, resulting in loss-of-function phenotypes (Subramanian et al., 2016). We have not used this method, which requires additional mutations (*tor1-1* to tolerate rapamycin, *fpr1* to remove an endogenous rapamycin-binding protein) to function, and which requires cytological assays to confirm effective nuclear depletion.

A second approach uses a plant-derived auxin-inducible degron to conditionally target a protein of interest for ubiquitin-mediated proteolysis (Nishimura, Fukagawa, Takisawa, Kakimoto, & Kanemaki, 2009; Nishimura & Kanemaki, 2014). This system has been used in a variety of organisms and has been modified specifically for use in budding yeast. It contains two components: a degron, based on the IAA17 transcription repressor, that is fused to the protein of interest, and an SCF E3 ligase component, TIR1, usually from rice (osTIR1). Auxins (usually indole acetic acid, IAA) promote the interaction of the IAA17-based tag with TIR1, leading to the rapid polyubiquitylation of the target protein and subsequent degradation by the proteasome, often in a matter of minutes in mitotically growing cells. Unfortunately, constructs and conditions that support rapid degradation of IAA17-tagged proteins in mitotic cells work poorly, if at all, in meiotic yeast cells. Efficient degradation appears to require high levels of osTIR1, and the original expression vectors developed by Nishimura and Kanemaki do not meet this need during meiosis. Two approaches have been used to address this problem. Tang et al. (2015) have constructed strains where osTIR1 expression is driven by the *CUP1* promoter, with osTIR1 expression being induced by adding $CuSO_4$ to cultures about 1 h before the addition of auxin. As an alternative, we have constructed strains where osTIR1 is constitutively expressed at high levels in both mitotic and meiotic cells, using either the *TEF1* or the *ZEO1* promoters; these strains can be used for auxin-inducible degradation of target proteins during both meiosis and the mitotic cell cycle, without the need for other factors (Fig. 1B). Construction of strains with IAA17-protein fusions, *pTEF1-osTIR1* or *pZEO1-osTIR1*, and use of these strains for meiotic depletion are described later.

2.2 Construction of Strains for Meiotic Protein Depletion

In the following sections, construction of meiotic depletion and of auxin-inducible degron yeast strains is described, using *SGS1* as an example.

2.2.1 CLB2 Promoter Replacement

The strategy used for promoter replacement is illustrated in Fig. 2 and starts with a plasmid (pMJ787, originally Amon plasmid 455, a gift from Angelika Amon) that contains a *kanMX* module for selecting transformants with G418, nt −1000 to +6 from the *CLB2* locus, and sequences encoding three tandem influenza hemagglutinin epitopes (3XHA). The *kanMX-pCLB2-3xHA* module is inserted at a locus by PCR amplification with primers corresponding to the edges of the module with an additional 40 nt of sequence from the target locus (see Section 2.3.3). While this homology is in theory sufficient to direct integration in the yeast genome (Manivasakam, Weber, McElver, & Schiestl, 1995), we have found that the yield of transformants can be substantially increased by including two additional DNA fragments that contain 150–200 nt of sequence flanking the site of insertion. The cotransforming DNA fragments contain sufficient homology to frequently recombine with each other (Kunes, Ma, Overbye, Fox, & Botstein, 1987), creating a *kanMX-pCLB2-3xHA* fragment with 150–200 nt of flanking homology that is able to integrate in the genome with greater efficiency (Fig. 2A). If the N-terminal 3xHA tag is found to interfere with protein function, the same plasmid can be used for amplification with an alternative set of primers to create a promoter replacement without the N-terminal HA tag (Fig. 2B).

2.2.2 IAA17 Tagging

The strategy used for tagging with the IAA17 auxin-response element has been described (Nishimura & Kanemaki, 2014); in our work, we have generally used C-terminal IAA17 tags with a 3xHA epitope tag between the gene's coding sequences and the IAA17 tag. This has the advantage of

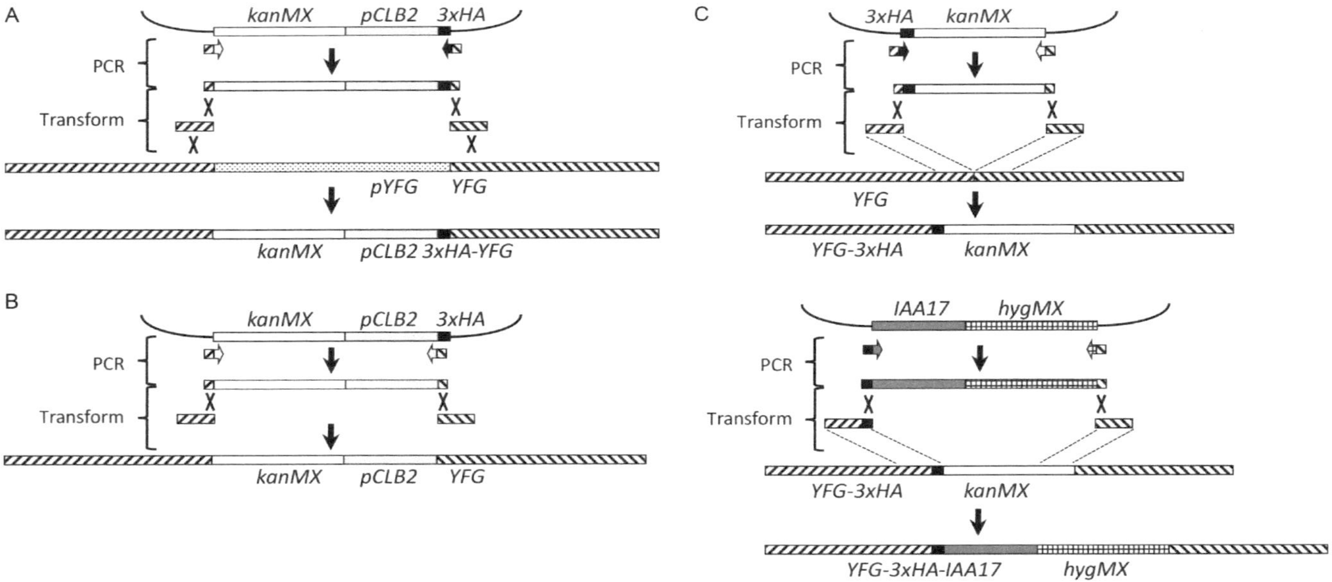

Fig. 2 Three-fragment transformation strategy for promoter replacement and IAA17 tagging. (A) Replacement of a target gene (*YFG*) promoter with the *CLB2* promoter and addition of an N-terminal HA tag (see text for details). A *kanMX-pCLB2-3xHA* fragment is amplified from plasmid DNA, using hybrid primers with 3′ homology to fragment boundaries and 5′ homology to the boundaries of the desired insertion site. This fragment is mixed with two fragments (150 nt or greater) that contain sequences at the boundaries of the insertion site, and transformed into yeast. The additional homology provided by the two fragments increases the recovery of integrants, which replace the endogenous promoter with *kanMX* and the *CLB2* promoter, and also add a 3xHA epitope tag at the N-terminus of the protein. (B) If desired, the 3xHA tag can be omitted by amplifying with a hybrid primer that contains 3′ homology to the right-hand *CLB2* promoter boundary, and 5′ homology to the right-hand boundary of the insertion site. (C) Strategy for creating a C-terminal hybrid 3xHA-IAA17 tag. First, a 3xHA epitope tag is inserted at the C-terminus of the gene linked to *kanMX*, using three-fragment transformation. The same strategy is used to replace the *kanMX* moiety with a *hygMX*-linked *IAA17* insert.

allowing protein expression from its native promoter, and of monitoring protein levels on Western blots (see Section 3.3). Construction is done in two steps (Fig. 2C). First, the protein is C-terminally tagged using pFA6a-3HA-kanMX6 (Longtine et al., 1998), using the same flanking homology fragment strategy as described earlier (Section 2.2.1). Second, the IAA17 tag is added to the C-terminal 3xHA tag, again using the flanking homology fragment strategy, using a derivative of pMK43 (Nishimura et al., 2009) in which the *kanMX* cassette is replaced by either a *hygMX* (pMJ894) or a *natMX* (pMJ893) drug-resistance cassette (Goldstein & McCusker, 1999).

2.2.3 *osTIR1* Expression

The second element of the auxin-inducible expression system, the rice-derived TIR1 F-box protein, was originally obtained as a yeast codon-optimized version whose expression is driven by the *GAL1* promoter (pYK6; Kubota, Nishimura, Kanemaki, & Donaldson, 2013). It appears that this protein must be expressed at high levels in order to function optimally, and the *GAL1* promoter is not compatible with such high levels of expression during meiosis. Using genome-wide meiotic expression data (Primig et al., 2000), we identified two genes, *TEF1* and *ZEO1*, that displayed uniformly high levels of expression both during the mitotic cell cycle and during meiosis; plasmids containing these promoters were constructed (Fig. 3) and integrated at the *URA3* locus.

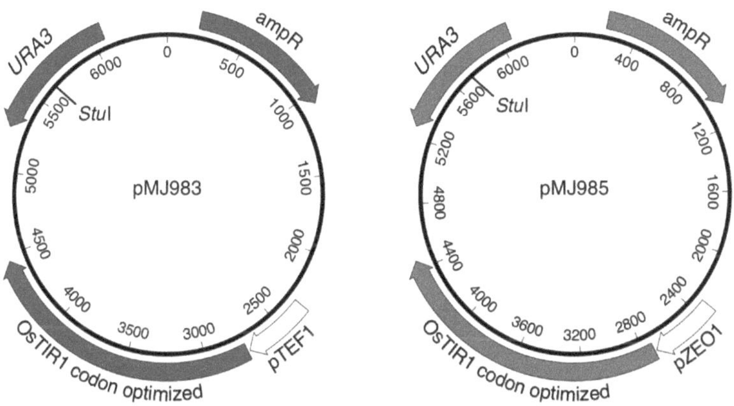

Fig. 3 Plasmids with *osTIR1* expressed from high-level, constitutive promoters. pMJ983 contains *TEF1* promoter sequences from −409 to −1 nt; pMJ985 contains *ZEO1* promoter sequences from −405 to −1 nt. The unique *StuI* site, used to direct integration to the *URA3* locus, is indicated.

2.3 Insertion of Constructs by Transformation

In the following section, transformation by electroporation, using the flanking homology fragment strategy (see Fig. 2), is described. We have found this approach to be suitable for strain SK1 derivatives; in other strains, other approaches may be used (Gietz & Woods, 2001).

2.3.1 Equipment

- High-capacity refrigerated tabletop or floor centrifuge with adaptors for 50-mL conical-bottom tubes
- Vortex mixer
- Electroporator capable of 1.5 kV, 200 Ω, 25 μF pulse (BioRad Micro-Pulser or equivalent)
- Electroporation cuvettes, 0.2 cm gap (BioRad 1652082 or equivalent). Keep at 4°C.
- Shaking incubator or water bath, set at 30°C.

2.3.2 Reagents

- YEPAD broth: 1% Bacto™ yeast extract, 2% Bacto™ peptone, 2% D-glucose, 0.004% adenine, pH 5.5 (adjusted with 1 N HCl). Prepare as a 1.1 × stock without glucose, and add glucose from a sterile 20% stock before use.
- YEPAD + 1 *M* sorbitol broth; make as above but include sorbitol at 1 *M*.
- 1 *M* sorbitol, filter-sterilize and store at 4°C
- Sonicated carrier DNA. We usually use either salmon or herring sperm DNA, dissolved in 10 m*M* TRIS, 1 m*M* EDTA, pH 7.5 (TE), sonicated until no longer viscous, phenol-extracted and ethanol precipitated, and resuspended in TE at 5 mg/mL. Chill to 4°C before use.
- Selective media, made with 1 *M* sorbitol.
- DNA fragments, made either by PCR or by digestion of plasmids, ideally at a concentration of 1–2 μg/μL (see Note 1). Chill to 4°C before use.

2.3.3 Procedure

1. Inoculate a single colony of the desired strain into 5 mL YEPAD; grow overnight with aeration at 30°C.
2. Wash three times with cold 1 *M* sorbitol by centrifuging 2 min at 3220 × *g* at 4°C, draining thoroughly.
3. After the last centrifugation, decant as much supernatant as possible, and resuspend the pellet by vortexing. The final yield should be about 30 μL of cells per 1 mL of original overnight culture.

4. Transfer 40 μL of cell suspension to a cold microfuge tube. At this point, cells can be frozen at −80°C.
5. If frozen cells are used, thaw cell suspension on ice.
6. Add 12.5 μg sonicated carrier DNA and 1–2 μg of each transforming DNA fragment.
7. Add the cell + DNA mix to a precooled electroporation cuvette (see Note 1).
8. Pulse-shock in electroporator at 1.5 kV, 200 Ω, 25 μF.
9. Immediately add 0.4 mL ice-cold 1 *M* sorbitol and mix.
10. If nutritional selection is used, cells can be immediately plated on the relevant dropout plates made with 1 *M* sorbitol (see Note 2).
11. If a drug-resistance selection is used, add the transformation mix to 4.5 mL of YEPAD + 1 *M* sorbitol and aerate at 30°C for 4–5 h before plating (see Note 2).
12. Incubate plates at 30°C for 2–3 days. Integrative transformants will be visible as larger colonies in a background of smaller microcolonies that contain abortive transformants.

3. SYNCHRONOUS MEIOSIS AND DNA, PROTEIN, AND CYTOLOGICAL ANALYSIS

A basic requirement for the real-time analysis of protein function during meiosis is the ability to generate populations of cells that undergo meiosis synchronously. Much of the work done in this regard has used strains of the SK1 genetic background (Alani, Cao, & Kleckner, 1987; Kane & Roth, 1974), where sporulation can be induced in liquid cultures with relative synchrony. The methods described later are the ones currently used by our lab to sporulate SK1, for isolation of DNA from meiotic cultures in a way that preserves recombination intermediates, for Western blots to monitor protein depletion, and for nuclear staining to monitor meiotic divisions.

3.1 Synchronous Sporulation

Most past and current studies induce SK1 strains (Kane & Roth, 1974) to sporulate by pregrowth in rich medium with acetate as the carbon source, to the point of transition between exponential growth and stationary phase. Cells are then shifted into starvation medium with acetate as the sole carbon source and are aerated vigorously to induce sporulation and meiosis. There is considerable variation between groups in terms of the composition of both presporulation and sporulation medium (Börner & Cha, 2015; Lichten, 2014;

Oh et al., 2009). The protocol below uses buffered presporulation medium to limit the alkalinization of cultures that occurs during growth in non-fermentable carbon sources. High pH prevents nutrient uptake (McCusker & Haber, 1977) and can result in premature entry into meiosis.

3.1.1 Equipment

- Rotating tissue culture drum (Bellco 7736-10115 or equivalent) in a 30°C incubator
- Glass culture tubes, 20 × 150 mm
- 2-L Erlenmeyer flasks (autoclaved)
- 250-mL centrifuge bottle (autoclaved)
- High-capacity tabletop or floor centrifuge with adaptors for 250-mL bottles
- 2.8-L triple-baffled Fernbach flask (Bellco 2551-02800 or equivalent)
- 2-L triple-baffled Erlenmeyer flasks (Bellco 2542-02000 or equivalent)
- Shaking incubator or shaking water bath capable of sustained operation at 375 rpm (see Note 3)
- Spectrophotometer with at least 1 cm distance between the cuvette and detector (see Note 4) and 1.5 mL semimicro disposable cuvettes, light path 10 mm

3.1.2 Reagents (See Note 5)

- YEPAD broth and plates: Broth—see Section 2.3.2. Plates—YEPAD plates are made the same way, but Bacto™ agar is added to 2% before autoclaving.
- Presporulation broth (SPS): 1% Bacto™ peptone, 0.5% Bacto™ yeast extract, 1% potassium acetate, 0.17% Bacto™ yeast nitrogen base without amino acids, 1% ammonium sulfate, 0.5% potassium hydrogen phthalate, pH 5.5 (adjusted with 10 N KOH).
- Sporulation medium (KAc): 1% potassium acetate, 0.001% polypropylene glycol 2000 (Sigma 202339, from a 1% stock suspension), supplemented as required by the specific auxotrophies of the strain being used. Supplements are added at 1/5 the concentration used to support vegetative growth in minimal media. Concentrations of common supplements used in our lab: histidine, 0.0004%; lysine, 0.0006%; arginine, 0.0004%; leucine, 0.0012%; uracil, 0.0002%. Add polypropylene glycol and supplements after autoclaving (see Note 5).
- 3-Indoleacetic acid (Auxin, Sigma I3750), 500 m*M* in DMSO. Make fresh the day of use.

3.1.3 Procedure (See Note 6)

1. (−4 days) Streak diploid strains to be used on YEPAD agar plates. Incubate 48 h at 30°C
2. (−2 days) Inoculate a single colony into 5 mL YEPAD broth culture in a 20 × 150 mm glass culture tube and incubate 24 h at 30°C on a roller drum.
3. (−1 day) Inoculate four SPS cultures (240 mL of SPS in a 2-L Erlenmeyer flask) with a range of dilutions of the overnight culture. We generally inoculate at 1/500, 1/1000, 1/1333, and 1/2000 dilutions. Incubate with vigorous shaking (300 rpm) for ~18 h at 30°C. Also, aliquot 400 mL supplemented KAc into a 2.8-L triple-baffled Fernbach flask and incubate at 30°C. Reserve 240 mL unsupplemented KAc for washing and incubate at 30°C.
4. About 18 h after initiating SPS cultures, measure culture OD_{600}. Grow cultures until one reaches an OD_{600} of 1.4 (~2×10^7 cells/mL).
5. Transfer culture to a 250-mL centrifuge bottle and centrifuge 3 min, 3220 × *g*. Decant the supernatant, drain briefly, and resuspend in 240 mL prewarmed KAc without supplements. Remove 35 mL sample for DNA extraction, 3 mL sample for protein extraction, and 1 mL sample for DAPI staining.
6. Centrifuge remaining culture as above and resuspend in a portion of the 400 mL supplemented KAc from the 2.8-L Fernbach flask, return to the flask, and incubate at 30°C with very vigorous shaking (375 rpm).
7. Take samples at time points as needed for DNA, protein, and DAPI staining, as in step 5.
8. For auxin-induced protein depletion: at desired time, split culture in half, into two 2-L baffled Erlenmeyer flasks, adding to one 500 m*M* auxin to a final concentration of 2 m*M* and an identical volume of DMSO to the other. Continue additions every hour to maintain auxin concentrations.
9. Alternatively, if greater sample volumes are required, start two SPS cultures for each dilution (step 3). On the day of the sporulation, pool the two cultures whose OD_{600} is closest to 1.4, split into two centrifuge bottles, and process as above (steps 5–6) using two 2.8-L Fernbach flasks. At the desired time, inoculate one flask with auxin and the other with DMSO as in step 8.

3.2 DNA Preparation by Combined CTAB Extraction and Psoralen Cross-Linking

Methods for the preparation of DNA for the analysis of meiotic recombination products and intermediates have been described elsewhere

(Ahuja & Börner, 2011; Allers & Lichten, 2000; Oh et al., 2009; Schwacha & Kleckner, 1994). Holliday junction (HJ)-containing intermediates are vulnerable to loss by branch migration, and three main strategies have been used to stabilize HJ-containing intermediates: psoralen cross-linking, which forms interstrand cross-links that limit branch migration; preparation in the presence of multivalent cations (including CTAB and hexamminecobalt(III) chloride, CoHex), which limits branch migration by stabilizing the stacked-X HJ conformation; and preparation in agarose plugs, which stabilizes HJs by limiting arm movement (Oh et al., 2009). Psoralen cross-linking has the advantage of allowing harsher downstream extraction procedures to be used, but has the disadvantage of requiring cross-link reversal if single-strand analyses are desired; in addition, cross-links at a restriction enzyme site block digestion, resulting in a low but unavoidable level of partial digestion products. Conversely, CTAB extraction is considerably more demanding technically, but yields DNA that requires no further processing for downstream analyses. We recently found that a combined CTAB extraction and psoralen cross-linking approach confers modest improvements in recombination intermediate yields (J. S. A. and M. L., unpublished), and describe it here. Methods for downstream analysis, including 1D- and 2D-gel electrophoresis, blotting, and probing, have been described elsewhere (Ahuja & Börner, 2011; Oh et al., 2009).

3.2.1 Equipment

- High-capacity tabletop or floor centrifuge with adaptors for 50-mL conical-bottom tubes and 5-mL round-bottom tubes
- Variable-speed microfuge
- Long-wavelength UV (366 nm) transilluminator, with output calibrated using a long-wave UV meter (UVP Blak-Ray, model J-221 or equivalent)
- Qubit 3.0 minifluorometer, Nanodrop 2000 microvolume spectrophotometer, or equivalent
- 50-mL conical-bottom polypropylene tubes (Falcon 352070 or equivalent)
- 5-mL round-bottom polypropylene tubes (Falcon 352063 or equivalent)
- 1.5-mL microfuge tubes
- 35-mm Petri dish (Falcon 353001 or equivalent)
- Fine-tip transfer pipettes (Samco 23220S)

3.2.2 Reagents

- 1 mg/mL psoralen (Trioxsalen, Sigma-Aldrich T6137) in 100% ethanol (see Note 7)
- 50 m*M* Tris–HCl, 50 m*M* EDTA, pH 8
- Storage solution: 1 *M* sorbitol, 50 m*M* potassium phosphate, 10 m*M* EDTA, 20% glycerol, pH 7.5. Filter-sterilize and store at 4°C
- Spheroplasting solution: 1 *M* sorbitol, 50 m*M* potassium phosphate, 10 m*M* EDTA, 5 m*M* hexamine cobalt chloride, pH 7.5. Filter-sterilize and store at RT
- Zymolyase 100T
- β-Mercaptoethanol
- CTAB extraction solution: 3% CTAB, 100 m*M* Tris, 25 m*M* EDTA, 2 *M* NaCl, 2% polyvinylpyrrolidone (avg. MW 40,000, PVP40), 20 m*M* CoHex. Preparation method is critical, as the final solution is too viscous to filter-sterilize, and components can precipitate if not added in the correct order

  For 100 mL:

  Weigh out 11.69 g NaCl; add 10 mL 1 *M* Tris–HCl pH 7.5 and 5 mL 0.5 *M* EDTA pH 8 and adjust to 40 mL total with water. Add 10 mL 10% PVP40 to dissolve the NaCl.

  Filter-sterilize in the following order, adding each component after the previous has completely passed through the filter:

  10 mL 200 m*M* CoHex
  10 mL 10% PVP40
  50 mL NaCl/Tris/EDTA/PVP40 from above
  30 mL 10% CTAB

  Store at 37°C. If a precipitate forms, discard.
- DNase-free RNase, 10 mg/mL
- Proteinase K, 20 mg/mL in 10 m*M* Tris, 20 m*M* $CaCl_2$, 50% glycerol, pH 7.5. Store aliquots at −20°C
- Chloroform:isoamyl alcohol, 24:1
- CTAB dilution solution: 1% CTAB, 50 m*M* Tris, 1 m*M* EDTA, 4 m*M* CoHex, pH 7.5. Filter-sterilize and store at RT
- 0.4 *M* NaCl, 10 m*M* Tris, 1 m*M* EDTA, 1 m*M* CoHex, pH 7.5. Filter-sterilize and store at 4°C
- 1.42 *M* NaCl, 10 m*M* Tris, 1 m*M* EDTA, 1 m*M* CoHex, pH 7.5. Filter-sterilize and store at 4°C
- 95% ethanol
- 70% ethanol, 0.3 m*M* CoHex

- 70% ethanol, 3 m*M* $MgCl_2$
- 10 m*M* Tris, 10 m*M* $MgCl_2$, 200 m*M* NaCl, pH 7.5. Filter-sterilize and store at 4°C
- 10 m*M* Tris, 2 m*M* $MgCl_2$, 50 μ*M* spermidine, pH 7.5. Filter-sterilize and store at 4°C

3.2.3 Procedure

1. Place a 35-mL aliquot of sporulation or presporulation culture in a 50-mL conical tube and centrifuge 2 min at 3220 × *g*. Drain well, resuspend in 1.5 mL psoralen solution, made fresh the same day by adding 1 mL of 1 mg/mL psoralen stock to 9 mL of 50 m*M* Tris, 50 m*M* EDTA. Vortex the psoralen working solution before each use to maintain suspension.
2. Place cells in a 35 mm Petri dish on long-wave UV transilluminator. Cross-link 10 min, 0.6 mW/cm^2, agitating plates every 2 min to resuspend settled cells (see Note 8).
3. Transfer cells to 1.5-mL microfuge tubes, centrifuge 2 min, 1200 × *g*.
4. Remove supernatant and resuspend in storage solution. Recentrifuge as above, remove supernatant, freeze pellet on dry ice, and store at −80°C.
5. Thaw frozen pellet on ice. While pellets are thawing, prepare zymolyase solution: 0.5 mg/mL zymolyase 100T, 1% β-mercaptoethanol in spheroplasting solution. Keep on ice until use.
6. Resuspend cell pellet in 0.5 mL zymolyase solution and transfer to a 5-mL round-bottom tube.
7. Incubate at room temperature for 2.5–3.5 min, monitoring degree of spheroplasting by microscope (see Note 9).
8. When 80% of cells are spheroplasted (see Note 9), centrifuge 2 min at 3200 × *g* at room temperature. Thoroughly remove supernatant. Gently resuspend cell pellet in 0.5 mL CTAB extraction solution by flicking or by pipetting up and down. Add 12.5 μL proteinase K solution, mix gently. Add 1 μL RNase solution, mix gently.
9. Incubate at 37°C for 15 min, mixing gently every 5 min.
10. Transfer to a 1.5-mL microfuge tube that contains 0.3 mL chloroform:isoamyl alcohol. Vortex vigorously for 10 s, shake, and vortex again for 5 s. Leave at room temperature for 2 min and vortex again for 5 s.
11. Centrifuge at top speed in microfuge for 3 min. Promptly remove upper aqueous layer, avoiding white interface layer, and transfer to a 5-mL round-bottom tube.

12. Layer 1.5 mL CTAB dilution solution on top, mix by inverting five times and then let sit at room temperature for 10 min. Invert again 20 times, until a precipitate forms. If a precipitate does not form, add another 0.5 mL CTAB dilution solution and repeat mixing and 10 min rest. The precipitate should sink to the bottom of the tube.
13. Using a fine-tip transfer pipette, carefully remove the supernatant. If the precipitate is loose and not all supernatant can be removed, pulse-spin *very briefly* in the microfuge to make the precipitate settle to the bottom of the tube. Immediately add 2 mL ice-cold 0.4 *M* NaCl, 10 m*M* Tris, 1 m*M* EDTA, 1 m*M* CoHex. Mix briefly until the pellet floats. Carefully remove the supernatant, avoiding drying the pellet. Repeat this wash.
14. Promptly add 0.5 mL ice-cold 1.42 *M* NaCl, 10 m*M* Tris, 1 m*M* EDTA, 1 m*M* CoHex, and shake tube gently until the pellet becomes translucent.
15. Transfer to a prechilled 1.5-mL microfuge tube and add 1 mL 95% ethanol. Invert gently until a precipitate forms. Allow precipitate to settle to the bottom of the tube. Pulse-spin if necessary. Remove supernatant with a fine-tip transfer pipette, add 1 mL 70% ethanol, 0.3 m*M* CoHex. Vortex briefly until pellet floats, then let settle. Pulse-spin if necessary. Remove supernatant.
16. If white CTAB particles are visible, repeat steps 14 and 15.
17. Completely remove ethanol and allow to air-dry briefly. The pellet should still be damp; overdrying will make resuspending difficult.
18. Add 0.1 mL ice-cold 10 m*M* Tris, 10 m*M* $MgCl_2$, 200 m*M* NaCl, soak pellet on ice for at least 30 min. Add 0.2 mL 95% ethanol, invert gently until precipitate forms. Allow precipitate to settle, pulse-spinning if necessary. Remove supernatant, wash pellet twice with 0.2 mL 70% ethanol, 3 m*M* $MgCl_2$ as in step 15.
19. Remove ethanol completely and air-dry briefly as in step 17.
20. Resuspend pellet in 50–100 μL 10 m*M* Tris, 2 m*M* $MgCl_2$, 50 μ*M* spermidine (see Note 10).

3.3 Protein Isolation and Western Blotting

Monitoring target protein levels during meiosis is important to determine the extent of depletion or induction. Western blots are used, probing either with antibodies specific to the protein of interest or using epitope-tagged constructs and probing with antiepitope antibodies. For Sgs1, Top3, and

Rmi1, we use N- or C-terminal 3xHA tags, detected with anti-HA mouse monoclonal antibody (clone 12CA5, Roche). Polyclonal antisera are used to detect Arp7 (sc-8961, Santa Cruz Biotechnology) as a loading control. Blots are either probed first with the protein of interest, stripped, and reprobed with anti-Arp7, or separate gels are run with identical loadings of the same sample and are separately probed. To allow gel-to-gel comparisons, a common sample of the same or related strain expressing the tagged protein (usually a mitotic culture) should be loaded on every gel.

Proteolysis is a major concern in isolating nuclear proteins from meiotic cells, due to the tight vacuole-nuclear association that forms during meiosis (Pan et al., 2000). The protocol below, adapted from Foiani, Marini, Gamba, Lucchini, and Plevani (1994), uses trichloroacetic acid (TCA) pre-treatment and lysis in the presence of TCA to minimize proteolysis.

3.3.1 Equipment

- High-capacity tabletop or floor centrifuge with adaptors for 5-mL round-bottom tubes
- Variable-speed microfuge
- 5-mL round-bottom polypropylene tubes (Falcon 352063 or equivalent)
- 1.5-mL screw-top microfuge tubes
- Acid-washed glass beads, 425–600 μm (Sigma G8772)
- Benchtop votexer
- Mini-Beadbeater (Biospec) or equivalent
- Block-type incubator or water bath, set to 95°C
- Vertical minigel box and transfer apparatus (XCel Mini-Cell and XCell II™ Blot Module, Thermo Fisher Scientific)
- Gel cassettes (Life Technologies NC2010)
- Electrophoresis and transfer apparatus (XCell II Blot Module, Invitrogen)
- Orbital platform shaker
- Invitrolon™ PVDF/Filter paper sandwich, 0.4 μm pore, 8.3 × 7.3 cm (Thermo Fisher Scientific, LC2005)
- CCD camera or scanning fluorimeter

3.3.2 Reagents

- 10% glycerol
- 20% and 5% TCA
- 1.5 *M* Tris–HCl, pH 8.8

- Sample buffer: 0.1 *M* Tris–HCl, 0.2% SDS, 5% β-mercaptoethanol, 10% glycerol, 0.01% bromophenol blue
- Protogel 29:1 acrylamide:bisacrylamide mix (National Diagnostics)
- Tetramethylethylenediamine (TEMED)
- 2 × Stacking buffer: 0.125 M Tris–HCl, 0.2% SDS, pH 6.8
- 2 × Resolving buffer: 0.75 M Tris–HCl, 0.2% SDS, pH 8.8
- 10% ammonium persulfate
- Precast 8%–16% Tris–Glycine Mini Gels, 12 well (Thermo Fisher Scientific, XP08162BOX)
- Running buffer: 0.192 *M* glycine, 25 m*M* Tris–HCl, 0.1%SDS, pH 8.3
- Transfer buffer: 0.096 *M* glycine, 12.5 m*M* Tris–HCl, 0.1% SDS, 20% methanol, pH 8.3. Chill to 4°C before use.
- Phosphate-buffered saline (PBS): 137 m*M* NaCl, 2.7 m*M* KCl, 10 m*M* Na_2HPO_4, 1.8 m*M* KH_2PO_4, pH 7.4.
- Biotin-free milk powder (I-Block™, Thermo Fisher Scientific, T2015)
- Tween 20
- Anti-HA mouse monoclonal antibody (clone 12CA5, Roche; 11583816001, used at 1:10,000 dilution)
- Anti-Arp7 goat polyclonal antibody (Santa Cruz Biotechnology, sc-8961, used at 1:1000 dilution)
- Alkaline phosphatase-conjugated goat-anti-mouse (Sigma A3562, used at 1:5000 dilution)
- Alkaline phosphatase-conjugated rabbit-anti-goat (Sigma A4187, used at 1:5000 dilution)
- Blocking buffer: PBS + 2 mg/mL biotin-free milk powder (I-block), 0.05% Tween 20. Add milk powder slowly to PBS until completely dissolved and then add Tween 20.
- Alkaline solution: 50 m*M* Tris–HCl, 100 m*M* NaCl, 1 m*M* $MgCl_2$, pH 9.5
- CDP-Star Chemiluminescent Substrate, 0.25 m*M* (Thermo Fisher Scientific T2147)
- Restore PLUS Western Blot Stripping Buffer (Thermo Fisher Scientific 46430)

3.3.3 Procedure

3.3.3.1 Sample Preparation

1. Collect 3 mL samples from a sporulating culture, centrifuge 2 min at 3220 × *g*, and drain. Resuspend in 1 mL 10% glycerol, transfer to 1.5 mL screw-top microfuge, and centrifuge in microfuge 2 min at top speed, drain, freeze on dry ice, and store at −80°C.

2. Thaw the pellet on ice, add 100 µL of 20% TCA, and vortex to resuspend.
3. Add glass beads such that there will be no free liquid on the top of the beads. Disrupt cells by shaking continuously on a Bead beater (preferred) or vortex at highest speed for 2 min.
4. Add 400 µL of 5% TCA to the broken cells + beads, vortex, and transfer the cloudy supernatant to a fresh 1.5-mL screw cap tube.
5. Add 200 µL of 5% TCA to beads, vortex, recover the supernatant, and add it to the previous supernatant.
6. Centrifuge the combined supernatants 5 min at top speed in the microfuge. Discard the clear supernatant, removing as much as is possible.
7. Add 80µL of sample buffer to the pellet and vortex vigorously. Much of the pellet is composed of cellular debris that will not go back into solution. If the sample turns yellow, add 5–20 µL Tris 1.5 *M* pH 8.8, until the blue color is restored.
8. Heat the sample at 95°C for 5 min. Centrifuge 3 min at top speed in the microfuge. Transfer supernatant containing the SDS-solubilized proteins to a new 1.5 mL screw cap tube.
9. Typically, 7.5 µL of extract is loaded per lane.

3.3.3.2 Gel Preparation and Electrophoresis

We use 6% gels to detect Sgs1, 10% gels to detect all other proteins. Alternatively, precast 8%–16% gradient gels can be used, but resolution will be somewhat compromised.

1. For a 10% 10 mL gel, mix 5 mL 2 × resolving buffer, 3.4 mL Protogel mix, 1.4 mL water, 150 µL 10% ammonium persulfate, 15 µL TEMED. For a 6% gel, 2 mL Protogel and 2.8 mL water are used. Pour acrylamide mix up to the last line on the cassette and immediately overlay with 100 µL of water. Allow to polymerize at least 30 min.
2. Prepare 5% stacking gel: mix 2.5 mL stacking buffer, 0.83 mL Protogel mix, 1.6 mL water, 75 µL 10% ammonium persulfate, 7.5 µL TEMED. Pour water off resolving gel and add stacking gel. Insert comb and allow 30 min to polymerize.
3. Remove comb and sealing tape and place gel in gel box. Add running buffer to cover wells, approximately 500 mL.
4. Heat samples to 95°C, 3 min. Load 5–7.5 µL sample per well.
5. Run gels at 180 V, room temperature, 1–1.5 h, until tracking dye reaches the bottom of the gel.

3.3.3.3 Western Blotting and Detection

1. Disassemble gel cassette; if stacking gels are used, remove the stacking layer. Place gel in 10 mL transfer buffer on orbital shaker, 10 min at 50 rpm.
2. Soak PVDF membrane in 10 mL 100% methanol for 1 min. Wash in 20 mL water on orbital shaker, 2 min at 50 rpm, and then place in 10 mL transfer buffer.
3. Soak filter paper from the membrane/filter kit and scotch pads from the transfer apparatus in transfer buffer in a glass beaker (or equivalent), using enough transfer buffer to completely submerse them.
4. Place two or three soaked scotch pads on the cathode side of the transfer apparatus, followed by the presoaked filter paper. Pour some buffer on the filter paper and place the gel over it. Pour some buffer on the gel, and place the PVDF membrane on it, followed by presoaked filter paper. Remove all air bubbles from the assembly and place two or three soaked scotch pads on top. Close the apparatus with the anode and place it in the gel tank. Pour transfer buffer inside the transfer apparatus until it covers the transfer sandwich. The outer chamber can be filled with either buffer or water.
5. Transfer at 15 V, 16 h at 4°C (for Sgs1) or at 25 V, 1.5 h at room temperature (for other proteins).
6. Disassemble transfer apparatus, place the membrane in 20 mL of I-block blocking buffer and incubate for at least 1 h at room temperature or overnight at 4°C on an orbital shaker.
7. Discard blocking buffer and add 20 mL of blocking buffer with the appropriate dilution of primary antibody. Incubate at least 3 h at room temperature or overnight at 4°C on an orbital shaker (recommended for Sgs1).
8. Discard antibody solution, add 20 mL blocking buffer, and wash 5 min at room temperature on an orbital shaker. Repeat for five additional washes.
9. Add 20 mL of blocking buffer with the appropriate dilution of secondary antibody and incubate 1 h at room temperature on an orbital shaker.
10. Discard secondary antibody solution and wash as in step 8.
11. Place the membrane in 10 mL alkaline buffer for 2 min. Remove the membrane, drain excess solution and place into 3 mL of CDP-star, making sure that membrane is complexly coated. Remove membrane, drain excess substrate, and wrap in plastic wrap or a developing folder.

12. Scan using a Fuji LAS3000 CCD camera or equivalent.
13. To strip blots, place membrane in 20 mL blocking buffer, wash 10 min at room temperature on an orbital shaker. Washed blots can be stored at 4°C in blocking buffer before stripping.
14. Remove blocking buffer, add 15 mL stripping buffer, and incubate at room temperature 15 min. Remove blot from stripping buffer and wash with 20 mL PBS, 1 min at room temperature at least three times. Wash the blot once with 20 mL blocking buffer, 1 min at room temperature. Remove the blot and block in 20 mL blocking buffer for at least 1 h. To reprobe, repeat steps 7–12.

3.4 Monitoring Nuclear Divisions by DAPI Staining

3.4.1 Equipment

- Variable-speed microfuge
- 1.5-mL microfuge tubes
- Prewashed glass slides
- Cover slips, 18 × 18 mm or 24 × 50 mm
- Epifluorescence microscope with blocks and filters for DAPI detection, with 60× or 100× objective

3.4.2 Reagents

- 95% ethanol
- 1 μg/mL 4′,6-diamidino-2-phenylindole dihydrochloride (DAPI). Store at 4°C in the dark

3.4.3 Procedure

1. Place a 1 mL culture sample in a microfuge tube and centrifuge 2 min at top speed.
2. Discard supernatant and resuspend cells in 50 μL water. Add 1 mL 95% ethanol (see Note 12).
3. Add 1 μL of 1 μg/mL DAPI. Incubate at room temperature, 10 min, protected from light.
4. Centrifuge 2 min at top speed; discard supernatant and resuspend in 1 mL water. Centrifuge 1 min at 7000 rpm; discard supernatant and repeat 1 mL water wash.
5. Discard supernatant, resuspend in 50 μL water. Store protected from light at 4°C.
6. Place 5 μL of cells on glass slide, cover with cover slip, and visualize using microscope (see Note 13).

4. NOTES

1. To prevent arcing, all DNA should be in either TE or nuclease-free water. In order to minimize sorbitol dilution, DNA should be as concentrated as possible. Briefly wiping electroporation cuvettes to remove condensation will also minimize arcing.
2. Transformant yields can be reduced if cells are plated at too high a concentration. We recommend spreading two selective plates with 100 μL of the transformation mix, and as many selective plates with 20 μL as convenient.
3. Use of a shaking incubator with cooling capacity (New Brunswick Innova 43 or equivalent) is recommended; incubators with limited ventilation or cooling can overheat.
4. Because light scattering, not absorbance, is measured here, a spectrophotometer with some distance between the cuvette and the detector should be used. We use a Pharmacia Ultrospec 2000, with ~7 cm between the cuvette and the detector.
5. All presporulation and sporulation media should be made fresh, preferably on the day before sporulation.
6. To minimize experiment-to-experiment variation in timing of events during sporulation, adherence to a fixed schedule, starting when diploids are struck on YEPAD plates, is highly recommended. For example, one such routine would streak diploids at 3:00 pm, incubate plates for exactly 48 h, inoculate YEPAD overnights at 3:00 pm, aerate exactly 24 h, and inoculate SPS presporulation cultures at 3:00 pm.
7. Psoralen is toxic; use appropriate precautions when handling.
8. Exposure time should be adjusted so that total exposure is 6 mW min/cm^2 (0.36 J/cm^2).
9. Earlier time points take less time to spheroplast than later time points, so the spheroplasting state should be monitored frequently. Cells are sufficiently spheroplated when 80%–90% of cells are phase dark; spheroplasting can also be checked by mixing cells with 20% SDS and examining their lysis under the microscope.
10. DNA pellets can be difficult to resuspend. We usually incubate pellets with buffer overnight at 4°C, followed by gentle mixing by gently pipetting the resuspension mix up and down ~20 times.
11. If nonfat milk powder is added too rapidly, it will form clumps and never dissolve. If blocking solution is urgently needed, the PBS solution can be warmed by microwaving (~1 min) beforehand.

12. Stain ethanol-fixed cells with DAPI and wash as soon as possible (i.e., the same day), as prolonged incubation in ethanol leads to nucleus distortion and mislocalization. Formaldehyde fixation followed by a brief ethanol permeabilization may be useful if it is impossible to stain cells on the same day.
13. To maximize the fraction of cells with nuclei in the same focal plane, use a small volume (~5 μL) of cell suspension and a large (20 × 40 mm) cover slip.

ACKNOWLEDGMENTS

All plasmids, plasmid sequences, and sequences of primers mentioned in this chapter are available from the authors upon request. We thank Angelika Amon and Masato Kanemaki for plasmids, and Anura Shodhan and Matan Cohen for comments. This work was supported by the Intramural Research Program of the NIH through the Center for Cancer Research at the National Cancer Institute.

REFERENCES

Ahuja, J. S., & Börner, G. V. (2011). Analysis of meiotic recombination intermediates by two-dimensional gel electrophoresis. *Methods in Molecular Biology (Clifton, NJ)*, *745*, 99–116 [Chapter 7]. https://doi.org/10.1007/978-1-61779-129-1_7.

Alani, E., Cao, L., & Kleckner, N. (1987). A method for gene disruption that allows repeated use of *URA3* selection in the construction of multiply disrupted yeast strains. *Genetics*, *116*(4), 541–545.

Allers, T., & Lichten, M. (2000). A method for preparing genomic DNA that restrains branch migration of Holliday junctions. *Nucleic Acids Research*, *28*(2), e6.

Bizzari, F., & Marston, A. L. (2011). Cdc55 coordinates spindle assembly and chromosome disjunction during meiosis. *The Journal of Cell Biology*, *193*(7), 1213–1228. https://doi.org/10.1083/jcb.201103076.

Blitzblau, H. G., & Hochwagen, A. (2013). ATR/Mec1 prevents lethal meiotic recombination initiation on partially replicated chromosomes in budding yeast. *eLife*, *2*, e00844. https://doi.org/10.7554/eLife.00844.018.

Börner, G. V., & Cha, R. S. (2015). Induction and analysis of synchronous meiotic yeast cultures. *Cold Spring Harbor Protocols*, *2015*(10), 908–913. https://doi.org/10.1101/pdb.prot085035.

Börner, G. V., Kleckner, N., & Hunter, N. (2004). Crossover/noncrossover differentiation, synaptonemal complex formation, and regulatory surveillance at the leptotene/zygotene transition of meiosis. *Cell*, *117*(1), 29–45.

Brito, I. L., Yu, H.-G., & Amon, A. (2010). Condensins promote coorientation of sister chromatids during meiosis I in budding yeast. *Genetics*, *185*(1), 55–64. https://doi.org/10.1534/genetics.110.115139.

Chan, A. C., Borts, R. H., & Hoffmann, E. (2009). Temperature-dependent modulation of chromosome segregation in *msh4* mutants of budding yeast. *PLoS One*, *4*(10), e7284. https://doi.org/10.1371/journal.pone.0007284.

Cho, R. J., Campbell, M. J., Winzeler, E. A., Steinmetz, L., Conway, A., Wodicka, L., et al. (1998). A genome-wide transcriptional analysis of the mitotic cell cycle. *Molecular Cell*, *2*(1), 65–73.

Chu, S., DeRisi, J., Eisen, M., Mulholland, J., Botstein, D., Brown, P. O., et al. (1998). The transcriptional program of sporulation in budding yeast. *Science (New York, NY)*, *282*(5389), 699–705.

Clyne, R. K., Katis, V. L., Jessop, L., Benjamin, K. R., Herskowitz, I., Lichten, M., et al. (2003). Polo-like kinase Cdc5 promotes chiasmata formation and cosegregation of sister centromeres at meiosis I. *Nature Cell Biology*, *5*(5), 480–485. https://doi.org/10.1038/ncb977.

Copsey, A., Tang, S., Jordan, P. W., Blitzblau, H. G., Newcombe, S., Chan, A. C.-H., et al. (2013). Smc5/6 coordinates formation and resolution of joint molecules with chromosome morphology to ensure meiotic divisions. *PLoS Genetics*, *9*(12), e1004071 https://doi.org/10.1371/journal.pgen.1004071.

De Muyt, A., Jessop, L., Kolar, E., Sourirajan, A., Chen, J., Dayani, Y., et al. (2012). BLM helicase ortholog Sgs1 is a central regulator of meiotic recombination intermediate metabolism. *Molecular Cell*, *46*(1), 43–53. https://doi.org/10.1016/j.molcel.2012.02.020.

Foiani, M., Marini, F., Gamba, D., Lucchini, G., & Plevani, P. (1994). The B subunit of the DNA polymerase alpha-primase complex in *Saccharomyces cerevisiae* executes an essential function at the initial stage of DNA replication. *Molecular and Cellular Biology*, *14*(2), 923–933.

Gangloff, S., McDonald, J. P., Bendixen, C., Arthur, L., & Rothstein, R. (1994). The yeast type I topoisomerase Top3 interacts with Sgs1, a DNA helicase homolog: A potential eukaryotic reverse gyrase. *Molecular and Cellular Biology*, *14*(12), 8391–8398.

Gietz, R. D., & Woods, R. A. (2001). Genetic transformation of yeast. *BioTechniques*, *30*, 816–831.

Goldstein, A. L., & McCusker, J. H. (1999). Three new dominant drug resistance cassettes for gene disruption in *Saccharomyces cerevisiae*. *Yeast (Chichester, England)*, *15*(14), 1541–1553. https://doi.org/10.1002/(SICI)1097-0061(199910)15:14<1541::AID-YEA476>3.0.CO;2-K.

Grandin, N., & Reed, S. I. (1993). Differential function and expression of *Saccharomyces cerevisiae* B-type cyclins in mitosis and meiosis. *Molecular and Cellular Biology*, *13*(4), 2113–2125.

Gray, S., Allison, R. M., Garcia, V., Goldman, A. S. H., & Neale, M. J. (2013). Positive regulation of meiotic DNA double-strand break formation by activation of the DNA damage checkpoint kinase Mec1(ATR). *Open Biology*, *3*(7), 130019. https://doi.org/10.1098/rsob.130019.

Haruki, H., Nishikawa, J., & Laemmli, U. K. (2008). The anchor-away technique: Rapid, conditional establishment of yeast mutant phenotypes. *Molecular Cell*, *31*(6), 925–932. https://doi.org/10.1016/j.molcel.2008.07.020.

Honigberg, S. M., & Esposito, R. E. (1994). Reversal of cell determination in yeast meiosis: Postcommitment arrest allows return to mitotic growth. *Proceedings of the National Academy of Sciences of the United States of America*, *91*(14), 6559–6563.

Jessop, L., & Lichten, M. (2008). Mus81/Mms4 endonuclease and Sgs1 helicase collaborate to ensure proper recombination intermediate metabolism during meiosis. *Molecular Cell*, *31*(3), 313–323. https://doi.org/10.1016/j.molcel.2008.05.021.

Jessop, L., Rockmill, B., Roeder, G. S., & Lichten, M. (2006). Meiotic chromosome synapsis-promoting proteins antagonize the anti-crossover activity of sgs1. *PLoS Genetics*, *2*(9), e155 https://doi.org/10.1371/journal.pgen.0020155.

Jin, H., Guacci, V., & Yu, H.-G. (2009). Pds5 is required for homologue pairing and inhibits synapsis of sister chromatids during yeast meiosis. *The Journal of Cell Biology*, *186*(5), 713–725. https://doi.org/10.1083/jcb.200810107.

Kane, S. M., & Roth, R. (1974). Carbohydrate metabolism during ascospore development in yeast. *Journal of Bacteriology*, *118*(1), 8–14.

Katis, V. L., Lipp, J. J., Imre, R., Bogdanova, A., Okaz, E., Habermann, B., et al. (2010). Rec8 phosphorylation by casein kinase 1 and Cdc7-Dbf4 kinase regulates cohesin cleavage by separase during meiosis. *Developmental Cell*, *18*(3), 397–409. https://doi.org/10.1016/j.devcel.2010.01.014.

Kaur, H., De Muyt, A., & Lichten, M. (2015). Top3-Rmi1 DNA single-strand decatenase is integral to the formation and resolution of meiotic recombination intermediates. *Molecular Cell*, *57*(4), 583–594. https://doi.org/10.1016/j.molcel.2015.01.020.

Kubota, T., Nishimura, K., Kanemaki, M. T., & Donaldson, A. D. (2013). The Elg1 replication factor C-like complex functions in PCNA unloading during DNA replication. *Molecular Cell*, *50*(2), 273–280.

Kunes, S., Ma, H., Overbye, K., Fox, M. S., & Botstein, D. (1987). Fine structure recombinational analysis of cloned genes using yeast transformation. *Genetics*, *115*(1), 73–81.

Larsen, N. B., & Hickson, I. D. (2013). *RecQ helicases: Conserved guardians of genomic integrity. In M. Spies (Ed.), DNA helicases and DNA motor proteins: Vol. 973* (pp. 161–184). New York, NY: Springer, New York. https://doi.org/10.1007/978-1-4614-5037-5_8.

Lee, B. H., & Amon, A. (2003). Role of polo-like kinase *CDC5* in programming meiosis I chromosome segregation. *Science (New York, NY)*, *300*(5618), 482–486. https://doi.org/10.1126/science.1081846.

Leung, W.-K., Humphryes, N., Afshar, N., Argunhan, B., Terentyev, Y., Tsubouchi, T., et al. (2015). The synaptonemal complex is assembled by a polySUMOylation-driven feedback mechanism in yeast. *The Journal of Cell Biology*, *211*(4), 785–793. https://doi.org/10.1083/jcb.201506103.

Li, P., Shao, Y., Jin, H., & Yu, H.-G. (2015). Ndj1, a telomere-associated protein, regulates centrosome separation in budding yeast meiosis. *The Journal of Cell Biology*, *209*, 247–259.

Lichten, M. (2014). Tetrad, random spore, and molecular analysis of meiotic segregation and recombination. *Methods in Molecular Biology (Clifton, NJ)*, *1205*, 13–28. https://doi.org/10.1007/978-1-4939-1363-3_2.

Lin, W., Jin, H., Liu, X., Hampton, K., & Yu, H.-G. (2011). Scc2 regulates gene expression by recruiting cohesin to the chromosome as a transcriptional activator during yeast meiosis. *Molecular Biology of the Cell*, *22*(12), 1985–1996. https://doi.org/10.1091/mbc.E10-06-0545.

Lin, W., Wang, M., Jin, H., & Yu, H.-G. (2011). Cohesin plays a dual role in gene regulation and sister-chromatid cohesion during meiosis in *Saccharomyces cerevisiae*. *Genetics*, *187*(4), 1041–1051. https://doi.org/10.1534/genetics.110.122358.

Longtine, M. S., McKenzie, A., Demarini, D. J., Shah, N. G., Wach, A., Brachat, A., et al. (1998). Additional modules for versatile and economical PCR-based gene deletion and modification in *Saccharomyces cerevisiae*. *Yeast (Chichester, England)*, *14*(10), 953–961. https://doi.org/10.1002/(SICI)1097-0061(199807)14:10<953::AID-YEA293>3.0.CO;2-U.

Manivasakam, P., Weber, S. C., McElver, J., & Schiestl, R. H. (1995). Micro-homology mediated PCR targeting in *Saccharomyces cerevisiae*. *Nucleic Acids Research*, *23*(14), 2799–2800. https://doi.org/10.1093/nar/23.14.2799.

McCusker, J. H., & Haber, J. E. (1977). Efficient sporulation of yeast in media buffered near pH 6. *Journal of Bacteriology*, *132*(1), 180–185.

Mullen, J. R., Kaliraman, V., Ibrahim, S. S., & Brill, S. J. (2001). Requirement for three novel protein complexes in the absence of the Sgs1 DNA helicase in *Saccharomyces cerevisiae*. *Genetics*, *157*(1), 103–118.

Nishimura, K., Fukagawa, T., Takisawa, H., Kakimoto, T., & Kanemaki, M. (2009). An auxin-based degron system for the rapid depletion of proteins in nonplant cells. *Nature Methods*, *6*(12), 917–922. https://doi.org/10.1038/nmeth.1401.

Nishimura, K., & Kanemaki, M. T. (2014). Rapid depletion of budding yeast proteins via the fusion of an auxin-inducible degron (AID). *Current Protocols in Cell Biology*, *64*, 20.9.1–16. https://doi.org/10.1002/0471143030.cb2009s64.
Oh, S. D., Jessop, L., Lao, J. P., Allers, T., Lichten, M., & Hunter, N. (2009). Stabilization and electrophoretic analysis of meiotic recombination intermediates in *Saccharomyces cerevisiae*. *Methods in Molecular Biology (Clifton, NJ)*, *557*, 209–234. https://doi.org/10.1007/978-1-59745-527-5_14.
Oh, S. D., Lao, J. P., Hwang, P. Y., Taylor, A. F., Smith, G. R., & Hunter, N. (2007). BLM ortholog, Sgs1, prevents aberrant crossing-over by suppressing formation of multichromatid joint molecules. *Cell*, *130*(2), 259–272. https://doi.org/10.1016/j.cell.2007.05.035.
Oh, S. D., Lao, J. P., Taylor, A. F., Smith, G. R., & Hunter, N. (2008). RecQ helicase, Sgs1, and XPF family endonuclease, Mus81-Mms4, resolve aberrant joint molecules during meiotic recombination. *Molecular Cell*, *31*(3), 324–336. https://doi.org/10.1016/j.molcel.2008.07.006.
Pan, X., Roberts, P., Chen, Y., Kvam, E., Shulga, N., Huang, K., et al. (2000). Nucleus-vacuole junctions in *Saccharomyces cerevisiae* are formed through the direct interaction of Vac8p with Nvj1p. *Molecular Biology of the Cell*, *11*(7), 2445–2457.
Primig, M., Williams, R. M., Winzeler, E. A., Tevzadze, G. G., Conway, A. R., Hwang, S. Y., et al. (2000). The core meiotic transcriptome in budding yeasts. *Nature Genetics*, *26*(4), 415–423. https://doi.org/10.1038/82539.
Rockmill, B., Fung, J. C., Branda, S. S., & Roeder, G. S. (2003). The Sgs1 helicase regulates chromosome synapsis and meiotic crossing over. *Current Biology: CB*, *13*(22), 1954–1962.
Schwacha, A., & Kleckner, N. (1994). Identification of joint molecules that form frequently between homologs but rarely between sister chromatids during yeast meiosis. *Cell*, *76*(1), 51–63.
Subramanian, V. V., Macqueen, A. J., Vader, G., Shinohara, M., Sanchez, A., Borde, V., et al. (2016). Chromosome synapsis alleviates Mek1-dependent suppression of meiotic DNA repair. *PLoS Biology*, *14*(2), e1002369 https://doi.org/10.1371/journal.pbio.1002369.
Tang, S., Wu, M. K., Zhang, R., & Hunter, N. (2015). Pervasive and essential roles of the Top3-Rmi1 decatenase orchestrate recombination and facilitate chromosome segregation in meiosis. *Molecular Cell*, *57*(4), 607–621. https://doi.org/10.1016/j.molcel.2015.01.021.
Vincenten, N., Kuhl, L.-M., Lam, I., Oke, A., Kerr, A. R., Hochwagen, A., et al. (2015). The kinetochore prevents centromere-proximal crossover recombination during meiosis. *eLife*, *4*, 923. https://doi.org/10.7554/eLife.10850.
Wach, A. (1996). PCR-synthesis of marker cassettes with long flanking homology regions for gene disruptions in *S. cerevisiae*. *Yeast (Chichester, England)*, *12*(3), 259–265. https://doi.org/10.1002/(SICI)1097-0061(19960315)12:3<259::AID-YEA901>3.0.CO;2-C.
Yu, H.-G., & Koshland, D. (2007). The aurora kinase Ipl1 maintains the centromeric localization of PP2A to protect cohesin during meiosis. *The Journal of Cell Biology*, *176*(7), 911–918. https://doi.org/10.1083/jcb.200609153.
Zhang, L., Wang, S., Yin, S., Hong, S., Kim, K. P., & Kleckner, N. (2014). Topoisomerase II mediates meiotic crossover interference. *Nature*, *511*(7511), 551–556. https://doi.org/10.1038/nature13442.

CHAPTER FIFTEEN

Quantitative Methods to Investigate the 4D Dynamics of Heterochromatic Repair Sites in *Drosophila* Cells

Christopher P. Caridi, Laetitia Delabaere, Harianto Tjong[2], Hannah Hopp[3], Devika Das[4], Frank Alber, Irene Chiolo[1]
University of Southern California, Los Angeles, CA, United States
[1]Corresponding author: e-mail address: chiolo@usc.edu

Contents

Abstract

Heterochromatin is mostly composed of long stretches of repeated DNA sequences prone to ectopic recombination during double-strand break (DSB) repair. In *Drosophila*, "safe" homologous recombination (HR) repair of heterochromatic DSBs relies on a striking relocalization of repair sites to the nuclear periphery. Central to understanding heterochromatin repair is the ability to investigate the 4D dynamics (movement in space and time) of repair sites. A specific challenge of these studies is preventing phototoxicity and photobleaching effects while imaging the sample over long periods of time, and with sufficient time points and *Z*-stacks to track repair foci over time. Here we describe an optimized approach for high-resolution live imaging of heterochromatic DSBs in *Drosophila* cells, with a specific emphasis on the fluorescent markers and imaging setup

[2] Present address: The Jackson Laboratory, Farmington, CT, United States.
[3] Present address: Keck School of Medicine, University of Southern California, Los Angeles, CA, United States.
[4] Present address: Mayo Clinic School of Medicine, AZ, United States.

Methods in Enzymology, Volume 601
ISSN 0076-6879
https://doi.org/10.1016/bs.mie.2017.11.033

used to capture the motion of repair foci over long-time periods. We detail approaches that minimize photobleaching and phototoxicity with a DeltaVision widefield deconvolution microscope, and image processing techniques for signal recovery post-imaging using SoftWorX and Imaris software. We present a method to derive mean square displacement curves revealing some of the biophysical properties of the motion. Finally, we describe a method in R to identify tracts of directed motions (DMs) in mixed trajectories. These approaches enable a deeper understanding of the mechanisms of heterochromatin dynamics and genome stability in the three-dimensional context of the nucleus and have broad applicability in the field of nuclear dynamics.

1. NUCLEAR DYNAMICS PLAY CRITICAL ROLES IN HETEROCHROMATIN REPAIR AND GENOME STABILITY

Double-strand breaks (DSBs) in pericentromeric heterochromatin (hereafter, "heterochromatin") are a major threat to genome stability (recently reviewed in Amaral, Ryu, Li, & Chiolo, 2017; Caridi, Delabaere, Zapotoczny, & Chiolo, 2017; Chiolo, Tang, Georgescu, & Costes, 2013). Heterochromatin comprises ~30% of fly and human genomes (Ho et al., 2014; Hoskins et al., 2007, 2015; Fig. 1A) and is mostly composed of repeated sequences prone to ectopic recombination during DNA repair (Amaral et al., 2017; Caridi et al., 2017; Chiolo et al., 2013; Peng & Karpen, 2008). In *Drosophila*, for example, about half of these sequences consist of simple "satellite" repeats (mostly tandem 5-base pair sequences) repeated for hundreds of kilobases to megabases, while the rest are mostly composed of scrambled clusters of transposable elements and about 250 isolated genes (Ho et al., 2014; Hoskins et al., 2007, 2015). While heterochromatin is a major component of the genome in multicellular eukaryotes, it is absent in budding yeast.

Despite the risk of aberrant recombination, homologous recombination (HR) is largely utilized for repairing heterochromatic DSBs in both flies (Chiolo et al., 2011; Janssen et al., 2016; Ryu, Bonner, & Chiolo, 2016; Ryu et al., 2015) and mammalian cells (Beucher et al., 2009; Tsouroula et al., 2016), and studies from our lab and others identified specialized pathways that promote heterochromatin repair while preventing aberrant recombination (reviewed in Amaral et al., 2017; Caridi et al., 2017; Chiolo et al., 2013). Working with *Drosophila* cells, we discovered that "safe" HR repair relies on a striking relocalization of heterochromatic repair sites to the nuclear periphery (Caridi et al., 2018; Chiolo et al., 2011; Ryu et al., 2016, 2015). Repair starts inside the heterochromatin domain—a

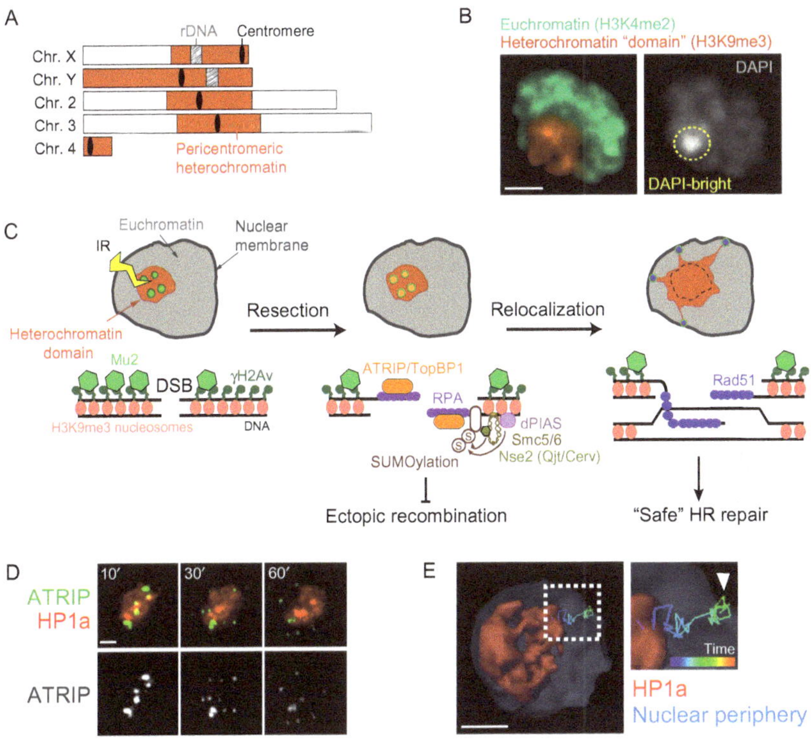

Fig. 1 Nuclear dynamics of heterochromatin repair mechanisms in *Drosophila* cells. (A) Schematic view of *Drosophila* chromosomes showing the position and extent of pericentromeric heterochromatin. (B) Immunofluorescence of a Kc cell showing heterochromatin organized as one distinct domain (*red*, H3K9me3) comprising the DAPI bright region (*yellow circle*). Euchromatin surrounds the heterochromatin domain (*green*, H3K4me2). Scale bar = 1 μm. (C) HR repair of heterochromatic DSBs starts inside the domain with H2Av phosphorylation and Mu2/Mdc1 recruitment. Resection also occurs inside the domain resulting in ATRIP and TopBP1 foci. Next, the heterochromatin domain expands and repair foci relocalize to the nuclear periphery to form Rad51 foci and continue HR. dPIAS, Nse2/Qjt, and Nse2/Cerv SUMO-ligases are required to block HR progression in the domain and prevent ectopic recombination. (D) Live imaging of one Kc cell expressing mCh-HP1a and EGFP-ATRIP shows ATRIP foci inside the heterochromatin domain at 10 min after IR, which leave the domain by 60 min after IR (Chiolo et al., 2011). (E) 4D reconstruction of one nucleus of a Kc cell expressing mCh-HP1a, mCh-LaminC, and GFP-ATRIP and, exposed to IR, shows an example of a heterochromatic repair focus that leaves the heterochromatin domain and reaches the nuclear periphery (Ryu et al., 2015).

distinct nuclear structure in flies (Chiolo et al., 2011; Riddle et al., 2011; Ryu et al., 2016, 2015; Fig. 1B)—and is temporarily halted after resection and ATRIP/TopBP1 focus formation (Chiolo et al., 2011; Ryu et al., 2016, 2015; Fig. 1C). This block to HR progression is mediated by SUMOylation via the SUMO E3 ligases dPIAS and Smc5/6 subunits Nse2/Qjt and Nse2/Cerv (Chiolo et al., 2011; Ryu et al., 2016, 2015; Fig. 1C). Next, the heterochromatin domain expands during relocalization (Chiolo et al., 2011). This likely reflects a global relaxation of the chromatin and facilitates damage signaling and/or nuclear dynamics. Repair continues with Rad51 recruitment at the nuclear periphery (Chiolo et al., 2011; Delabaere & Chiolo, 2016; Ryu et al., 2015; Fig. 1D), revealing a tight regulation of repair progression in space and time. Inactivating the relocalization pathway leads to aberrant recombination between heterochromatic sequences, chromosomal aberrations, and heterochromatin instability (Caridi et al., 2018; Chiolo et al., 2011; Ryu et al., 2016, 2015), revealing the importance of these dynamics for genome integrity. In mouse cells, where heterochromatin is organized in several "chromocenters," these domains expand during repair (Ayoub, Jeyasekharan, Bernal, & Venkitaraman, 2008; Burgess, Burman, Kruhlak, & Misteli, 2014; Tsouroula et al., 2016) and repair sites leave the domains before Rad51 recruitment and HR progression (Chiolo et al., 2013; Jakob et al., 2011; Tsouroula et al., 2016). These observations suggest highly conserved strategies for "safe" heterochromatin repair. Relocalization likely prevents ectopic recombination by isolating the damaged site and its homologous sequences (on the homologous chromosome or the sister chromatid) away from similar sequences on ectopic chromosomes, before strand invasion (Amaral et al., 2017; Caridi et al., 2017; Chiolo et al., 2011, 2013; Ryu et al., 2016, 2015).

In addition, a major body of work in yeast and mammalian cells revealed the dynamic nature of chromatin in both damaged and undamaged regions (reviewed in Amaral et al., 2017; Caridi et al., 2017). Repair sites are highly dynamic during homology search for HR repair (Cho, Dilley, Lampson, & Greenberg, 2014; Dion, Kalck, Horigome, Towbin, & Gasser, 2012; Miné-Hattab & Rothstein, 2012; Neumann et al., 2012; Saad et al., 2014) or during relocalization of relatively rare classes of DSBs to specific subnuclear compartments. For example, DSBs induced in rDNA of yeast and human cells leave the nucleolus during repair (Torres-Rosell et al., 2007; van Sluis & McStay, 2015). Further, persistent DSBs, eroded telomeres, subtelomeric breaks, and collapsed forks relocalize to the nuclear periphery for repair (Chung et al., 2015; Churikov et al., 2016; Horigome et al., 2014, 2016; Kalocsay, Hiller, & Jentsch, 2009; Khadaroo et al., 2009; Nagai et al., 2008; Oza, Jaspersen, Miele, Dekker, & Peterson, 2009; Su, Dion, Gasser, & Freudenreich, 2015; Swartz, Rodriguez, & King, 2014; Therizols et al., 2006; recently reviewed in Amaral et al., 2017; Caridi et al., 2017). Interestingly, SUMOylation appears to drive the relocalization of repair sites in different contexts, revealing conserved pathways for the regulation of nuclear dynamics during DSB repair (Churikov et al., 2016; Horigome et al., 2016; Kalocsay et al., 2009; Khadaroo et al., 2009; Nagai et al., 2008; Oza et al., 2009; Su et al., 2015; Torres-Rosell et al., 2007). Structural components in the cytoplasm appear to influence nuclear dynamics via SUN–KASH protein complexes traversing the nuclear membrane (Lottersberger, Karssemeijer, Dimitrova, & de Lange, 2015; Spichal et al., 2016). In addition, the entire genome becomes more dynamic in response to DSBs, albeit to a lesser extent relative to repair sites (Krawczyk et al., 2012; Lottersberger et al., 2015; Miné-Hattab & Rothstein, 2012; Seeber, Dion, & Gasser, 2013). This global mobilization of the genome correlates with increased susceptibility to micrococcal nuclease digestion in response to DSB induction (Ziv et al., 2006) and possibly results from a global decrease in histone levels following DNA damage (Hauer et al., 2017; Kruhlak et al., 2006), release of chromatin-associated components (Ziv et al., 2006), or loss of anchoring to nuclear structures (Agmon, Liefshitz, Zimmer, Fabre, & Kupiec, 2013; Dion, Kalck, Seeber, Schleker, & Gasser, 2013; Strecker et al., 2016).

Understanding the molecular mechanisms involved in the relocalization of DSBs in heterochromatin and other DNA sequences requires the ability to analyze the 4D dynamics of repair sites. This can be accomplished by live imaging of repair components, given that many repair factors form cytologically visible foci upon recruitment to DSBs (Costes, Chiolo, Pluth, Barcellos-Hoff, & Jakob, 2010; Delabaere et al., 2017; Haaf, Golub, Reddy, Radding, & Ward, 1995; Lisby, Barlow, Burgess, & Rothstein, 2004; Lisby & Rothstein, 2004; Liu, Li, Lee, & Maizels, 1999; Maser, Monsen, Nelms, & Petrini, 1997; Ryu et al., 2015; Scully et al., 1997). For example, the repair component Mu2/Mdc1 associates with the phosphorylated form of the histone variant H2Av (Chiolo et al., 2011; Dronamraju & Mason, 2009; Stucki et al., 2005) (γH2Av, corresponding to mammalian γH2AX (Fernandez-Capetillo, Lee, Nussenzweig, & Nussenzweig, 2004; Madigan, Chotkowski, & Glaser, 2002)), which marks DSBs, and mediates the recruitment of other HR proteins (Chapman & Jackson, 2008; Goldberg et al., 2003; Lou et al., 2006; Wang, Matsuoka, Carpenter, & Elledge, 2002). Thus, Mu2/Mdc1 foci can be used as a marker of repair sites throughout early and late steps of HR repair (Chiolo et al., 2011). ATRIP and TopBP1 are recruited to resected DSBs (Mordes, Glick, Zhao, & Cortez, 2008; Zou & Elledge, 2003); thus foci of these proteins mark resection (Chiolo et al., 2011). In late stages of HR, Rad51 promotes the search for a homologous template and strand invasion

(Sung, 1994), and Rad54 stabilizes Rad51-mediated strand invasion intermediates (Petukhova, Stratton, & Sung, 1998; Petukhova, Van Komen, Vergano, Klein, & Sung, 1999). GFP-tagging of these components can thus be used as a marker for later repair steps (Chiolo et al., 2011) (Fig. 1C).

To investigate the spatial and temporal dynamics of heterochromatin repair, mEGFP-tagged repair components are monitored relative to mCherry (mCh)-tagged HP1a as a marker for the heterochromatin domain. Early studies revealed that Mu2/Mdc1, ATRIP, and TopBP1 foci form inside the heterochromatin domain (Fig. 1D; Chiolo et al., 2011), while Rad51 and Rad54 foci form after relocalization (Chiolo et al., 2011). Repair foci leave the domain primarily between 10 and 30 min after DSB induction with ionizing radiation (IR) (Caridi et al., 2018; Chiolo et al., 2011; Ryu et al., 2016, 2015), and several foci reach the nuclear periphery in the first hour after IR (Caridi et al., 2018; Ryu et al., 2015) (Fig. 1E). Thus, investigating the 4D dynamics of heterochromatic repair foci requires live-cell imaging for at least 1 h after IR. A specific challenge of these studies is imaging the sample over long periods of time and with sufficient time points and *Z*-stacks to enable focus tracking, while limiting phototoxicity and photobleaching effects.

The use of *Drosophila* cultured cells greatly facilitates these experiments. In addition to the existence of a distinct heterochromatin domain, *Drosophila* cells are maintained at room temperature and ambient CO_2 concentrations (Cherbas & Gong, 2014), which minimizes stress from environmental changes during cell culturing, sample processing, and live imaging. Further, these cells are mostly in S and G2 phases of the cell cycle (Chiolo et al., 2011), which is ideal for studying HR repair. Finally, efficient RNAi procedures facilitate the investigation of the mechanisms involved in heterochromatin repair with genetic approaches (Zhou, Mohr, Hannon, & Perrimon, 2013).

Here, we describe a procedure for monitoring the spatial and temporal dynamics of heterochromatic DSBs in *Drosophila* cells following IR (Fig. 2), including: (i) the generation of stable cell lines expressing fluorescent-tagged repair and heterochromatin marks; (ii) a technique to image the same field before and after IR; (iii) the setup used to minimize light exposure with a DeltaVision deconvolution system (Applied Precision/GE Healthcare); and (iv) postimage processing done with SoftWorX (Applied Precision/GE Healthcare), which maximizes the recovery of information from low-exposure experiments while correcting for modest photobleaching. Additionally, we describe the workflow we implemented to track repair foci with

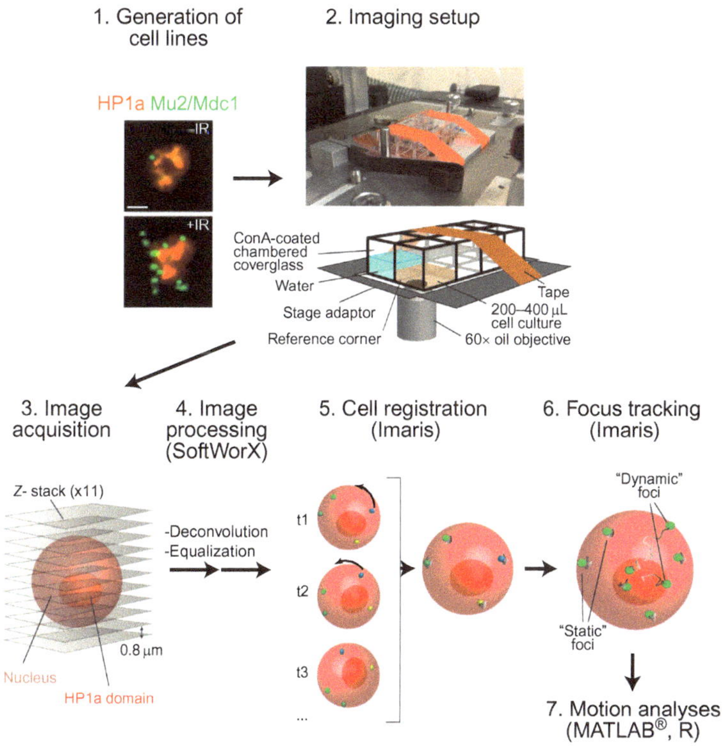

Fig. 2 Pipeline for image acquisition, processing, and analysis of focus dynamics in *Drosophila* heterochromatin. A stable cell line homogeneously expressing fluorescent-tagged protein (e.g., mCH-HP1a and mEGFP-Mu2/Mdc1) is placed in a well of a ConA-coated chambered coverslide, which is secured to the adaptor placed on the microscope stage. For long movies, evaporation of the media is limited by adding water to the surrounding wells. For tracking experiments, which require frequent imaging and long kinetics, photobleaching and phototoxicity are reduced by: (i) selecting bright fluorescent tags, which reduce the exposure time required to detect repair foci and heterochromatin; (ii) using the 2×2 binning option of the CoolsnapHQ2 camera, which dramatically increases the light collected by the camera per time unit; (iii) underexposing the sample and recovering most image details with postimage processing. Individual nuclei are registered with Imaris to correct for modest rotational and translational shifts, using "static" foci as a reference. "Dynamic" foci are then tracked using Imaris and the biophysical properties of the motion are characterized using customized scripts in MATLAB® and R.

Imaris (Bitplane) and derive mean square displacement (MSD) curves with a MATLAB® (Mathworks) script. Finally, we describe a new method we developed in R to identify directed motions within trajectories characterized by different types of motions. Together, these techniques enable descriptive and quantitative analyses of the spatial and temporal dynamics of heterochromatin repair, which cannot be accomplished with fixed cell studies or biochemical approaches. Similar approaches have also broad applicability for the study of nuclear dynamics of repair foci in other contexts, from yeast to mammalian cells.

2. METHODS

Successful focus tracking and 4D analyses of focus motion over long-time periods require the optimization of different steps, including the selection of the brightest and most photoresistant fluorescent tags for live-cell imaging, the use of microscopes that mitigate the risk of cell damage, and the identification of imaging conditions and postimaging denoising approaches to recover image details while minimizing light exposure. Cell mobilization and nucleus registration approaches are also needed to isolate focus dynamics from cell dynamics. Once the positional data are collected, quantitative analyses enable the understanding of the biophysical properties of focus motion, including identifying directed motions.

2.1 Live-Cell Imaging of *Drosophila* Cells

2.1.1 Generation of Stable Cell Lines Expressing Fluorescent-Tagged Proteins

Live-cell imaging experiments greatly benefit from using stably transfected Kc167 (Kc) cells maintained as exponentially growing cultures. Kc cells are preferred over S2 cells because they adhere better to the substrate and maintain a more stable karyotype in the population. Live imaging and tracking experiments are facilitated by generating cell lines with homogeneous signals across the cell population, and a careful selection of tags for live imaging.

2.1.1.1 Cell Maintenance

Kc cells are maintained in Schneider's medium (Sigma) supplemented with 10% FBS (Gemini) and 2% Antibiotic-Antimycotic (Gibco) at 27°C. Schneider's media is an optimal choice for live imaging given the low autofluorescence. If cells are grown in media with high autofluorescence (e.g., SF-900 II, Gibco), we recommended shifting the cells to Schneider's media just before live cell imaging. Cells are kept in the exponential phase by splitting the culture every 3–5 days to maintain a concentration of $1.5–9 \times 10^6$ cells/mL. Cells do not grow equally well when seeded at a density below 1.5×10^5 cells/mL. For information about *Drosophila* cell maintenance, see the Drosophila Genomics Resource Center (DGRC) website and Yang and Reth (2012).

2.1.1.2 Selecting Fluorescent Tags for Live-Cell Imaging

Live-cell imaging experiments are critically dependent on selecting the best combination of fluorescent tags for each protein of interest to maximize signal recovery during the entire kinetic. We have successfully used several fluorescent tags for live imaging of *Drosophila* cells (i.e., EGFP, GFP, mEGFP, mCitrin, mCerulean, mTurquoise2, mCherry, Aquamarine). However, mEGFP or EGFP tags' signals are among the brightest and most resistant to photobleaching and have been the best choice for 4D tracking of repair foci that requires frequent imaging over long periods of time (Ryu et al., 2015). mCherry (mCh) is also quite resistant to photobleaching and can be used to detect very abundant proteins or large nuclear structures in the same experiments, such as the heterochromatin domain (e.g., via expression of mCh-HP1a) or nuclear periphery components (e.g., via expression of mCh-LaminC) (Chiolo et al., 2011; Ryu et al., 2016, 2015).

2.1.1.3 Generating Cell Lines That Express Tagged Proteins

Generating *Drosophila* cell lines that express fluorescent markers for DSB repair and heterochromatin is facilitated by using agents that deliver high transfection efficiency, like Cellfectin (Invitrogen), Transit-Insect (Mirus Bio LLC), Transit-2020 (Mirus Bio LLC), or Effectene (Qiagen) (Fig. 3). Transfections are done following manufacturer's procedures in six-well plates and typically using 2.5 μg of plasmids expressing each tagged protein of interest. Transfections with up to three plasmids result in most cells expressing all plasmids. For live imaging experiments requiring transient transfection, Cellfectin is preferred because it forms few if any precipitates during transfection. Transgene expression can be tested 3–4 days after transfection.

Using stable cell lines facilitates acquiring significantly more data in a short time, as every imaged field will contain ∼20–50 transfected cells that can be analyzed for focus dynamics. Additionally, some repair components and nuclear architecture markers are not detectable after transient transfection, in which case stable cell lines are required. This needs to be empirically determined. To generate stable cell lines, 1 μg of plasmid carrying the selection cassette (i.e., pCoPuro (Iwaki, Figuera, Ploplis, & Castellino, 2003) or pCoHygro (Invitrogen)) is added to the plasmid mix during transfection, and cells are split in media containing the selection agent (i.e., 150 μg/mL

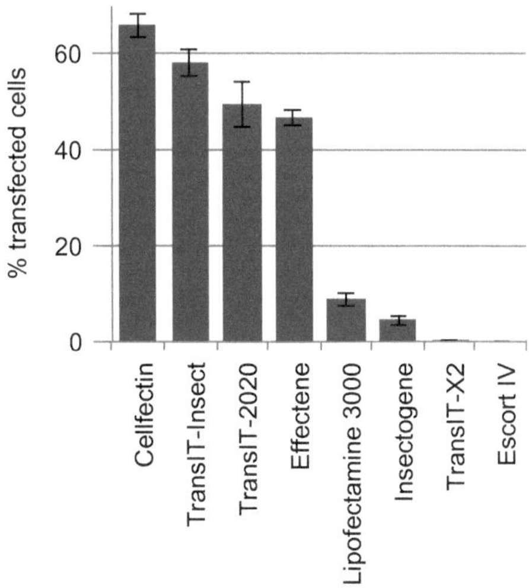

Fig. 3 High transfection efficiency is obtained with Cellfectin. High transfection efficiency facilitates the generation of stable Kc cell lines expressing fluorescent markers of the heterochromatin domain and damage foci. To optimize this step, different transfection agents were tested following manufacturers' instructions. Cells were transfected with 2.5 μg of a midiprep of pCopia-mCh-HP1a, and the average number of cells expressing HP1a was estimated 72 h after transfection. The highest transfection efficiency obtained after several optimization steps is shown for each reagent, revealing the highest transfection efficiency using Cellfectin. Error bars show ±SEM.

hygromycin or 300 μg/mL puromycin) 3–5 days after transfection. For the following weeks, cells are split as needed to maintain them in the exponential growth phase, and recovery from the selection is typically observed after 4–5 weeks. Longer selection times render the signal more homogeneous across the culture but might also result in reduced signal intensity. Thus, for tracking experiments we recommend using the cells shortly after establishing a stable line.

Notes: We mostly use the Copia promoter to constitutively induce a protein of interest at low levels (typically similar to endogenous levels), although the metallothionein promoter (induced by $CuSO_4$) and the heat-shock promoter (induced by a 30-min temperature shift at 37°C) have also been used successfully. Given the potential effects of temperature shifts on gene expression, nuclear architecture, chromatin dynamics, or repair pathways (Li et al., 2015; Seong, Li, Shimizu, Nakamura, & Ishii, 2011; Velichko, Petrova, Kantidze, & Razin, 2012), and the potential secondary effects of expressing nonphysiological levels of the proteins, specific controls need to be applied when using these promoters. Additional considerations apply to multicolor imaging. Imaging more than two channels increases the risk of phototoxicity and the time required for imaging each field. These might particularly affect tracking experiments, which require many time points and frequent imaging. Additionally, transfections with four plasmids (three plasmids expressing the tagged proteins plus the plasmid carrying a selection marker) are not very efficient, which complicates the generation of stable cell lines expressing all three fluorescent markers. If three-color imaging is required, it is recommended to express two (or three) of the tagged proteins in a multicistronic vector (Gonzalez et al., 2011). Alternatively, the selection marker can be cloned into one of the three vectors, or multiple tagged components can be integrated in the genome with sequential transfections.

2.1.2 Immobilization of Cells for Microscopy

A key step for 4D analysis of nuclear dynamics is immobilizing the cells on the substrate. This minimizes rotational and translational movements of the cells, reducing the need for postimaging corrections of the movement and dramatically improving the analysis of nuclear dynamics. We tested Concanavalin A (ConA) (Type VI or IV, Sigma), Polylysine (Sigma), and Cell-Tak (Corning), and found that ConA Type VI produces the best results.

Procedure for coverslide coating with ConA:

2.1.2.1 Prepare a solution of 1 mg/mL ConA in water. Stir for about 1 h, until mostly dissolved. Filter the solution with a 0.22-μm pore-size filter.

2.1.2.2 Add ~100 μL of ConA solution to each well of the eight-well chambered imaging coverslide (Nunc™ Lab-Tek™ II Chambered Coverglass, Thermo Scientific). The solution should form a thin layer coating the surface. Let the solution dry in the hood or in a 30°C incubator. For best results, repeat the coating two more times and use the coated coverslides within 1 week of preparation.

2.1.3 Temperature Regulation During Live Imaging

Maintaining a stable temperature while imaging is important to ensure consistency throughout the time course as well as across different experiments. To achieve stable temperature conditions, we use an environmental chamber mounted around the microscope that maintains the sample at 25°C during the experiment.

Note: For movies lasting a few hours, it is also important to limit evaporation of the growth media by adding water to empty wells surrounding the well with the cells (Fig. 2). For even longer movies, it is helpful to place a slightly damp paper towel in the empty chambers to further limit evaporation.

2.1.4 Image Acquisition Setup With a DeltaVision Deconvolution Microscope

Depending on the proteins being analyzed and the microscope available, imaging parameters need to be empirically optimized to minimize photobleaching and phototoxicity. For imaging the dynamics of heterochromatic repair sites, we typically utilize cells stably expressing mCh-HP1a and a repair protein tagged with mEGFP (e.g., Mu2/Mdc1, ATRIP, or Rad51 (Caridi et al., 2018; Chiolo et al., 2011; Ryu et al., 2016, 2015)). The following procedure is optimized for cells expressing mCh-HP1a and mEGFP-Mu2/Mdc1 and imaging with a DeltaVision Elite deconvolution inverted microscope equipped with: white light LED (rated 100 Lumens at 350 mA); seven-color InsightSSI solid-state illumination system; PlanApo 60 × oil objective with N.A. 1.42; Ultimate Focus module; a Coolsnap HQ2 camera; and controlled by SoftWorX software (v. 6.1.3). The DeltaVision System is optimized for low-light imaging and image processing through deconvolution, which enables an excellent recovery of image details in underexposed samples (see Section 2.2.1). This makes this system an excellent choice for the experiments described here. Spinning disk microscopes or widefield microscopes used in combination with external image processing software have also been used successfully for similar experiments (see Hediger, Taddei, Neumann, & Gasser, 2004; Meister, Gehlen, Varela, Kalck, & Gasser, 2010 for an overview of different imaging technologies and Cho et al., 2014; Dimitrova, Chen, Spector, & de Lange, 2008; Dion et al., 2012; Lottersberger et al., 2015; Miné-Hattab & Rothstein, 2012; Su et al., 2015 for examples of their application to focus tracking experiments).

2.1.4.1 Split cells to a density of 2×10^6 cells/mL 48 h before the experiment.

2.1.4.2 On the day of the experiment, transfer 200–400 μL of cells into one well of the chambered coverslide, and let the cells settle for 10–15 min before imaging. Meanwhile, set the temperature of the environmental chamber of the microscope to 25°C.

2.1.4.3 On the microscope, manually position the 60 × objective and the dichroic filter for GFP/mCh imaging.

2.1.4.4 Place immersion oil on the objective lens. Note that the immersion oil needs to be optimized based on objective, temperature conditions, and coverslip thickness to minimize spherical aberration while maximizing contrast in the images. With 1.5 mm chambered coverslide, we use an immersion oil refractive index = 1.512 (GE Healthcare Life Sciences). Follow the microscope manufacturer's instructions for this step.

2.1.4.5 Place the chambered coverslide on the stage adaptor. When imaging multiple fields, the movement of the stage can shift the coverglass on the stage adaptor, making it difficult to return to the same field of cells. To prevent this issue, secure the coverslide tightly using tape (Fig. 2) before placing it on the microscope stage. Adjust the stage level to bring the sample into focus.

2.1.4.6 In an effort to minimize light exposure during imaging, thus reducing photobleaching and phototoxicity, set the Coolsnap HQ2 camera at 2 × 2 binning. This results in less resolution but higher intensity collected per pixel. With this setting, we set an image size of 512 × 512 px, to collect the largest possible field of view.

2.1.4.7 Select the fields of interest. Illumination intensity is adjusted to the minimum level sufficient to see the sample, thus minimizing photobleaching of the signal while choosing the fields (e.g., 10% excitation intensity, or 10% T, for mCh and GFP). Suitable fields should contain an even distribution of cells as a monolayer, in addition to intense and homogeneous signals for mCh and GFP-tagged proteins. The number of fields that can be imaged in a single experiment is limited by the time required for imaging each field and the time interval between time points. We typically image four to five fields for each experiment. Once each field is selected, save its coordinates using "*Mark Point*" option in the "*Point List*" section in SoftWorx. The list of selected fields will appear in the "*Point List*" window.

2.1.4.8 Optimize the imaging path to visit all the fields using "*Optimize List*" followed by "*Compact List*" commands in the "*Point List*" section. This will minimize the time required to visit all of the selected fields, enabling imaging of more fields for each experiment.

2.1.4.9 Because cells are imaged before and after IR, it is essential to be able to return to the same fields after removing the chambered coverslide from the microscope for IR exposure. However, removal and repositioning of the sample might result in a slight shift

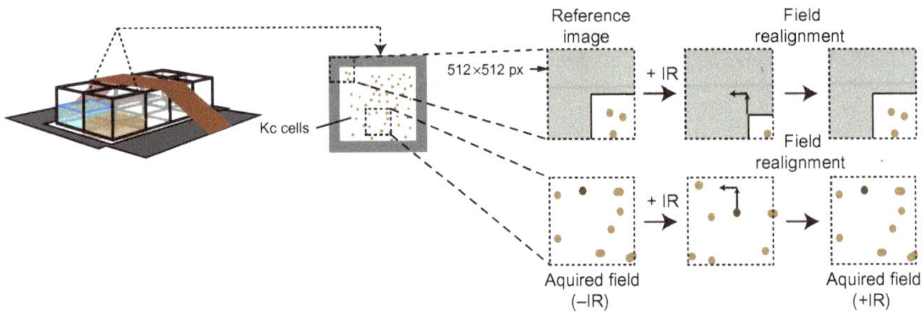

Fig. 4 Field realignment post IR. Focus tracking requires cell imaging before and after IR, but repositioning of the chambered coverslide on the stage after IR frequently results in a slight shift of the stage. This can be corrected using a reference field (e.g., a corner of the well). The reference field is manually recentered after IR, and the extent of correction applied enables determination of the *X* and *Y* shift. Corresponding corrections are applied to all fields of interest.

of the stage (Fig. 4). To identify the same field of cells, select a "landmark" field to use as a reference (such as a corner of the slide), and save its coordinates in addition to those of the selected fields of cells (Fig. 4).

2.1.4.10 Select the time of exposure for the sample based on the need to detect sufficient signal with the minimum exposure. For example, we use 10% T and a target intensity of 200 and 600 counts (approximately 5 and 15 ms) for mCh-HP1a and mEGFP-Mu2/Mdc1, respectively. This results in underexposing the image, but most image details can be recovered postimaging by optimized deconvolution and photobleaching correction algorithms available in SoftWorX (Fig. 5 and Section 2.2.1 of this protocol).

2.1.4.11 Adjust image acquisition settings to image the sample across its entire thickness with a minimum number of *Z*-stacks. This requires some optimization depending on the thickness of the sample. For *Drosophila* cells, image 11 *Z*-stacks at 0.8 μm distance between the *Z*-stacks. While a relatively high distance between *Z*-stacks reduces the resolution in this dimension, this is an excellent compromise between spatial resolution required for tracking and the need to minimize photobleaching and phototoxicity.

2.1.4.12 List the fields of interest (i.e., the points saved in the "*Point List*" section) in the "*Design/Run Experiment,*" "*Points*" tab, "*Visit Point List*" section. Activate the "*Ultimate Focus*" options with five

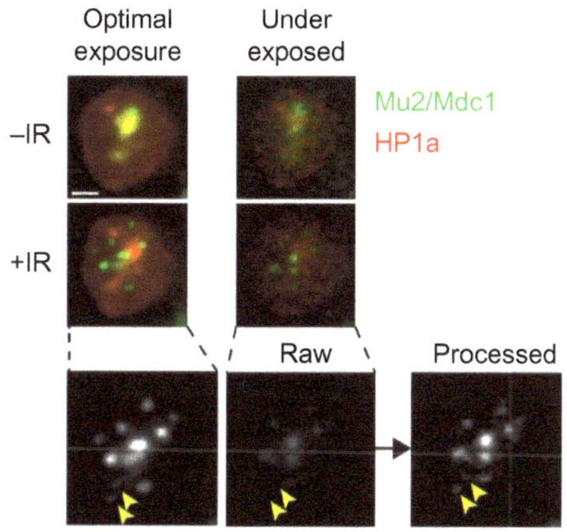

Fig. 5 Image details in underexposed samples are recovered with postimage processing. Underexposure of the cells enables the long and frequent image collection required for focus tracking while minimizing cell damage and photobleaching. The application of deconvolution and equalization algorithms enables the recovery of most details in underexposed images. In the example shown, a Kc cell stably expressing mEGFP-Mu2 and mCh-HP1a shows an overall enrichment of Mu2/Mdc1 signals inside the heterochromatin domain before IR (−IR), and damage foci associated with the heterochromatin domain at 10 min after IR (+IR) (Ryu et al., 2015). The signal has been collected first in underexposed conditions (5 ms for mCh and 20 ms for GFP, with 10% T), then in optimal imaging conditions (12 ms for mCh and 20 ms for GFP, with 100% T), for both time points as indicated. The noisy signal of raw underexposed images is significantly improved by postimage processing (deconvolution and equalization), revealing even the weak signals associated with small foci (*arrowheads*). Images are maximum intensity projections of one nucleus. Scale bar = 1 μm.

iterations to assure that the cell focus is maintained throughout the experiment. This option corrects for axial drifts with an infrared laser-based system that detects the position of the coverslip relative to the sample. This is preferred over the "*Image-based autofocus before imaging*" option which causes photodamage while refocusing the image. If the "*Unlimited Focus*" module is not available on the system, manual adjustments of the focus or automated image-based autofocusing might be required throughout the kinetic.

2.1.4.13 Begin image capturing using "*Run*"-"*Start Scan*" (green arrowhead in the "*Design/Run Experiment*" window). The acquired image set corresponds to the "untreated" time point (−IR, Figs. 1–4).

2.1.4.14 Once all frames have been imaged, carefully remove the chambered coverslide without disturbing the stage position or the cells. Apply additional immersion oil onto the objective lens as needed.

2.1.5 IR Treatment, Field Realignment, and Image Acquisition After IR

This procedure enables the acquisition of several selected fields of cells after IR for each experiment, providing a large number of cells for analyses. Importantly, a direct investigation of cellular responses to IR requires the ability to image the same fields before and after IR, which is accomplished as described later.

2.1.5.1 We routinely expose the sample to 1.7 or 5 Gy IR, which corresponds to 14.5 or 44 s in our X-ray irradiator (X-RAD iR-160, Precision X-Ray, stage level 30), respectively. The timer is started half way through IR exposure. We use 5 Gy to track ATRIP foci (Ryu et al., 2015), and 1.7 Gy for Mu2/Mdc1 foci. Mu2/Mdc1 foci are typically more numerous than ATRIP foci, so reducing the intensity of IR minimizes overlapping signals (e.g., deriving from focus clustering; Chiolo et al., 2011, 2013), reducing the risk of ambiguous tracks. Notably, we did not detect major differences in MSD values between 1.7 and 5 Gy in our cells.

2.1.5.2 After exposure to IR, carefully place the chambered coverslide back on the microscope stage. Position the sample on the reference field (e.g., the corner of the chambered coverslide) using its saved coordinates (Fig. 4). Compare to the untreated images. If the stage shifted, move back the reference field until the original position and record the extent of the shift required for this realignment. Readjust the position of each field of cells accordingly. For example, if the reference field is shifted +300 μm along the *X*-axis and +400 μm along the *Y*-axis, move each field of cells −300 and −400 μm along the *X*- and *Y*-axes, respectively, and save the new position (Fig. 4). For each field, also readjust the *Z*-axis to ensure that the sample is in focus and save the new position.

2.1.5.3 Adjust the "*Image Capturing*" parameters of the "*Experiment Setup*" option to collect images at 40 s time intervals for 1 h movies (91 images total). In SoftWorX, these options are under the "*Time-lapse*" tab in the "*Design/Run Experiment*" window. Click "*Start Scan*" to begin the imaging of the time points after IR (+IR, Fig. 2). Record the time point from IR at which the imaging starts (this is 3–5 min after IR in our experiments). Let the system acquire all the time points.

Notes: The imaging protocol described has been optimized for live imaging and tracking for MSD analysis of focus motion in response to IR. However, similar imaging approaches have broader applicability. For example, damage can be induced with hydroxyurea (1–2 m*M*) or methyl-methanesulfonate (0.0033%–0.1%). If cells are treated with chemicals, it is recommended to image untreated cells under control conditions to check for cellular toxicity coming from the solvent or imaging procedures. Longer movies are typically collected with longer time intervals between images to minimize photobleaching and phototoxicity. For example, we acquire images every 80 s for 2 h movies and every 30–60 min for 24 h movies. Finally, it is necessary to optimize imaging parameters (number of *Z*-stacks, intensity of light during exposure, frequency of time points) for each cell line, tagged component, and treatment. We recommend testing the effect of the imaging approach per se on DSB formation (e.g., Mu2/Mdc1 focus number (Caridi et al., 2018)) and cell ability to divide (e.g., by following the cell overnight and capturing cell division by imaging with 40-min time intervals (Chiolo et al., 2011)). Noninvasive imaging approaches should not interfere with cell division or induce DSB formation (Meister et al., 2010).

2.2 Image Processing and Focus Tracking

2.2.1 Image Processing With SoftWorX

The low light exposure conditions used in this protocol enable imaging over long time periods while limiting phototoxicity and photobleaching, but they also result in low signal-to-noise ratio in the collected images. A critical step in this analysis is the application of algorithms (deconvolution and equalization) that recover most image details postimaging, enabling precise identification and tracking of repair foci throughout the entire kinetic (Fig. 5). Equalization compensates for photobleaching effects by normalizing the signal intensity of each time point to a reference time point. This greatly facilitates automated detection and tracking of foci, which relies on the average signal intensity across the kinetic. Deconvolution is the mathematical correction of the distortion of an image resulting from diffraction and aberration of light passing through the optics of a microscope (Scalettar, Swedlow, Sedat, & Agard, 1996). This distortion can be quantified by establishing the shape of a single point object imaged with a microscope, or point spread function (PSF) (Scalettar et al., 1996). Deconvolution algorithms use the PSF to correct for the distortions, resulting in image deblurring, noise reduction, resolution improvement, and contrast enhancement (Scalettar et al., 1996). In addition to facilitating automated focus detection by increasing resolution, deconvolution corrects for image distortions in the axial direction, dramatically reducing noise in the tracks coming from this dimension. In the DeltaVision microscope, equalization and deconvolution algorithms are fully integrated with the system and optimized for the specific microscope setup. However, similar results can be obtained with other software (e.g., Huygens Software Suite for deconvolution or Imaris for equalization).

2.2.1.1 Combine "−IR" and "+IR" files using the "*Image Fusion*" function in SoftWorX with the "*Combine time points for like wavelengths*" option.

2.2.2.1 Deconvolve the fused images in SoftWorX using five iterations of the "*Conservative*" protocol to improve contrast and resolution. In our experience, this is sufficient to improve image sharpness while limiting potential artifacts.

2.2.3.1 Correct the deconvolved file for modest photobleaching using the "*Equalize Time Points*" option.

2.2.4.1 Having individual cells cropped facilitates further image analysis with Imaris. For this purpose, crop selected cells using the "*Save File*" option in SoftWorX by selecting the field section of interest. Cells better suited for the registration step described next need to fit the following criteria: (i) the cell should appear nearly static throughout the kinetic; (ii) at least four "*static*" foci (i.e., foci that do not significantly move in the nucleus) are present throughout the kinetic (Fig. 2)—these will be used for registration; (iii) GFP and mCherry signals remain visible throughout the kinetics.

2.2.2 Cell Registration With Imaris

After cropping the selected cells, we utilize Imaris (v. 7.7.1 with XT module) to correct for minor cell/nucleus movement ("registration") and to track foci for motion analysis. Registration is performed by tracking all of the foci (usually 4–12) that remain largely static throughout the kinetic and by correcting cell drift using those as a reference (Fig. 2). Using 7+ foci as a reference typically results in a better registration.

2.2.2.1 File Cropping

Remove the first time point (UNT) at this stage of the analysis by using the "*Crop Time*" function in Imaris. The first time point does not contain any IR-induced focus, so it cannot be used for registration. Save the cropped file with a new name to use this for further analysis. Keeping the original file is also important, as it contains information about which repair foci were already present before IR. Those foci can be used for registration but not for tracking of IR-induced foci and MSD analyses.

2.2.2.2 Optional: Noise Reduction

Automatic focus tracking is greatly improved by reducing background signals and small focus vibrations along the *Z*-stack. Background signals are reduced using Imaris, by applying the "*Baseline Subtraction*" option in the "*Thresholding*" function of "*Image Processing*" to the channel with the foci. In addition, minor vibrations along the *Z*-stack are reduced by applying the "*Smooth Time*" function of "*Image Processing*" with a filter width of 1 to all time points. Together, these steps facilitate the subsequent automated tracking of foci.

2.2.2.3 Automated Focus Tracking

Focus tracking is done using the Imaris "*Spot Detection Tool*," and the tracked foci are used to register the nucleus. Apply the following steps to generate the tracks, clicking the right pointing blue arrow to proceed through each step:

i. Generate a new "Spot" in Imaris. Select the magic wand icon, and click on "*Rebuild*";

ii. Select the "*Track Spots Over Time*" box;

iii. Select the "*Source Channel*" corresponding to the wavelength at which foci were imaged, using the dropdown menu;

iv. In the "*Spot Detection*" section, select 0.2 μm as "*Estimated XY Diameter.*" This value reliably detects most DNA repair foci we examined. However, depending on the proteins being studied, this parameter may require empirical adjustments;

v. The algorithm will place spheres corresponding to all detected foci. In the "*Filters*" section select "*Quality.*" Adjust the lowest threshold to a point at which the faintest foci are reliably distinguished from the background;

vi. In the "*Add/Delete (Cursor Intersects with)*" section, using the dropdown menu, select the "*Specific Channel*" corresponding to the wavelength at which foci were imaged;

vii. In the "*Algorithm*" section, select the "*Autoregressive Motion.*" In the "*Parameters*" section change the "M*ax Distance*" to 0.5 μm and the "*Max Gap Size*" to 3. Check the box labeled "Fill gaps with all detected objects";

viii. We apply two filters in the "*Classify Tracks*" section. Select "*Track Duration*" and adjust the lower threshold to eliminate tracks that only last a few time points. Add the "*Track Length*" filter and adjust the lower threshold to further remove short tracks. Click the right pointing orange double arrow icon to finalize the track detection.

2.2.2.4 Manual Editing of Tracks

Sometimes the tracks generated by Imaris include large jumps to unrelated foci, especially when those are in close spatial proximity relative to the focus of interest, in which case tracks detected automatically require manual adjustments. To edit a track, select the corresponding spot and the "*Edit Tracks*" icon. Select the time point that requires editing, delete it, manually recreate a new spot, and connect it to the preexistent track. Edit each track as necessary to assure that each focus is correctly identified throughout the kinetic.

2.2.2.5 Selection of Tracks for Registration

Identify foci that remain relatively stationary throughout the time course. These are typically euchromatic foci that originate outside the HP1a domain or foci that were already present in the nuclei before IR (e.g., spontaneous damage). We find that registration works better using at least seven foci, although four foci are sometimes sufficient. Foci characterized by extensive motion will affect the registration process and should be excluded at this stage.

2.2.2.6 Registration

Highlight all suitable tracks and click the "*Correct Drift*" button below the tracks window. In the "*Drift Correction Options*," select "*Translational And Rotational Drift*." For the "*Result Dataset Size*" select "*Include Entire Result*." Confirm that the "*Correct objects' positions*" box is selected. Then click "OK." Imaris will register the nucleus based on the selected tracks, which will compensate for any minor translational and rotational motion of the nucleus during the experiment. Save the resulting file with a new name. This will be used for further focus tracking (Fig. 2).

2.2.3 Focus Tracking With Imaris

Tracking DNA damage foci is similar to the registration process, except that a new "spot" is generated for each tracked focus.

2.2.3.1 Optional

To minimize artificial vibrations of the nucleus resulting from the registration process, reapply the "*Smooth time*" function of "*Image Processing*" with a filter width of 1 to all the time points.

2.2.3.2 Focus Tracking

Visually identify a focus for tracking. Repeat steps 2.2.2.3 and 2.2.2.4, except that most tracks are filtered out using both upper and lower thresholds for "*Quality*" and "*Track Duration*" filters; only the focus of interest remains tracked. Repeat this step as many times as necessary to track all foci under investigation.

2.2.3.3 Nuclear Periphery Detection

The nuclear periphery is identified by creating a volume that corresponds to the diffuse nuclear signal generated by background mCh-HP1a or GFP-Mu2/Mdc1 signals. Using the "*Automatic Creation*" function, select the channel corresponding to HP1a or Mu2/Mdc1, and manually adjust smoothness and threshold to create a volume fitting the nuclear signal. Alternatively, a specific marker for the nuclear periphery (e.g., mCh-LaminC) can be used (Ryu et al., 2015).

2.2.4 4D Image Rendering

2.2.4.1 Optional

4D rendering of individual tracks can be done in Imaris to facilitate the analysis and display of each track (i.e., Fig. 1D). 4D rendering is obtained by generating a volume corresponding to the HP1a domain. Using the "*Automatic Creation*" function, select the channel corresponding to the HP1a domain and manually adjust smoothness and threshold to create a volume fitting the HP1a signal. The volume corresponding to the nuclear periphery (defined as described in Section 2.2.3.3) can be displayed at this step as well. Finally, select the focus track of interest, and deselect the green and red channels of the original image before saving the image.

2.3 Analysis of Focus Dynamics

2.3.1 Mean Square Displacement (MSD) Analysis

MSD analyses, which plot the average squared distance traveled by a focus at progressively increasing time intervals, provide quantitative measurements of the dynamic properties of focus motion (Dion & Gasser, 2013; Saxton & Jacobson, 1997; Spichal & Fabre, 2017). MSD values are

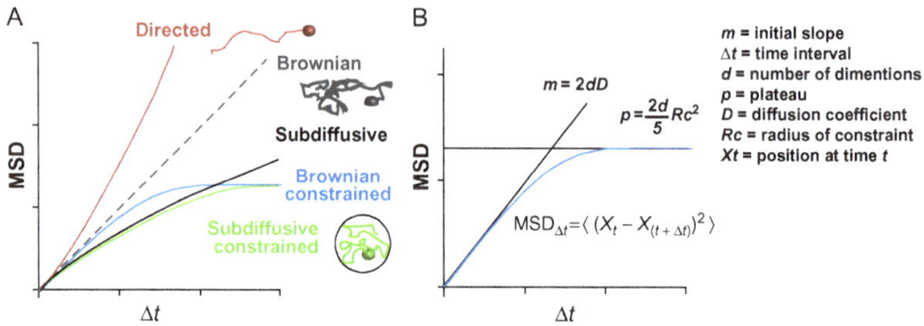

Fig. 6 MSD curves. MSD values are derived from the mean of the squared displacement calculated over increasing time intervals, Δt. (A) Examples of MSD plots corresponding to directed motion, Brownian motion, subdiffusive motion, and subdiffusive or Brownian motions via a constrained space are indicated. (B) For Brownian or subdiffusive motions, containment radius and diffusion coefficient are calculated using the plateau of the curve or the initial slope, as indicated (see text for details).

calculated at multiple time intervals to generate a curve for each track, and these curves are averaged to describe the behavior of a population of foci. The shape of the MSD curve reveals whether the motion of a particle is Brownian, subdiffusive or directed (Dion & Gasser, 2013; Saxton & Jacobson, 1997; Spichal & Fabre, 2017). MSD curves with increasing slope reflect DM, while linear MSD graphs indicate Brownian motion (Dion & Gasser, 2013; Meister et al., 2010; Saxton & Jacobson, 1997; Spichal & Fabre, 2017; Fig. 6A). However, given that the chromatin behaves as a polymer, and other constraints to the movement exist (e.g., chromatin compaction, molecular crowding, and anchoring to nuclear structures; reviewed in Dion & Gasser, 2013; Spichal & Fabre, 2017), chromatin motion is typically subdiffusive rather than Brownian, resulting in flattened MSD curves (Amitai, Seeber, Gasser, & Holcman, 2017; Marshall et al., 1997; Spichal & Fabre, 2017; Fig. 6A). In addition, subdiffusive (or Brownian) motion occurring in a constrained space (e.g., the nucleus or subnuclear domains) is characterized by MSD graphs that reach a plateau (Dion & Gasser, 2013; Saxton & Jacobson, 1997; Spichal & Fabre, 2017), and this is proportional to the radius of constraint (i.e., the radius of the volume explored by the focus; Fig. 6B). MSD curves enable calculating the radius of constraint and the diffusion coefficient as follows.

2.3.1.1 Extracting Positional Data

For each tracked spot, select the "*Statistics*" tab and click on the "*Position*" information in the dropdown menu of the "*Detailed*" tab. Click on the floppy disk icon to save the data as an excel file. The file contains three columns: posX, posY, and posZ, corresponding to the coordinates of the focus at each time point. To use the MSD script provided as Supplementary Information 1: "MSD_script" (available on https://doi.org/10.1016/bs.mie.2017.11.033), add a column before those three and name it "t." Number each time point with increasing number starting from "001" for the first time point, add the corresponding number at the beginning of each file name, and save this as a comma-separated values comma-separated values (.csv) file editable in Excel.

2.3.1.2 MSD Calculation

To derive MSD curves, positional data obtained from each track are processed with a MATLAB® script that generates MSD values for each focus. Open the MSD script in MATLAB® and point to the folder containing the .csv files in line 2 (e.g., "*myInputFolder*"). Enter the information of the destination folder in line 3 (e.g., "*myOutputFolder*"). Adjust the number of time points to be analyzed by editing the *timelapse_length* variable in line 8 (e.g., 92 time points in our experiments). This change will be reflected in the MSD table size and calculation. MSDs were calculated as described in Dion and Gasser (2013), Meister et al. (2010), Miné-Hattab and Rothstein (2012), and Spichal and Fabre (2017); for each position x_t (characterized by XYZ coordinates) at time t, and time interval Δt:

$$\mathrm{MSD}_{\Delta t} = \left\langle \left(x_t - {}^t x_{(t+\Delta t)}\right)^2 \right\rangle$$

Run the script. Output files will be saved in the destination folder. Note that multiple files (i.e., positional data from a large number of foci) can be read at once. As long as the file name is preceded by a progressively increasing number (i.e., 001, 002, etc.), the output will be a .csv file, with columns corresponding to MSD values for each focus and the focus number indicated at the bottom of each column. Rows correspond to increasing Δt (i.e., $\Delta t = 1$ in row 1, $\Delta t = 2$ in row 2, etc.). Combine all MSD data in one Excel file. Calculate and plot the average MSD value and standard errors.

2.3.1.3 Calculation of the Radius of Constraint and the Diffusion Coefficient for a Population of Foci

In order to calculate the plateau of the MSD curve, derive the curve that fits the data using MATLAB®'s "*curve fitting tools*" with this equation:

$$y = cst + A\left(1 - e^{-\frac{(x - x_t)}{\delta}}\right)$$

The plateau p of the curve, corresponding to ($p = cst + A$), enables to calculate the radius of constraint Rc as:

$$Rc = \sqrt{5p/2d}$$

where d is the number of dimensions (three in this case). The diffusion coefficient D is derived from:

$$D = m/2d$$

where m is the initial slope of the MSD curve.

2.3.2 Detection of Directed Motions (DMs) in Mixed Trajectories

When DM occurs in a trajectory characterized by different types motion (e.g., preceded and/or followed by subdiffusive or constrained motions), also called mixed trajectory, an MSD analysis applied to the entire kinetic might mask the presence of DMs. This is even more likely when MSD values across different foci are averaged, using DMs initiated at different time points for each focus. Here we describe a method to identify time intervals characterized by DMs in these mixed trajectories, for each focus track. MSD curves for individual tracks are analyzed through a script in R (Supplementary Information 2: "DM_script.R" (available on https://doi.org/10.1016/bs.mie.2017.11.033)) that returns the time intervals characterized by MSD graphs with increasing slope.

2.3.2.1 DM Identification

Open the script in R (https://cran.r-project.org) or R studio (https://www.rstudio.com) and point to the folder containing one of the .csv files obtained in Section 2.3.1.1, in line 202 (i.e., "*list.files*"). The script derives MSD curves starting from each time point of the trajectory, and with progressively increasing Δt (with $\Delta t > 10$ time intervals). MSD curves are then smoothed using the *lowess* function with default parameters (https://www.stat.ethz.ch/R-manual/R-devel/library/stats/html/lowess.html) and segmented in two or more lines with increasing slopes. Points of slope change along the curve are detected using a *change point* analysis (Zeileis, Kleiber, Krämer, & Hornik, 2003) (*breakpoints* function with parameter fraction = 0.3; i.e., every regression fit must use at least 30% of the total data points). The ratio of slopes between two lines is used to identify DM, i.e., time intervals when the

MSD graph shows progressively increasing slopes. A few parameters can be modified to optimize the detection of DMs; the current values robustly detect DMs in our datasets, but adjustments might be required for different data:

i. "*minratio*" is the minimum ratio between lines derived from MSD plots for considering the graph characterized by increasing slope, and the current value is set at 3.5. Higher values of *minratio* increase the stringency of detection of DMs (Note: a *minratio* higher than 2 is required to minimize the detection of false positives).

ii. "*r2min*" is the minimum coefficient determination of the fitted line. This is set at 0.99 and higher values of this parameter increase the stringency of detection of DMs.

iii. "*minslope*" is the minimum slope of the regression line at which data points are considered (current setting: 0.012). Lower values result in more foci with limited motions included in the analysis.

Run the script with the selected parameters. Time intervals characterized by DM appear in the "*Console*" work space and the corresponding graph is shown in the "*Plots*" section. If parameters are set correctly, those graphs show an upward curvature indicating DMs. The first and last time point of each time interval characterized by DM is saved in a file generated with file extension ".out" defined in the "*outsuffix*" variable and is shown as time intervals. Data are compiled to show the number and duration of DMs across the kinetics. Those typically correspond to time intervals characterized by DM in a mixed trajectory. The detection of time intervals characterized by DM can be validated by running the MSD script described in Section 2.3.1 on the positional data corresponding to each time interval.

ACKNOWLEDGMENTS

This work was supported by NIH R21ES021541, NIH R01GM117376, the Mallinckrodt Foundation Award, and NSF Career 1751197 to I.C. We would like to thank S. Keagy for insightful comments on the chapter, B. Spatola for his help on an early version of the chapter, T. Nellimoottil for his help with the MSD script, T. Ryu for her analysis of Imaris filtering options and ConA optimization experiments, K. Chrzanowski for his input on Imaris tools, and the Chiolo Lab for helpful discussions. We also thank R. Rothstein, J. Miné-Hattab, and M. Smith for their input on data analyses.

DISCLOSURES

The authors have nothing to disclose.

REFERENCES

Agmon, N., Liefshitz, B., Zimmer, C., Fabre, E., & Kupiec, M. (2013). Effect of nuclear architecture on the efficiency of double-strand break repair. *Nature Cell Biology, 15*(6), 694–699. https://doi.org/10.1038/ncb2745.

Amaral, N., Ryu, T., Li, X., & Chiolo, I. (2017). Nuclear dynamics of heterochromatin repair. *Trends in Genetics, 33*(2), 86–100. https://doi.org/10.1016/j.tig.2016.12.004.

Amitai, A., Seeber, A., Gasser, S. M., & Holcman, D. (2017). Visualization of chromatin decompaction and break site extrusion as predicted by statistical polymer modeling of single-locus trajectories. *Cell Reports, 18*(5), 1200–1214. https://doi.org/10.1016/j.celrep.2017.01.018.

Ayoub, N., Jeyasekharan, A. D., Bernal, J. A., & Venkitaraman, A. R. (2008). HP1-beta mobilization promotes chromatin changes that initiate the DNA damage response. *Nature, 453*(7195), 682–686. https://doi.org/10.1038/nature06875.

Beucher, A., Birraux, J., Tchouandong, L., Barton, O., Shibata, A., Conrad, S., et al. (2009). ATM and Artemis promote homologous recombination of radiation-induced DNA double-strand breaks in G2. *The EMBO Journal, 28*(21), 3413–3427. https://doi.org/10.1038/emboj.2009.276.

Burgess, R. C., Burman, B., Kruhlak, M. J., & Misteli, T. (2014). Activation of DNA damage response signaling by condensed chromatin. *Cell Reports, 9*(5), 1703–1717. https://doi.org/10.1016/j.celrep.2014.10.060.

Caridi, C., D'Agostino, C., Ryu, T., Zapotoczny, G., Li, X., Delabaere, L., et al. (2018). *Directed motions of heterochromatic DSBs require nuclear actin filaments and myosins,* Under revision.

Caridi, P. C., Delabaere, L., Zapotoczny, G., & Chiolo, I. (2017). And yet, it moves: Nuclear and chromatin dynamics of a heterochromatic double-strand break. *Philosophical Transactions of the Royal Society of London. Series B, Biological Sciences, 372*(1731), 20160291. https://doi.org/10.1371/journal.pgen.1002473.

Chapman, J. R., & Jackson, S. P. (2008). Phospho-dependent interactions between NBS1 and MDC1 mediate chromatin retention of the MRN complex at sites of DNA damage. *EMBO Reports, 9*(8), 795–801. https://doi.org/10.1038/embor.2008.103.

Cherbas, L., & Gong, L. (2014). Cell lines. *Methods, 68*(1), 74–81. https://doi.org/10.1016/j.ymeth.2014.01.006.

Chiolo, I., Minoda, A., Colmenares, S. U., Polyzos, A., Costes, S. V., & Karpen, G. H. (2011). Double-strand breaks in heterochromatin move outside of a dynamic HP1a domain to complete recombinational repair. *Cell, 144*(5), 732–744. https://doi.org/10.1016/j.cell.2011.02.012.

Chiolo, I., Tang, J., Georgescu, W., & Costes, S. V. (2013). Nuclear dynamics of radiation-induced foci in euchromatin and heterochromatin. *Mutation Research, 750*(1–2), 56–66. https://doi.org/10.1016/j.mrfmmm.2013.08.001.

Cho, N. W., Dilley, R. L., Lampson, M. A., & Greenberg, R. A. (2014). Interchromosomal homology searches drive directional ALT telomere movement and synapsis. *Cell, 159*(1), 108–121. https://doi.org/10.1016/j.cell.2014.08.030.

Chung, D. K., Chan, J. N., Strecker, J., Zhang, W., Ebrahimi-Ardebili, S., Lu, T., et al. (2015). Perinuclear tethers license telomeric DSBs for a broad kinesin- and NPC-dependent DNA repair process. *Nature Communications, 6,* 7742. https://doi.org/10.1038/ncomms8742.

Churikov, D., Charifi, F., Eckert-Boulet, N., Silva, S., Simon, M. N., Lisby, M., et al. (2016). SUMO-dependent relocalization of eroded telomeres to nuclear pore complexes controls telomere recombination. *Cell Reports, 15*(6), 1242–1253. https://doi.org/10.1016/j.celrep.2016.04.008.

Costes, S. V., Chiolo, I., Pluth, J. M., Barcellos-Hoff, M. H., & Jakob, B. (2010). Spatiotemporal characterization of ionizing radiation induced DNA damage foci and their relation to chromatin organization. *Mutation Research*, *704*(1–3), 78–87. https://doi.org/10.1016/j.mrrev.2009.12.006.

Delabaere, L., & Chiolo, I. (2016). ReiNF4rcing repair pathway choice during cell cycle. *Cell Cycle*, *15*(9), 1182–1183. https://doi.org/10.1080/15384101.2016.1159108.

Delabaere, L., Ertl, H. A., Massey, D. J., Hofley, C. M., Sohail, F., Bienenstock, E. J., et al. (2017). Aging impairs double-strand break repair by homologous recombination in Drosophila germ cells. *Aging Cell*, *16*, 320–328. https://doi.org/10.1111/acel.12556.

Dimitrova, N., Chen, Y. C., Spector, D. L., & de Lange, T. (2008). 53BP1 promotes non-homologous end joining of telomeres by increasing chromatin mobility. *Nature*, *456*(7221), 524–528. https://doi.org/10.1038/nature07433.

Dion, V., & Gasser, S. M. (2013). Chromatin movement in the maintenance of genome stability. *Cell*, *152*(6), 1355–1364. https://doi.org/10.1016/j.cell.2013.02.010.

Dion, V., Kalck, V., Horigome, C., Towbin, B. D., & Gasser, S. M. (2012). Increased mobility of double-strand breaks requires Mec1, Rad9 and the homologous recombination machinery. *Nature Cell Biology*, *14*(5), 502–509. https://doi.org/10.1038/ncb2465.

Dion, V., Kalck, V., Seeber, A., Schleker, T., & Gasser, S. M. (2013). Cohesin and the nucleolus constrain the mobility of spontaneous repair foci. *EMBO Reports*, *14*(11), 984–991. https://doi.org/10.1038/embor.2013.142.

Dronamraju, R., & Mason, J. M. (2009). Recognition of double strand breaks by a mutator protein (MU2) in Drosophila melanogaster. *PLoS Genetics*, *5*(5), e1000473. https://doi.org/10.1371/journal.pgen.1000473.

Fernandez-Capetillo, O., Lee, A., Nussenzweig, M., & Nussenzweig, A. (2004). H2AX: The histone guardian of the genome. *DNA Repair (Amst)*, *3*(8–9), 959–967. https://doi.org/10.1016/j.dnarep.2004.03.024.

Goldberg, M., Stucki, M., Falck, J., D'Amours, D., Rahman, D., Pappin, D., et al. (2003). MDC1 is required for the intra-S-phase DNA damage checkpoint. *Nature*, *421*(6926), 952–956. https://doi.org/10.1038/nature01445.

Gonzalez, M., Martin-Ruiz, I., Jimenez, S., Pirone, L., Barrio, R., & Sutherland, J. D. (2011). Generation of stable Drosophila cell lines using multicistronic vectors. *Scientific Reports*, *1*, 75. https://doi.org/10.1038/srep00075.

Haaf, T., Golub, E. I., Reddy, G., Radding, C. M., & Ward, D. C. (1995). Nuclear foci of mammalian Rad51 recombination protein in somatic cells after DNA damage and its localization in synaptonemal complexes. *Proceedings of the National Academy of Sciences of the United States of America*, *92*(6), 2298–2302.

Hauer, M. H., Seeber, A., Singh, V., Thierry, R., Sack, R., Amitai, A., et al. (2017). Histone degradation in response to DNA damage enhances chromatin dynamics and recombination rates. *Nature Structural & Molecular Biology*, *24*(2), 99–107. https://doi.org/10.1038/nsmb.3347.

Hediger, F., Taddei, A., Neumann, F. R., & Gasser, S. M. (2004). Methods for visualizing chromatin dynamics in living yeast. *Methods in Enzymology*, *375*, 345–365.

Ho, J. W., Jung, Y. L., Liu, T., Alver, B. H., Lee, S., Ikegami, K., et al. (2014). Comparative analysis of metazoan chromatin organization. *Nature*, *512*(7515), 449–452. https://doi.org/10.1038/nature13415.

Horigome, C., Bustard, D. E., Marcomini, I., Delgoshaie, N., Tsai-Pflugfelder, M., Cobb, J. A., et al. (2016). PolySUMOylation by Siz2 and Mms21 triggers relocation of DNA breaks to nuclear pores through the Slx5/Slx8 STUbL. *Genes & Development*, *30*(8), 931–945. https://doi.org/10.1101/gad.277665.116.

Horigome, C., Oma, Y., Konishi, T., Schmid, R., Marcomini, I., Hauer, M. H., et al. (2014). SWR1 and INO80 chromatin remodelers contribute to DNA double-strand break perinuclear anchorage site choice. *Molecular Cell*, *55*(4), 626–639. https://doi.org/10.1016/j.molcel.2014.06.027.

Hoskins, R. A., Carlson, J. W., Kennedy, C., Acevedo, D., Evans-Holm, M., Frise, E., et al. (2007). Sequence finishing and mapping of Drosophila melanogaster heterochromatin. *Science*, *316*(5831), 1625–1628. https://doi.org/10.1126/science.1139816.

Hoskins, R. A., Carlson, J. W., Wan, K. H., Park, S., Mendez, I., Galle, S. E., et al. (2015). The release 6 reference sequence of the Drosophila melanogaster genome. *Genome Research*, *25*(3), 445–458. https://doi.org/10.1101/gr.185579.114.

Iwaki, T., Figuera, M., Ploplis, V. A., & Castellino, F. J. (2003). Rapid selection of Drosophila S2 cells with the puromycin resistance gene. *BioTechniques*, *35*(3), 482–484, 486.

Jakob, B., Splinter, J., Conrad, S., Voss, K. O., Zink, D., Durante, M., et al. (2011). DNA double-strand breaks in heterochromatin elicit fast repair protein recruitment, histone H2AX phosphorylation and relocation to euchromatin. *Nucleic Acids Research*, *39*(15), 6489–6499. https://doi.org/10.1093/nar/gkr230.

Janssen, A., Breuer, G. A., Brinkman, E. K., van der Meulen, A. I., Borden, S. V., van Steensel, B., et al. (2016). A single double-strand break system reveals repair dynamics and mechanisms in heterochromatin and euchromatin. *Genes & Development*, *30*(14), 1645–1657. https://doi.org/10.1101/gad.283028.116.

Kalocsay, M., Hiller, N. J., & Jentsch, S. (2009). Chromosome-wide Rad51 spreading and SUMO-H2A.Z-dependent chromosome fixation in response to a persistent DNA double-strand break. *Molecular Cell*, *33*(3), 335–343. https://doi.org/10.1016/j.molcel.2009.01.016.

Khadaroo, B., Teixeira, M. T., Luciano, P., Eckert-Boulet, N., Germann, S. M., Simon, M. N., et al. (2009). The DNA damage response at eroded telomeres and tethering to the nuclear pore complex. *Nature Cell Biology*, *11*(8), 980–987. https://doi.org/10.1038/ncb1910.

Krawczyk, P. M., Borovski, T., Stap, J., Cijsouw, T., ten Cate, R., Medema, J. P., et al. (2012). Chromatin mobility is increased at sites of DNA double-strand breaks. *Journal of Cell Science*, *125*(Pt. 9), 2127–2133. https://doi.org/10.1242/jcs.089847.

Kruhlak, M. J., Celeste, A., Dellaire, G., Fernandez-Capetillo, O., Müller, W. G., McNally, J. G., et al. (2006). Changes in chromatin structure and mobility in living cells at sites of DNA double-strand breaks. *The Journal of Cell Biology*, *172*(6), 823–834.

Li, L., Lyu, X., Hou, C., Takenaka, N., Nguyen, H. Q., Ong, C. T., et al. (2015). Widespread rearrangement of 3D chromatin organization underlies polycomb-mediated stress-induced silencing. *Molecular Cell*, *58*(2), 216–231. https://doi.org/10.1016/j.molcel.2015.02.023.

Lisby, M., Barlow, J. H., Burgess, R. C., & Rothstein, R. (2004). Choreography of the DNA damage response: Spatiotemporal relationships among checkpoint and repair proteins. *Cell*, *118*(6), 699–713.

Lisby, M., & Rothstein, R. (2004). DNA damage checkpoint and repair centers. *Current Opinion in Cell Biology*, *16*(3), 328–334. https://doi.org/10.1016/j.ceb.2004.03.011.

Liu, Y., Li, M., Lee, E. Y., & Maizels, N. (1999). Localization and dynamic relocalization of mammalian Rad52 during the cell cycle and in response to DNA damage. *Current Biology*, *9*(17), 975–978.

Lottersberger, F., Karssemeijer, R. A., Dimitrova, N., & de Lange, T. (2015). 53BP1 and the LINC complex promote microtubule-dependent DSB mobility and DNA repair. *Cell*, *163*(4), 880–893. https://doi.org/10.1016/j.cell.2015.09.057.

Lou, Z., Minter-Dykhouse, K., Franco, S., Gostissa, M., Rivera, M. A., Celeste, A., et al. (2006). MDC1 maintains genomic stability by participating in the amplification of ATM-dependent DNA damage signals. *Molecular Cell*, *21*(2), 187–200. https://doi.org/10.1016/j.molcel.2005.11.025.

Madigan, J. P., Chotkowski, H. L., & Glaser, R. L. (2002). DNA double-strand break-induced phosphorylation of Drosophila histone variant H2Av helps prevent radiation-induced apoptosis. *Nucleic Acids Research*, *30*(17), 3698–3705.

Marshall, W. F., Straight, A., Marko, J. F., Swedlow, J., Dernburg, A., Belmont, A., et al. (1997). Interphase chromosomes undergo constrained diffusional motion in living cells. *Current Biology*, 7(12), 930–939.

Maser, R. S., Monsen, K. J., Nelms, B. E., & Petrini, J. H. (1997). hMre11 and hRad50 nuclear foci are induced during the normal cellular response to DNA double-strand breaks. *Molecular and Cellular Biology*, *17*(10), 6087–6096.

Meister, P., Gehlen, L. R., Varela, E., Kalck, V., & Gasser, S. M. (2010). Visualizing yeast chromosomes and nuclear architecture. *Methods in Enzymology*, *470*, 535–567. https://doi.org/10.1016/S0076-6879(10)70021-5.

Miné-Hattab, J., & Rothstein, R. (2012). Increased chromosome mobility facilitates homology search during recombination. *Nature Cell Biology*, *14*(5), 510–517. https://doi.org/10.1038/ncb2472.

Mordes, D. A., Glick, G. G., Zhao, R., & Cortez, D. (2008). TopBP1 activates ATR through ATRIP and a PIKK regulatory domain. *Genes & Development*, *22*(11), 1478–1489. https://doi.org/10.1101/gad.1666208.

Nagai, S., Dubrana, K., Tsai-Pflugfelder, M., Davidson, M. B., Roberts, T. M., Brown, G. W., et al. (2008). Functional targeting of DNA damage to a nuclear pore-associated SUMO-dependent ubiquitin ligase. *Science*, *322*(5901), 597–602. https://doi.org/10.1126/science.1162790.

Neumann, F. R., Dion, V., Gehlen, L. R., Tsai-Pflugfelder, M., Schmid, R., Taddei, A., et al. (2012). Targeted INO80 enhances subnuclear chromatin movement and ectopic homologous recombination. *Genes & Development*, *26*(4), 369–383. https://doi.org/10.1101/gad.176156.111.

Oza, P., Jaspersen, S. L., Miele, A., Dekker, J., & Peterson, C. L. (2009). Mechanisms that regulate localization of a DNA double-strand break to the nuclear periphery. *Genes & Development*, *23*(8), 912–927. https://doi.org/10.1101/gad.1782209.

Peng, J. C., & Karpen, G. H. (2008). Epigenetic regulation of heterochromatic DNA stability. *Current Opinion in Genetics & Development*, *18*(2), 204–211. https://doi.org/10.1016/j.gde.2008.01.021.

Petukhova, G., Stratton, S., & Sung, P. (1998). Catalysis of homologous DNA pairing by yeast Rad51 and Rad54 proteins. *Nature*, *393*(6680), 91–94. https://doi.org/10.1038/30037.

Petukhova, G., Van Komen, S., Vergano, S., Klein, H., & Sung, P. (1999). Yeast Rad54 promotes Rad51-dependent homologous DNA pairing via ATP hydrolysis-driven change in DNA double helix conformation. *The Journal of Biological Chemistry*, *274*(41), 29453–29462.

Riddle, N. C., Minoda, A., Kharchenko, P. V., Alekseyenko, A. A., Schwartz, Y. B., Tolstorukov, M. Y., et al. (2011). Plasticity in patterns of histone modifications and chromosomal proteins in Drosophila heterochromatin. *Genome Research*, *21*(2), 147–163. https://doi.org/10.1101/gr.110098.110.

Ryu, T., Bonner, M. R., & Chiolo, I. (2016). Cervantes and Quijote protect heterochromatin from aberrant recombination and lead the way to the nuclear periphery. *Nucleus*, 7(5), 485–497. https://doi.org/10.1080/19491034.2016.1239683.

Ryu, T., Spatola, B., Delabaere, L., Bowlin, K., Hopp, H., Kunitake, R., et al. (2015). Heterochromatic breaks move to the nuclear periphery to continue recombinational repair. *Nature Cell Biology*, *17*(11), 1401–1411. https://doi.org/10.1038/ncb3258.

Saad, H., Gallardo, F., Dalvai, M., Tanguy-le-Gac, N., Lane, D., & Bystricky, K. (2014). DNA dynamics during early double-strand break processing revealed by non-intrusive imaging of living cells. *PLoS Genetics*, *10*(3), e1004187. https://doi.org/10.1371/journal.pgen.1004187.

Saxton, M. J., & Jacobson, K. (1997). Single-particle tracking: Applications to membrane dynamics. *Annual Review of Biophysics and Biomolecular Structure*, *26*, 373–399. https://doi.org/10.1146/annurev.biophys.26.1.373.

Scalettar, B. A., Swedlow, J. R., Sedat, J. W., & Agard, D. A. (1996). Dispersion, aberration and deconvolution in multi-wavelength fluorescence images. *Journal of Microscopy*, *182*(Pt. 1), 50–60.

Scully, R., Chen, J., Ochs, R. L., Keegan, K., Hoekstra, M., Feunteun, J., et al. (1997). Dynamic changes of BRCA1 subnuclear location and phosphorylation state are initiated by DNA damage. *Cell*, *90*(3), 425–435.

Seeber, A., Dion, V., & Gasser, S. M. (2013). Checkpoint kinases and the INO80 nucleosome remodeling complex enhance global chromatin mobility in response to DNA damage. *Genes & Development*, *27*(18), 1999–2008. https://doi.org/10.1101/gad.222992.113.

Seong, K. H., Li, D., Shimizu, H., Nakamura, R., & Ishii, S. (2011). Inheritance of stress-induced, ATF-2-dependent epigenetic change. *Cell*, *145*(7), 1049–1061. https://doi.org/10.1016/j.cell.2011.05.029.

Spichal, M., Brion, A., Herbert, S., Cournac, A., Marbouty, M., Zimmer, C., et al. (2016). Evidence for a dual role of actin in regulating chromosome organization and dynamics in yeast. *Journal of Cell Science*, *129*(4), 681–692. https://doi.org/10.1242/jcs.175745.

Spichal, M., & Fabre, E. (2017). The emerging role of the cytoskeleton in chromosome dynamics. *Frontiers in Genetics*, *8*, 60. https://doi.org/10.3389/fgene.2017.00060.

Strecker, J., Gupta, G. D., Zhang, W., Bashkurov, M., Landry, M. C., Pelletier, L., et al. (2016). DNA damage signalling targets the kinetochore to promote chromatin mobility. *Nature Cell Biology*, *18*(3), 281–290. https://doi.org/10.1038/ncb3308.

Stucki, M., Clapperton, J. A., Mohammad, D., Yaffe, M. B., Smerdon, S. J., & Jackson, S. P. (2005). MDC1 directly binds phosphorylated histone H2AX to regulate cellular responses to DNA double-strand breaks. *Cell*, *123*(7), 1213–1226. https://doi.org/10.1016/j.cell.2005.09.038.

Su, X. A., Dion, V., Gasser, S. M., & Freudenreich, C. H. (2015). Regulation of recombination at yeast nuclear pores controls repair and triplet repeat stability. *Genes & Development*, *29*(10), 1006–1017. https://doi.org/10.1101/gad.256404.114.

Sung, P. (1994). Catalysis of ATP-dependent homologous DNA pairing and strand exchange by yeast RAD51 protein. *Science*, *265*(5176), 1241–1243.

Swartz, R. K., Rodriguez, E. C., & King, M. C. (2014). A role for nuclear envelope-bridging complexes in homology-directed repair. *Molecular Biology of the Cell*, *25*(16), 2461–2471. https://doi.org/10.1091/mbc.E13-10-0569.

Therizols, P., Fairhead, C., Cabal, G. G., Genovesio, A., Olivo-Marin, J. C., Dujon, B., et al. (2006). Telomere tethering at the nuclear periphery is essential for efficient DNA double strand break repair in subtelomeric region. *The Journal of Cell Biology*, *172*(2), 189–199. https://doi.org/10.1083/jcb.200505159.

Torres-Rosell, J., Sunjevaric, I., De Piccoli, G., Sacher, M., Eckert-Boulet, N., Reid, R., et al. (2007). The Smc5-Smc6 complex and SUMO modification of Rad52 regulates recombinational repair at the ribosomal gene locus. *Nature Cell Biology*, *9*(8), 923–931.

Tsouroula, K., Furst, A., Rogier, M., Heyer, V., Maglott-Roth, A., Ferrand, A., et al. (2016). Temporal and spatial uncoupling of DNA double strand break repair pathways within mammalian heterochromatin. *Molecular Cell*, *63*(2), 293–305. https://doi.org/10.1016/j.molcel.2016.06.002.

van Sluis, M., & McStay, B. (2015). A localized nucleolar DNA damage response facilitates recruitment of the homology-directed repair machinery independent of cell cycle stage. *Genes & Development*, *29*(11), 1151–1163. https://doi.org/10.1101/gad.260703.115.

Velichko, A. K., Petrova, N. V., Kantidze, O. L., & Razin, S. V. (2012). Dual effect of heat shock on DNA replication and genome integrity. *Molecular Biology of the Cell*, *23*(17), 3450–3460. https://doi.org/10.1091/mbc.E11-12-1009.

Wang, B., Matsuoka, S., Carpenter, P. B., & Elledge, S. J. (2002). 53BP1, a mediator of the DNA damage checkpoint. *Science*, *298*(5597), 1435–1438. https://doi.org/10.1126/science.1076182.

Yang, J., & Reth, M. (2012). Drosophila S2 Schneider cells: A useful tool for rebuilding and redesigning approaches in synthetic biology. *Methods in Molecular Biology*, *813*, 331–341. https://doi.org/10.1007/978-1-61779-412-4_20.

Zeileis, A., Kleiber, C., Krämer, W., & Hornik, K. (2003). Testing and dating of structural changes in practice. *Computational Statistics and Data Analysis*, *44*, 109–123.

Zhou, R., Mohr, S., Hannon, G. J., & Perrimon, N. (2013). Inducing RNAi in Drosophila cells by transfection with dsRNA. *Cold Spring Harbor Protocols*, *2013*(5), 461–463. https://doi.org/10.1101/pdb.prot074351.

Ziv, Y., Bielopolski, D., Galanty, Y., Lukas, C., Taya, Y., Schultz, D. C., et al. (2006). Chromatin relaxation in response to DNA double-strand breaks is modulated by a novel ATM- and KAP-1 dependent pathway. *Nature Cell Biology*, *8*(8), 870–876. https://doi.org/10.1038/ncb1446.

Zou, L., & Elledge, S. J. (2003). Sensing DNA damage through ATRIP recognition of RPA-ssDNA complexes. *Science*, *300*(5625), 1542–1548. https://doi.org/10.1126/science.1083430.

CHAPTER SIXTEEN

Analysis of Meiotic Double-Strand Break Initiation in Mammals

Kevin Brick[*,2], **Florencia Pratto**[*,2], **Chi-Yu Sun**[†], **Rafael D. Camerini-Otero**[*,1], **Galina Petukhova**[†,1]
[*]Genetics & Biochemistry Branch, NIDDK, National Institutes of Health, Bethesda, MD, United States
[†]Uniformed Services University of the Health Sciences, Bethesda, MD, United States
[1]Corresponding authors: e-mail address: rc10d@nih.gov; galina.petukhova@usuhs.edu

Contents

Abstract

The repair of programmed DNA double-strand breaks (DSBs) physically tethers homologous chromosomes in meiosis to allow for accurate segregation through meiotic cell divisions. This process, known as recombination, also results in the exchange of alleles between parental chromosomes and contributes to genetic diversity. In mammals, meiotic DSBs occur predominantly in a small fraction of the genome, at sites known as hotspots. Studies of the formation and repair of meiotic DSBs in mammals are challenging, because few cells undergo meiotic DSB formation at a given time. To better understand the initiation and control of meiotic recombination in mammals, we have devised a

[2] Equal contributions.

Methods in Enzymology, Volume 601
ISSN 0076-6879
https://doi.org/10.1016/bs.mie.2017.11.037

highly sensitive method to map the sites of meiotic DSBs genome wide. Our method first isolates DNA bound to DSB repair proteins and then specifically sequences the associated single-stranded DNA. This protocol has generated the first meiotic DSB maps in several mammals and the only map of meiotic DSBs in humans.

1. INTRODUCTION

Meiotic recombination predominantly occurs at recombination hotspots, discrete kilobase-size regions of the genome where the recombination rate can be thousands of times higher than in adjacent regions (Arnheim, Calabrese, & Tiemann-Boege, 2007; Paigen & Petkov, 2010). The first human hotspot was described in 1982 (Orkin et al., 1982), and until 2005, only a handful of individual hotspots had been studied in detail. In 2005, the first genome-wide maps of recombination hotspots were created by inferring recombination events from patterns of linkage disequilibrium in human populations (Myers, Bottolo, Freeman, McVean, & Donnelly, 2005). Over 30,000 recombination hotspots were identified by this method, opening the door to better understanding the process of meiotic recombination. Building on this work, a major breakthrough was made in 2010, when three labs independently identified the PRDM9 protein as responsible for targeting meiotic double-strand breaks (DSBs) in the human and mouse genomes (Baudat et al., 2010; Myers et al., 2010; Parvanov, Petkov, & Paigen, 2010). PRDM9 directs DSBs to specific hotspots through binding to specific DNA sequences (Baudat et al., 2010; Billings et al., 2013; Grey et al., 2011) and subsequently trimethylating lysine 4 of the histone H3 (Hayashi, Yoshida, & Matsui, 2005). The DNA DSB formation machinery is recruited to hotspot locations to initiate recombination (reviewed by Baudat, Imai, & de Massy, 2013; Bolcun-Filas & Schimenti, 2012; Hunter, 2015; Keeney, Lange, & Mohibullah, 2014), and the ends of meiotic DSBs are nucleolitically processed to form long single-stranded DNA (ssDNA) overhangs. ssDNA overhangs are bound by recombinases, RAD51 and DMC1, that facilitate the repair of DSBs via homologous recombination. This repair produces either crossover or noncrossover products, depending on the specific mechanism used. Although PRDM9 defines practically all recombination hotspots in mice (Brick, Smagulova, Khil, Camerini-Otero, & Petukhova, 2012) and humans (Pratto et al., 2014), many more PRDM9 binding sites are present in the genome than there are hotspots. This indicates that factors other than PRDM9 will shape the recombination landscape and highlights the importance of experimentally determining hotspot locations genome wide.

Here we describe an approach to map recombination initiation hotspots—the sites of meiotic DSBs (Khil, Smagulova, Brick, Camerini-Otero, & Petukhova, 2012). We first capture DNA bound by DMC1 or RAD51 using chromatin immunoprecipitation (ChIP). Next, we apply a bespoke protocol to enrich for ssDNA and then sequence these ssDNA fragments by high-throughput sequencing (Khil et al., 2012) (Fig. 1). Since our approach involves direct physical detection of recombinase-bound ssDNA ends, it is broadly applicable to any sexually reproducing species. The precision of hotspot mapping is independent of polymorphism density (which affects crossover-mapping-based methods), and since our method detects

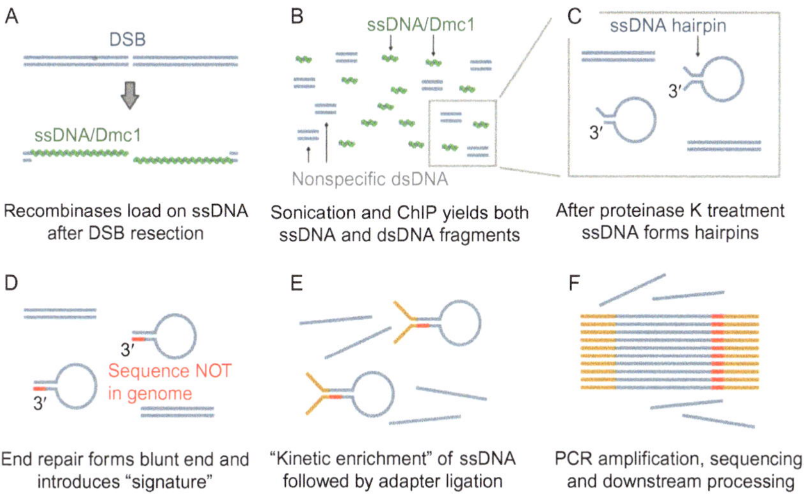

Fig. 1 ChIP-SSDS schematic. (A) ChIP-SSDS captures ssDNA bound by meiotic recombinases to map DSB hotspot locations. (B) After fixation and sonication, DMC1-bound ssDNA is immunoprecipitated using an anti-DMC1 antibody. (C) Deproteinized ssDNA will spontaneously form hairpins, mediated by intramolecular microhomologies. (D) The 3′ ends of hairpins are trimmed, then extended, using the 5′ end of the ssDNA molecule as a template. This both introduces a "signature" that allows downstream identification of ssDNA-derived sequencing reads and ensures that hairpins have double-stranded ends. (E) The DNA is denatured and then allowed to return to room temperature. In the reaction time frame, ssDNA hairpins will reform because of fast intramolecular reannealing; however, dsDNA will not renature. Sequencing adapters are then ligated to blunt-ended DNA, which will primarily occur at ssDNA-derived hairpins. (F) DNA with ligated adapters is then amplified and sequenced. A computational pipeline then specifically uses the "signature" sequence (D) to identify ssDNA.

the earliest intermediates in DSB repair, it can identify hotspots independent of downstream repair pathway decisions. Furthermore, since this method surveys recombination in a population of meiocytes, it yields accurate quantitative estimates of hotspot usage that are not achievable by other techniques.

Single-stranded DNA sequencing (SSDS) has been predominantly used in mice (Brick et al., 2012; Davies et al., 2016; Grey et al., 2017; Smagulova et al., 2011; Smagulova, Brick, Pu, Camerini-Otero, & Petukhova, 2016) but has recently allowed the generation of the first maps of recombination hotspots in individual human genomes (Pratto et al., 2014). In addition to mapping meiotic DSBs, SSDS may be useful for detecting DSBs in somatic tissues, or for detecting other ssDNA species, such as replication intermediates and certain viruses.

2. ChIP AND SSDS

2.1 Reagents

Paraformaldehyde (PFA) (Sigma, P6148)
5 *M* NaCl (Sigma, S5150-1L)
SDS (Sigma)
1 *M* Tris–Cl, pH 6.5 at 25°C (KD Medical)
1 *M* Tris–Cl, pH 8.0 at 25°C (KD Medical)
KOH (Sigma)
Triton X-100 (Sigma)
0.5 *M* EDTA (Sigma)
0.1 *M* EGTA (Sigma)
Proteinase Inhibitor cocktail (Roche, 11 836 153 001)
UltraPure DNase/RNase-Free Distilled Water (Invitrogen, 10977-023)
PBS (KD Medical)
TE (KD Medical)
8 *M* LiCl (Sigma, L7026)
IGEPAL-CA630 (Sigma, 18896)
Sodium deoxycholate monohydrate (Sigma, D5670)
$NaHCO_3$ (Sigma, S5761)
Dynabeads Protein G (Invitrogen, Cat no: 10004D)
Proteinase K, 20 mg/mL (Thermo Fisher Scientific # 25530049)
Anti-DMC1 antibody (Abcam, Ab 11054)
MinElute PCR Cleanup Kit (Qiagen, 28004)
TruSeq Nano DNA LT Library Preparation Kit (Illumina)

T4 DNA polymerase (3 U/μL, New England Biolabs, M0203)
DNA Polymerase I Large (Klenow) Fragment (5 U/μL, New England Biolabs, M0210)
T4 Polynucleotide Kinase (PNK) (10 U/μL, New England Biolabs, M0201)
Klenow fragment ($3' \rightarrow 5'$ *exo*-) (5 U/μL with 10 × NEBuffer 2, New England Biolabs, M0212)
Quick ligation kit (New England Biolabs, M2200)
T4 DNA Ligase Reaction Buffer (New England Biolabs, B0202)
dATP and dNTPs (New England Biolabs)
Qubit dsDNA HS Assay Kit (Thermo Fisher Scientific # Q32851) or Quant-iT PicoGreen dsDNA Assay Kit (Thermo Fisher Scientific, P7589)
AMPureXP beads (Beckman Coulter)

2.2 Key Equipment and Consumables

DNA LoBind 1.5 mL tubes (Eppendorf 022431021) and low retention filtered tips (USA Scientific)
Slide-A-Lyzer G2 Dialysis Cassette 10k MWCO, 3 mL (Thermo Scientific, 87730)
DynaMagTM-2 Magnet (Invitrogen, 12321D)
15 mL Dounce homogenizer with a tight fit pestle
70 mm cell strainer (BD Biosciences, 352350)
Thermomixer (Eppendorf)
PCR machine
Qubit Fluorometer (Thermo Fisher Scientific # Q33216) to use with the Qubit dsDNA HS Assay Kit or any fluorometer to use with PicoGreen kit
Bioruptor UCD 200 (Diagenode) and 15 mL polystyrene Falcon tubes
Library Quantification Kit (KAPA Biosystems # KK4824)

2.3 Buffers

Note: all buffers are prepared with deionized water and passed through 0.22-μm filter.
10% SDS
1 N KOH
Quenching buffer, store at 4°C:
1.25 *M* Glycine

Cell wash buffer 1, store at 4°C:

0.25% Triton X-100, 10 m*M* EDTA, 0.5 m*M* EGTA, 10 m*M* Tris–Cl, pH 8

Cell wash buffer 2, store at 4°C:

200 m*M* NaCl, 1 m*M* EDTA, 0.5 m*M* EGTA, 10 m*M* Tris–Cl, pH 8

Lysis buffer, store at 4°C:

1% SDS, 10 m*M* EDTA, 50 m*M* Tris–Cl, pH 8

Note: Warm to room temperature to allow SDS to redissolve before use.

Protease Inhibitors 100× store at −20°C:

One tablet of Protease Inhibitor Cocktail (Roche, 11836153001) dissolved in 500 μL of UltraPure water.

ChIP buffer, store at 4°C:

16.7 m*M* Tris–HCl, pH 8.0, 167 m*M* NaCl, 1.2 m*M* EDTA, 0.01% SDS, 1.1% Triton X-100

Note: 1 L of this buffer is freshly prepared at day 1.

ChIP wash 1, store at 4°C:

20 m*M* Tris–HCl pH 8, 150 m*M* NaCl, 2 m*M* EDTA, 1% Triton X-100, 0.1% SDS

ChIP wash 2, store at 4°C:

20 m*M* Tris–HCl pH 8, 500 m*M* NaCl, 2 m*M* EDTA, 1% Triton X-100, 0.1% SDS

ChIP wash 3, store at 4°C:

10 m*M* Tris–HCl pH 8, 0.25 *M* LiCl, 1 m*M* EDTA, 1% deoxycholic acid, 1% IGEPAL-CA630

1 *M* $NaHCO_3$, store at −20°C:

Note: Warm to room temperature to allow $NaHCO_3$ to redissolve before use.

2.4 Day 1: Chromatin Preparation and ChIP

2.4.1 Prepare the Following

18.5% PFA (5 mL)

Note: Refer to safety data sheet from the supplier while handling PFA and dispose accordingly.

- Weigh 0.925 g PFA and transfer into a 50-mL Falcon tube containing 4.8 mL H_2O and 35 μL 1 N KOH.
- Put the tube into 500 mL glass beaker filled with water to approximately 200 mL. Bring the water in the beaker to boiling using a microwave oven. Mix PFA suspension to dissolve the PFA powder. Repeat if necessary.
- Cool down the solution, keep on ice.

1% PFA (10 mL per sample)

- Add 540 μL of the 18.5% PFA to 9.46 mL of 1 × PBS
- Keep at room temperature

2.4.2 Experimental Procedure

1. Dissect testes, place in a Petri dish with cold 1 × PBS, transfer to a dry dish, remove tunica albuginea, and gently pull apart testis with fine forceps.
2. Transfer testes to 10 mL of 1% PFA in a 15-mL Falcon tube and rock for 10 min at room temperature on a rocker.
3. Immediately add 1 mL of quenching buffer, rock for 5 min at room temperature, and place the sample on ice.
4. Transfer solution with the fixed tissue to a 15-mL autoclaved Dounce homogenizer, and homogenize the tissue with 10 strokes of the pestle.
   Note: Depending on the tissue used, aggregates may remain in the suspension.
5. Pass the cell suspension through a 70-μm cell strainer into a 50-mL Falcon tube.
6. Transfer to a 15-mL Falcon tube and spin at 900 × *g* for 5 min at 4°C. Remove supernatant.
7. Resuspend the pellet in 10 mL of 1 × PBS. Spin at 900 × *g* for 5 min at 4°C and remove supernatant.
8. Resuspend in 10 mL of cell wash buffer 1. Spin at 900 × *g* for 5 min at 4°C and remove supernatant.
9. Resuspend in 10 mL of cell wash buffer 2. Spin at 900 × *g* for 5 min at 4°C and remove supernatant.
10. Resuspend in 1.5 mL of lysis buffer and add 15 μL of the 100 × protease inhibitor stock solution.
11. Transfer supernatant to a 15-mL polystyrene Falcon tube.
    Note: Polystyrene tubes are required for optimal sonication.
12. Sonicate chromatin for 15 min in the Bioruptor with the following parameters:
    15 s ON/45 s OFF (setting: high)
    Note: Adjust the time if necessary aiming for a fragment size ~1000 bp.
13. While waiting for sonication to complete, place a dialysis chamber in 1 L of cold ChIP buffer in the cold room. Spin at low speed using a stir bar.

14. Leave sonicated chromatin at room temperature until SDS is redissolved (about 1 min)
15. Transfer the chromatin to a 1.5-mL Eppendorf tube and spin at 12,000 × *g* for 10 min at 4°C.
16. Transfer the supernatant to a fresh 15-mL Falcon tube and add 1.5 mL of fresh ChIP buffer (total volume = 3 mL).
17. Transfer the 3 mL of sonicated chromatin to the dialysis chamber (from step 13) and dialyze for 5 h at 4°C with constant but slow stirring.
18. Transfer the dialyzed solution into two 1.5 mL tubes and keep 50 μL of the solution to check sonication efficiency.
19. Add 15 μL of 100 × of protease inhibitor stock solution to each tube.
20. Add 12 μg of DMC1 antibody per tube (24 μg total).
21. Rotate the tubes slowly overnight at 4°C.
22. Check sonication efficiency:
    - Add 250 μL H_2O to the 50 μL aliquot from step 18
    - Add NaCl to 0.2 *M*
    - Boil for 15 min to reverse crosslinking
    - Cool down to room temperature, add 1 μL of 10 mg/mL RNase A and incubate for 10 min at 37°C
    - Clean up the DNA with Qiagen MinElute kit using the protocol provided with the kit
    - Determine the size of sonicated DNA on 1% agarose gel

2.5 Day 2: ChIP (Continued)

2.5.1 Prepare the Following

Prepare elution buffer (1 mL)

- 800 μL ddH_2O
- 100 μL 1 *M* $NaHCO_3$ (bring to room temperature to dissolve particulates)
- 100 μL 10% SDS

2.5.2 Experimental Procedure

1. Prepare and add the Protein G Dynabeads
   a. Aliquot 150 μL of the bead slurry, separate the beads on the magnet, and remove supernatant.
   b. Wash the beads three times with the ChIP buffer.
   c. Resuspend the beads in ~100 μL of the ChIP sample from day 1, step 21 and transfer back to the rest of the ChIP sample, splitting between the two sample tubes.
   d. Rotate for 2 h at 4°C.

2. Briefly (1–2 s) spin the tubes to collect solution from the lid, separate the beads on the magnet, remove and save the supernatant (store at −20°C).
3. Wash the beads/chromatin complex by sequentially suspending the beads in 1.2 mL of each of the cold buffers in the order listed below. After each suspension, rotate for 5 min in the cold room, briefly spin, then collect beads on the magnet, and carefully dispose of the supernatant.
   a. ChIP wash 1
   b. ChIP wash 2
   c. ChIP wash 3
   d. TE buffer (two washes)
4. Add 150 μL of elution buffer to each tube and mix by gentle flicking.
5. Incubate in a thermomixer at 65°C, shaking at 800 RPM for 30 min.
6. Separate the beads on the magnet and combine the supernatant from both tubes into a fresh 1.5-mL Eppendorf tube for a total volume of 300 μL.
   Note: Be careful not to discard the supernatant at this step.
7. To the 300 μL sample, add 12 μL of 5 *M* NaCl and incubate at 65°C overnight (18–20 h) to reverse DNA–protein cross-links.

2.6 Day 3: DNA Cleanup

2.6.1 Experimental Procedures

1. Digest the protein
   Add 6 μL of 0.5 *M* EDTA, 12 μL of 1 *M* Tris–HCl pH 6.5 and 5 μL of Proteinase K (20 mg/mL) to the reverse-cross-linked chromatin and incubate at 45°C for 1 h.
2. Clean up the DNA with Qiagen MinElute kit using the following protocol:
   a. Transfer the sample to a 15-mL Falcon tube.
   b. Add 2.4 mL (seven volumes) of Buffer PB and mix well.
   c. Transfer 0.7 mL of the mixture to a MinElute column, incubate at room temperature for 1 min and centrifuge for 1 min.
   d. Discard flow-through and place the MinElute column back into the same tube.
   e. Repeat steps c and d until the entire mixture is passed through the column.
   f. Add 750 μL of Buffer PE to the MinElute column with the bound DNA, incubate at room temperature for 1 min, and centrifuge for 1 min at full speed.

   g. Discard flow-through and place the MinElute column back in the same tube.
   h. Repeat steps f and g once.
   i. Centrifuge the column for an additional 1 min at maximum speed.
   j. Remove residual ethanol around the ring of the column using a pipette.
   k. Place the MinElute column in a clean 1.5-mL microcentrifuge tube.
   l. Add 12 μL of Buffer EB to the center of the membrane, incubate at room temperature for 1 min, and centrifuge for 1 min at full speed. DNA will be in the eluate.
3. Use Qubit kit (Invitrogen) or PicoGreen kit (Thermo Fisher Scientific) to measure the resulting DNA concentration of the samples.
   Note: typical concentration for anti-DMC1 ChIP samples is within the range of 1–3 ng/μL.
4. Store samples at −20°C or proceed with library preparation.

2.7 Day 4: Library Preparation

The protocol for library preparation is based on Illumina protocols but has some important modifications that ensure optimal library yield from ssDNA. Standard Illumina protocols may not result in a satisfactory library.
All reactions are performed in 1.5 mL Eppendorf tubes. The exception is the PCR reaction, which uses a 0.2-mL tube.

2.7.1 Experimental Procedures

1. End-repair

End-Repair Premix

10 × T4 ligation buffer	5 μL
10 m*M* dNTPs	2 μL
T4 DNA polymerase (3 U/μL)	1 μL
Klenow fragment (1 U/μL; see Note 1)	1 μL
T4 PNK (10 U/μL)	1 μL
H_2O	30 μL
Total	40 μL

Note 1: Original Klenow concentration is 5 U/μL, prepare a 1/5 dilution every time.

Note 2: Current Illumina protocols perform this step at higher temperatures. This is NOT recommended for SSDS library preparation.

a. Mix 40 μL of the end-repair premix with 10 μL of the ChIP DNA from day 3, step 4.

b. Incubate at 20°C in a water bath for 30 min.

2. Purification of DNA using Qiagen MinElute PCR cleanup

Add 350 μL (seven reaction volumes) of Buffer PB to the reaction mix and follow the purification protocol provided with the kit. Elute DNA with 12 μL of Buffer EB.

3. A-tailing

A-Tailing Premix

10 × NEBuffer 2	5 μL
Klenow fragment 3′ → 5′*exo-* (5 U/μL)	1 μL
1 m*M* dATP	10 μL
H_2O	24 μL
Total	40 μL

a. Mix 40 μL of the A-tailing premix with 10 μL of the end-repaired DNA from step 2.

b. Incubate at 37°C for 30 min.

4. Purification of DNA using Qiagen MinElute PCR Cleanup Kit

Add 350 μL (seven reaction volumes) of Buffer PB to the reaction mix and follow the purification protocol provided with the kit. Elute DNA with 12 μL of Buffer EB.

5. Kinetic enrichment for ssDNA

Note: This is the key step in the protocol to enrich for ssDNA.

a. Incubate the eluted DNA from step 4 at 95°C for 3 min in the Thermomixer.

b. Spin the sample briefly and allow to return to room temperature.

6. Adapter ligation

Ligation Setup

DNA	10 μL
Quick DNA ligase buffer (2 ×)	15 μL
H_2O	3 μL

Continued

Ligation Setup

Diluted TruSeq adaptor (see Note 1)	1 μL
Quick T4 DNA ligase (see Note 2)	1 μL
Total	30 μL

Note 1: The stock TruSeq adapter concentration is 15 μ*M*. This should be diluted 1:10 in Buffer EB, and further diluted 1:10. 1 μL of this 1:100 dilution is used for ligation.
Note 2: The T4 DNA ligase is added last.
Note 3: Current Illumina protocols perform this step at higher temperatures. This is NOT recommended for SSDS library preparation.
Incubate at 20°C for 30 min in a water bath.

7. Purification of DNA using Qiagen MinElute PCR Cleanup Kit
Add 150 μL (five reaction volumes) of Buffer PB to the reaction mix and follow the purification protocol provided with the kit. Elute DNA with 12 μL of Buffer EB.
8. Library amplification

Amplification Premix

Illumina TruSeq PCR Master Mix (2×)	25 μL
Illumina TruSeq PCR primer cocktail	5 μL
H_2O	10 μL
Total	40 μL

a. Mix 40 μL of the amplification premix with the 10 μL of adapter-ligated DNA from step 7.
b. Run the following program on a PCR machine:
12–15 cycles of:
- 98°C 30 s
- 98°C 10 s
- 60°C 30 s
- 72°C 30 s

Followed by:
- 72°C 5 min
- 4°C hold

9. Purification of DNA using AMPureXP beads
   a. After PCR, add 0.9 × volumes of AMPureXP beads.
      Note: 0.9 × volumes is chosen to optimize the removal of adapter dimers (~125 bp) but to keep ssDNA-derived fragments (mean = 205–245 bp). This should be carefully calibrated if different beads are used.
   b. Incubate the PCR tube at room temperature for 10 min to bind DNA to the beads.
   c. Collect the beads on a magnet and discard supernatant.
   d. With the tube still in the magnet, add 200 μL of 80% ethanol.
   e. Incubate at room temperature for 1 min.
   f. Remove and discard ethanol.
   g. Repeat steps d–f once.
   h. Dry the beads at room temperature for 3–5 min or until all the ethanol has evaporated.
   i. Remove the tube from the magnet and resuspend the beads in 20 μL of elution buffer (Qiagen EB buffer).
   j. Incubate at room temperature for at least 2 min.
   k. Collect the beads on a magnet and transfer supernatant to a new Eppendorf tube.
   l. Measure DNA concentration using Qubit or PicoGreen kits.
   m. Store samples at −20°C.
10. Library quality check
    To obtain optimal cluster density for sequencing, it is important to accurately quantify the DNA concentration of sequencing libraries.
    Perform the following:
    a. Determine the concentration of the library using qPCR (KAPA Library Quantification Kit).
    b. Check the length distribution of DNA fragments by running an aliquot of the DNA library on the Agilent Technologies 2100 Bioanalyzer using a high sensitivity DNA chip. Typical bioanalyzer profiles for good SSDS libraries are shown in Fig. 2.
    c. Calculate the molarity of the library according to concentration and fragment size distribution.
11. Perform paired-end sequencing of SSDS library
    Note 1: Computational parsing of ssDNA uses 36 bp from read 1 and 40 bp from read 2. Thus, longer sequencing reads are not required.

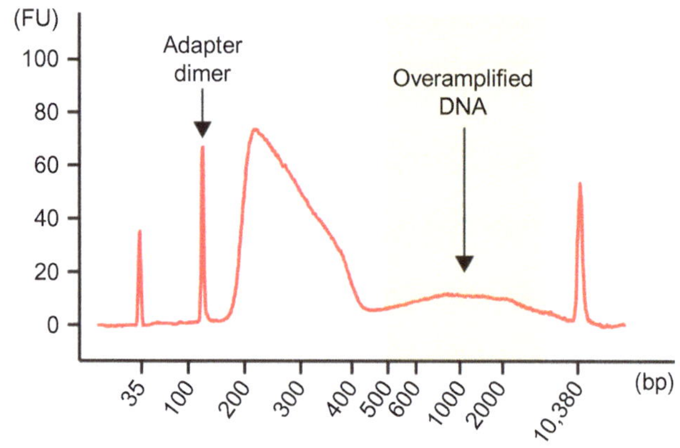

Fig. 2 Bioanalyzer output depicting the fragment size distribution in a typical SSDS library. The proportions of adapter dimers and overamplified DNA may vary considerably between samples. Libraries with a large excess of either these populations may reduce the yield of ssDNA. This library was not purified using beads (day 4: step 9) so that the adapter dimer peak could be clearly seen.

Note 2: SSDS libraries form secondary structures that appear to adversely affect clustering and sequencing on Illumina platforms. We recommend loading the Illumina flow cell at double the recommended molarity to obtain near optimal cluster density. Running samples in rapid run mode of the Illumina HiSeq 2500 has produced the best results.

3. COMPUTATIONAL ANALYSIS: IDENTIFICATION OF DSB HOTSPOTS FROM SSDS DATA

Most of the software tools for the SSDS data analysis are commonly used for NGS analyses. For more advanced users, it is possible to build a custom pipeline that recapitulates the steps below. The protocol begins with fastq files provided by the sequencing facility. Reads in these fastq files will be aligned to the reference genome, ssDNA will be identified, and then DSB hotspots are called.

3.1 Materials

Computer with Ubuntu Linux operating system
Linux user with sudo privileges

8 GB of RAM and preferably multiple CPU cores

Note: Instructions are provided for Ubuntu Linux. Installation can be performed with minor modifications on other Linux distributions or Apple Macs. The SSDS pipelines will not run on Windows PCs.

3.2 Required Programs/Packages

3.2.1 Linux Packages

bioperl
default-jre
git
libxml2-dev
libcurl4-openssl-dev
python-setuptools
python-pip
python-numpy
python-scipy
r-base-core
zlib1g-dev

3.2.2 Perl Packages

Bio::DB::Sam
File::Temp
Getopt::Long
List::Util
Make::Build
Math::Round
Statistics::Descriptive
Switch

3.2.3 R Packages

RCurl
XML

3.2.4 R Packages (Bioconductor)

Rtracklayer
ShortRead

3.2.5 Other Software

macs_2.1.0.20150731

3.3 SSDS Alignment Pipeline

3.3.1 Introduction

The following instructions are to be carried out from the Linux terminal. In Ubuntu, the terminal can be accessed from the applications menu or by pressing "Ctrl + Alt + T" key combination. In this protocol, terminal commands are displayed in monospaced font and are preceded by the ">" symbol. The ">" should not be included in commands.

For example, the following line

```
>apt-get install testPackage
```

should be interpreted as:

Type "apt-get install testPackage" at the Linux terminal and then hit "Enter"

Note: The Linux terminal is case sensitive; therefore all commands must be typed exactly as specified.

3.3.2 Installation

Install git if required:

```
>sudo apt-get install git
```

Use git to clone the SSDS pipeline on your computer:

```
>git clone https://github.com/kevbrick/SSDSpipeline.git
```

The easiest way to install the SSDS alignment pipeline is to simply run the configuration script located in the folder downloaded from github:

```
>cd SSDSpipeline
```

The configure script will install the SSDS alignment pipeline to /home/$USER/SSDS_pipeline_1.0.0.

To install the pipeline to the default locations:

```
>sudo ./configure.sh
```

Alternatively, you can specify the installation and/or genomes folders:

```
>sudo ./configure.sh -i /share/SSDS -g /share/SSDSgenomes
```

The configuration script will install all the required dependencies and test the pipeline.

If the configuration process successfully completes, the pipeline has been installed. Please skip Sections 3.3.3.1–3.3.3.3. If the configuration process fails, please follow the manual instructions outlined in Sections 3.3.3.1–3.3.3.3. These instructions also serve as a guide for users of other operating systems.

3.3.3 Manual Installation

3.3.3.1 Set Environment Variables

SSDSPIPELINEPATH	Location of run_SSDS_pipeline binary
SSDSPICARDPATH	Picard location (recommend: $SSPIPELINEPATH/ picard-tools-2.3.0)
SSDSFASTXPATH	Path to fastx_trimmer binary (recommend: $SSPIPELINEPATH)
SSDSSAMTOOLSPATH	Path to samtools binary (recommend: $SSPIPELINEPATH)
SSDSGENOMESPATH	Genomes top level folder
SSDSTMPPATH	Temporary folder location
PERL5LIB	$SSDSPIPELINEPATH must be added to the Perl path

It is best to define these environment variables in your user configuration file (i.e., ~/.bashrc). Add the following lines to ~/.bashrc:

```
export SSDSPIPELINEPATH=/XXX/YYY/SSDSpipeline
export SSDSPICARDPATH=/XXX/YYY/SSDSpipeline/picard-tools-2.3.0
export SSDSFASTXPATH=/XXX/YYY/SSDSpipeline
export SSDSSAMTOOLSPATH=/XXX/YYY/SSDSpipeline
export SSDSGENOMESPATH=/XXX/YYY/SSDSpipeline/genomes
export SSDSTMPPATH=/tmp
export PERL5LIB=$PERL5LIB:$SSDSPIPELINEPATH
```

Initialize variables in .bashrc file.

```
>source ~/.bashrc
```

3.3.3.2 Install Dependencies

The following programs are required for the SSDS pipeline:

Java runtime.

Get java runtime (root privileges required):

```
>sudo apt-get install default-jre
```

The following Perl packages are required for the SSDS pipeline:

Bio::DB::Sam

File::Temp

Getopt::Long
Math::Round
Statistics::Descriptive
Switch
Make::Build
Install Perl modules as follows (root privileges required):

```
>export PERL_MM_USE_DEFAULT=1
>sudo cpan File::Temp
>sudo cpan Getopt::Long
>sudo cpan Make::Build
>sudo cpan Math::Round
>sudo cpan Statistics::Descriptive
>sudo cpan Switch
>sudo apt-get install zlib1g-dev
>sudo apt-get install bioperl
>cd $SSDSPIPELINEPATH/Bio-SamTools-1.43
>perl INSTALL.pl
```

3.3.3.3 Run the Unit Tests

```
>cd $SSDSPIPELINEPATH/unitTest/
>sh runTest.sh
```

This will test the addGenome.pl script and the SSDS pipeline using either fastq, fastq.gz, or bam files as input. Successful completion of the tests will result in the following output:

```
Test genome was added and indexed successfully !!
SSDS pipeline from FASTQ successful !!
SSDS pipeline from FASTQ.GZ successful !!
```

The SSDS pipeline has been installed successfully and can be run.

Note 1: The SSDS pipeline has been tested and used on UCSC genomes. Therefore, chromosome names in the genome.fa file should follow the "chr##" nomenclature. Chromosome names that lack a "chr" preface or that use letter-based designations may not work.

Note 2: Specific versions of picard, samtools, and fastx_trimmer are included in the SSDS pipeline repository. We recommend using these versions as other versions may not be compatible with the SSDS pipeline. Expert users can tweak the pipeline scripts to use different versions of these programs if desired.

3.3.4 Indexing the Reference Genome With BWA

The genomes folder is used for both alignment and for sorting BAM files. Each reference genome must be in a unique subfolder.

The $SSDSPIPELINEPATH/addGenome.pl script will add a genome and generate all the required files. It takes a genome fasta file as input.

```
>perl $SSDSPIPELINEPATH/addGenome.pl --fa (genome FASTA file) --name (genome name; ie. mm10) --g (genome assembly; ie. mm10) --s (species; ie. mus_musculus)
```

This will index the genome and create a folder structure in $SSDSGENOMESPATH. The name provided to the addGenome.pl script is the genome name that will be used for running the pipeline (--g argument).

Alternatively, the genome folder can be populated manually using steps 1–4 below:

1. Create the folder. The folder name will be used as the genome name when running the SSDS pipeline:
```
>mkdir $SSDSGENOMESPATH/genome_name/
```
2. Copy the genome fasta file to the genomes folder:
```
>cp genome.fa $SSDSGENOMESPATH/genome_name/genome.fa
```
3. Create the genome fasta file index (makes fasta.fai file):
```
>samtools faidx $SSDSGENOMESPATH/genome_name/genome.fa
```
4. Index the genome for the BWA version required:
   The genome folders structure is as follows (folder names must be exact):
```
>mkdir $SSDSGENOMESPATH/genome_name/BWAIndex
>mkdir $SSDSGENOMESPATH/genome_name/BWAIndex/version0.5.x
>mkdir $SSDSGENOMESPATH/genome_name/BWAIndex/version0.7.10
```
   Once this structure has been built, index the genome:
```
>cd $SSDSGENOMESPATH/genome2use/BWAIndex/version0.5.x
>ln -s ../../genome.fa .
>$SSDSPIPELINEPATH/BWA_0.5.x index -a bwtsw genome.fa
>cd $SSDSGENOMESPATH/genome2use/BWAIndex/version0.7.10
>ln -s ../../genome.fa .
>$SSDSPIPELINEPATH/BWA_0.7.12 index -a bwtsw genome.fa
```
   Create genome dictionary file:
```
>java -jar $SSDSPICARDPATH/picard.jar CreateSequenceDictionary
  R=$SSDSGENOMESPATH/genome_name/genome.fa
  O=$SSDSGENOMESPATH/genome_name/genome.dict GENOME_ASSEMBLY=XXXX
  SPECIES=mySpecies VALIDATION_STRINGENCY=LENIENT
```

3.3.5 Align Reads to Reference Genome With the SSDS Pipeline

Starting from fastq/fastq.gz files, the pipeline can be run with the following command:

```
>$SSDSPIPELINEPATH/run_ssDNAPipeline \
--g {genome name} \
--n {number of threads} \
--fq1 {path to fastq / fastq.gz for read 1} \
--fq2 {path to fastq / fastq.gz for read 2} \
--sample {sample name} \
--lane {lane number; or ANY number} \
--date {date in DDMMYY format} \
--outdir {output folder}
```

Starting from a paired-end BAM file, the pipeline can be run with the following command:

```
>$SSDSPIPELINEPATH/run_ssDNAPipeline \
--g {genome name} \
--n {number of threads} \
--bam {path to paired end bam file} \
--sample {sample name} \
--lane {lane number; or ANY number} \
--date {date in DDMMYY format} \
--outdir {output folder}
```

The arguments for the SSDS pipeline can be accessed from the command line:

```
>$SSDSPIPELINEPATH/run_ssDNAPipeline -h
```

Alternatively, they are listed as follows:

Argument	Synopsis (* = required)	Detail
Note: either --bam OR --fq1 & --fq2 arguments are required		
--bam	*Input BAM file	Full path to a paired-end BAM file
--fq1	*Input fastq for read 1	Full path to a fastq / fastq.gz file for read 1
--fq2	*Input fastq for read 2	Full path to a fastq / fastq.gz file for read 2
--g	*Genome name	Name must match a folder in $SSDSGENOMESPATH

--n	Number of threads (for alignment step)	Default = 12
--r1BP	Trim size for read 1 (bp)	Integer value (default = 36)
--r2BP	Trim size for read 2 (bp)	Integer value (default = 40)
--splitSz	Large BAM files are split for ssDNA detection	A lower value reduces the memory footprint (default = 20,000,000)
--outdir	Output folder	Defaults to BAM/FASTQ folder
--sample	Sample name for read group	—
--date	Date for read group	Date in DDMMYY format
--lane	Lane for read group	—
--BWAVers	BWA version to use	0.5 or 0.7 (default)
--v	Use verbose mode	—
--h/help	Show help	—

3.3.6 Outputs From the SSDS Pipeline

The SSDS pipeline aligns Illumina sequencing reads to a reference genome and then generates output files for each of five subcategories of DNA types (for more detail, see Khil et al., 2012):

ssDNA type 1: high confidence ssDNA
ssDNA type 2: low confidence ssDNA
dsDNA: low confidence dsDNA
dsDNA strict: higher confidence dsDNA
unclassified: other ambiguous DNA

Importantly, these designations represent the most likely source of DNA, given the alignment and detectable inverted terminal repeat structure. These designations are not 100% correct: for example, ssDNA with perfect short homology between the start and end of a fragment will be classified as dsDNA. It is very unlikely that dsDNA will ever be classified as ssDNA type 1.

Sequencing reads of each type are output to BAM files. In ssDNA BAM files, the ITRs have been removed, but details of the ITR structure are retained as tag fields:

Microhomology length: uh:i:{number}
Mismatches within microhomology: mm: i:{number}
Distance of microhomology to fragment end (offset): os:i:{number}
ITR length: it: i:{number}
Fragments, representing the entire sequence from the start of the first-end read to the end of the second-end read, are output to BED files. Columns of the BED files correspond to:
Chromosome
Fragment start
Fragment end
Read 1 Quality Score_Read 2 Quality Score
ITR length_Microhomology length_Offset length
Fragment strand

3.4 Identification of DSB Hotspots

3.4.1 Installation

Install git if required:

```
>sudo apt-get install git
```

Use git to clone the callHotspots pipeline on your computer:

```
>git clone https://github.com/kevbrick/callHotspotsSSDS.git
```

3.4.2 Configure the Pipeline

The easiest way to configure the SSDS hotspot calling pipeline is to simply run the configuration script located in the folder downloaded from github:

```
>cd callHotspotsSSDS
```

The configure script will install the SSDS hotspot calling pipeline to /home/$USER/SSDS_callHotspots_1.0.0.
To install the pipeline to the default locations:

```
>sudo ./configure.sh
```

Alternatively, you can specify the installation folder:

```
>sudo ./configure.sh -i /share/SSDS_callHotspots_1.0.0
```

The configuration script will install all the required dependencies and test the pipeline.
If the configuration process successfully completes, the pipeline has been installed. Please skip Sections 3.4.3.1–3.4.3.4. If the configuration process fails, please follow the manual instructions outlined in Sections 3.4.3.1–3.4.3.4. These instructions also serve as a guide for users of other operating systems.

3.4.3 Manual Installation

3.4.3.1 Copy callHotspotsSSDS Folder to Desired Location

For the callHotspotsSSDS pipeline to be available for all users on a system, it is recommended to copy the entire contents of the callHotspotsSSDS folder to /usr/share:

```
>cp -r ./callHotspotsSSDS /usr/share
```

This is a recommended, but NOT a required step.

3.4.3.2 Install Dependencies

The following programs are required to run the callHotspots pipeline:
R (>version 2.10)
MACS (version 2.1.0.20150731)
Pip
First, install bash libraries (root privileges required):

```
>apt-get install -y r-base-core libxml2-dev libcurl4-openssl-dev python-setuptools python-pip python-numpy python-scipy
```

To install R packages, start R from the command line:

```
>R
```

From the R prompt, run:

```
>install.packages("RCurl", repos="http://cran.rstudio.com")
>install.packages("XML", repos="http://cran.rstudio.com")
>source("https://bioconductor.org/biocLite.R")
>biocLite("rtracklayer", repos="http://cran.rstudio.com")
>biocLite("ShortRead", repos="http://cran.rstudio.com")
```

MACS can be installed from the command line as follows (root access is required):

```
>sudo pip install --root=/XXX/YYY/callHotspotsSSDS/macs_2.1.0.20150731 -U MACS2==2.1.0.20150731
```

The following Perl packages are required for the callHotspots pipeline:
File::Temp
Getopt::Long
List::Util
Math::Round
Statistics::Descriptive
Install Perl modules as follows (root privileges required):

```
>export PERL_MM_USE_DEFAULT=1
>sudo cpan File::Temp
>sudo cpan Getopt::Long
>sudo cpan Math::Round
>sudo cpan Statistics::Descriptive
>sudo cpan List::Util
```

3.4.3.3 Set Environment Variables

CHSPATH	Location of pipeline
CHSNCISPATH	Path to NCIS R script (recommend: $CHSPATH/NCIS)
CHSBEDTOOLSPATH	Path to bedtools binaries (recommend: $CHSPATH/bedtools)
CHSTMPPATH	Temporary folder location
PERL5LIB	$CHSPATH must be added to the Perl path
CHSMACSPATH	Path to macs2 binary
PYTHONPATH	Path to macs2 python libraries must be added

It is best to define these environment variables in the bash configuration file (~/.bashrc) for each user. /XXX/YYY/ is a placeholder for the installation path of the callHotspotsSSDS folder (i.e., /usr/share/).

Add the following lines:

```
export CHSPATH=/XXX/YYY/callHotspotsSSDS
export CHSNCISPATH=/XXX/YYY/callHotspotsSSDS/NCIS
export CHSBEDTOOLSPATH=/XXX/YYY/callHotspotsSSDS/bedtools
export CHSTMPPATH=/tmp
export PERL5LIB=$PERL5LIB:/XXX/YYY/callHotspotsSSDS
export PATH=$PATH:/XXX/YYY/callHotspotsSSDS
export CHSMACSPATH=/XXX/YYY/callHotspotsSSDS/macs2.1.0.20150731/usr/local/bin
export PYTHONPATH=$PYTHONPATH:/XXX/YYY/callHotspotsSSDS/macs2.1.0.20150731/usr/local/lib/python2.7/dist-packages
```

Initialize these variables from .bashrc:

```
>source ~/.bashrc
```

3.4.3.4 Run the Unit Tests to Ensure That the Pipeline Works

This will test the hotspot calling and strength estimation scripts using test data. This test performs peak calling in a 30-Mb region.

```
>cd $CHSPATH/unitTest/
>sh runTest.sh
```

Successful completion of the tests will result in the following output:

Success !! Hotspot calling complete.

The CHS pipeline has been installed successfully and can be run.

Note: Specific versions of MACS, bedtools, and NCIS are included in the callHotspots pipeline repository. We recommend using these versions as other versions may not be compatible. Expert users can tweak the pipeline scripts to use different versions of these programs if desired.

3.4.4 Call Hotspots

To identify the locations of DSB hotspots from a DMC1 ChIP-SSDS experiment, we compare treatment and control experiments. This requires:
ssDNA_type_1 fragments BED file from DMC1-SSDS experiment
ssDNA_type_1 fragments BED file from an input or IgG-SSDS experiment
The peak calling pipeline is run using the following syntax:

```
>$CHSPATH/run_callHotspotsPipeline \
  --t {BED file : type 1 ssDNA from dmc1 experiment}
  --c {BED file : type 1 ssDNA from input/IgG experiment} \
  --gSz {estimated size of mappable genome}
  --name {prefix for output file names} \
  --out {output folder}
```

The arguments for the hotspot calling pipeline can be accessed from the command line:

```
>$CHSPATH/run_callHotspotsPipeline -h
```

Alternatively, they are listed as follows:

Argument	Synopsis (* = required)	Detail
--t	*Treatment BED file	Type 1 ssDNA BED file
--c	*Control BED file	Type 1 ssDNA BED file
--gSz	*Effective genome size	Estimated size of mappable genome
--gName	*Genome/species name (can be used instead of --gSz)	Use precomputed effective genome size For human: hg19 | hg38 | hg | hs | human For mouse: mm9 | mm10 | mm | mouse For rat: rn | rat
--name	*Name prefix for output files	—
--out	Output folder	Default = current folder
--blist	Blacklist file	BED file of genomic regions with sequencing biases that result in spurious peak calls, i.e., for mouse (mm10 genome): $CHSPATH/mm10_hotspot_blacklist.bed

Continued

Argument	Synopsis (* = required)	Detail
--tuniq	Treatment BED file with ONLY unique fragments	Optional, but not recommended
--cuniq	Control BED file with ONLY unique fragments	Optional, but not recommended
--debug	DEBUG mode	Builds scripts but does not execute. *Note*: This is a logical argument, so does not take any value (pass as --debug)
--q30	Q30 mode	Use only fragments where both reads have a q-score ≥ 30. This is useful for removing reads that map with low confidence (i.e., to high copy repeats) *Note*: This is a logical argument, so does not take any value (pass as --q30)
--nc	Do not call peaks	Do not run peak calling *Note*: This is a logical argument, so does not take any value (pass as --debug)
--h/help	Show help	

3.4.5 Outputs From the Hotspot Calling Pipeline

The hotspot calling pipeline generates output files of DSB hotspot locations.

Filename	Detail
$name_peaks.bedgraph	Recentered hotspots with strength estimate (BEDGRAPH file) Column 1: chromosome Column 2: hotspot start Column 3: hotspot end Column 4: hotspot strength *Note*: Hotspots are recentered as described in Brick et al. (2012). Briefly, the hotspot center is defined as the midpoint of the forward and reverse strand ssDNA coverage
$name_peaks.tab	Recentered hotspots table: Column 1: chromosome Column 2: hotspot start Column 3: hotspot end Column 4: hotspot strength Column 5: strength as percentage of total Column 6: strength as rank Column 7: signal fragments Column 8: noise fragments

The pipeline will also output other files that should only be used for debugging purposes:

Filename	Detail
$name.NCISout	NCIS output: Treatment:Control ratios
$name_model.r	MACS output: R script to visualize MACS peak model for these data
$name_peaks.xls	MACS output: raw peak calls (Excel file)
$name_summits.bed	MACS output: raw peak summits (BED file)
$name_peaks.narrowPeak	MACS output: raw peak calls (BED file)
$name_peaks.bed	MACS output: raw peak calls (BED file)
$name.macs2callpeak. YYMMDD_NNNNN·OUT	STDOUT from MACS peak calling
$name.macs2callpeak. YYMMDD_NNNNN.ERR	STDERR from MACS peak calling

REFERENCES

Arnheim, N., Calabrese, P., & Tiemann-Boege, I. (2007). Mammalian meiotic recombination hot spots. *Annual Review of Genetics*, *41*, 369–399. https://doi.org/10.1146/annurev.genet.41.110306.130301.

Baudat, F., Imai, Y., & de Massy, B. (2013). Meiotic recombination in mammals: Localization and regulation. *Nature Reviews. Genetics*, *14*, 794–806. https://doi.org/10.1038/nrg3573.

Baudat, F., et al. (2010). PRDM9 is a major determinant of meiotic recombination hotspots in humans and mice. *Science*, *327*, 836–840. https://doi.org/10.1126/science.1183439.

Billings, T., et al. (2013). DNA binding specificities of the long zinc-finger recombination protein PRDM9. *Genome Biology*, *14*, R35. https://doi.org/10.1186/gb-2013-14-4-r35.

Bolcun-Filas, E., & Schimenti, J. C. (2012). Genetics of meiosis and recombination in mice. *International Review of Cell and Molecular Biology*, *298*, 179–227. https://doi.org/10.1016/B978-0-12-394309-5.00005-5.

Brick, K., Smagulova, F., Khil, P., Camerini-Otero, R. D., & Petukhova, G. V. (2012). Genetic recombination is directed away from functional genomic elements in mice. *Nature*, *485*, 642–645. https://doi.org/10.1038/nature11089.

Davies, B., et al. (2016). Re-engineering the zinc fingers of PRDM9 reverses hybrid sterility in mice. *Nature*, *530*, 171–176. https://doi.org/10.1038/nature16931.

Grey, C., et al. (2011). Mouse PRDM9 DNA-binding specificity determines sites of histone H3 lysine 4 trimethylation for initiation of meiotic recombination. *PLoS Biology*, *9*. e1001176, https://doi.org/10.1371/journal.pbio.1001176.

Grey, C., et al. (2017). In vivo binding of PRDM9 reveals interactions with noncanonical genomic sites. *Genome Research*, *27*, 580–590. https://doi.org/10.1101/gr.217240.116.

Hayashi, K., Yoshida, K., & Matsui, Y. (2005). A histone H3 methyltransferase controls epigenetic events required for meiotic prophase. *Nature*, *438*, 374–378. https://doi.org/10.1038/nature04112.

Hunter, N. (2015). Meiotic recombination: The essence of heredity. *Cold Spring Harbor Perspectives in Biology, 7*. https://doi.org/10.1101/cshperspect.a016618.
Keeney, S., Lange, J., & Mohibullah, N. (2014). Self-organization of meiotic recombination initiation: General principles and molecular pathways. *Annual Review of Genetics, 48*, 187–214. https://doi.org/10.1146/annurev-genet-120213-092304.
Khil, P. P., Smagulova, F., Brick, K. M., Camerini-Otero, R. D., & Petukhova, G. V. (2012). Sensitive mapping of recombination hotspots using sequencing-based detection of ssDNA. *Genome Research, 22*, 957–965. https://doi.org/10.1101/gr.130583.111.
Myers, S., Bottolo, L., Freeman, C., McVean, G., & Donnelly, P. (2005). A fine-scale map of recombination rates and hotspots across the human genome. *Science, 310*, 321–324. https://doi.org/10.1126/science.1117196.
Myers, S., et al. (2010). Drive against hotspot motifs in primates implicates the PRDM9 gene in meiotic recombination. *Science, 327*, 876–879. https://doi.org/10.1126/science.1182363.
Orkin, S. H., et al. (1982). Linkage of beta-thalassaemia mutations and beta-globin gene polymorphisms with DNA polymorphisms in human beta-globin gene cluster. *Nature, 296*, 627–631.
Paigen, K., & Petkov, P. (2010). Mammalian recombination hot spots: Properties, control and evolution. *Nature Reviews. Genetics, 11*, 221–233. https://doi.org/10.1038/nrg2712.
Parvanov, E. D., Petkov, P. M., & Paigen, K. (2010). Prdm9 controls activation of mammalian recombination hotspots. *Science, 327*, 835. https://doi.org/10.1126/science.1181495.
Pratto, F., et al. (2014). DNA recombination. Recombination initiation maps of individual human genomes. *Science, 346*, 1256442, https://doi.org/10.1126/science.1256442.
Smagulova, F., Brick, K., Pu, Y., Camerini-Otero, R. D., & Petukhova, G. V. (2016). The evolutionary turnover of recombination hot spots contributes to speciation in mice. *Genes & Development, 30*, 266–280. https://doi.org/10.1101/gad.270009.115.
Smagulova, F., et al. (2011). Genome-wide analysis reveals novel molecular features of mouse recombination hotspots. *Nature, 472*, 375–378. https://doi.org/10.1038/nature09869.

FURTHER READING

Hansen, L., Kim, N. K., Mariño-Ramírez, L., & Landsman, D. (2011). Analysis of biological features associated with meiotic recombination hot and cold spots in Saccharomyces cerevisiae. *PLoS One, 6*, e29711. https://doi.org/10.1371/journal.pone.0029711.

Kolibri 360 GmbH
Ferdinand-Jühlke-Straße 7
99095 Erfurt, Deutschland
produktsicherheit@kolibri360.de